Dahlem Workshop Reports
Life Sciences Research Report 23
Biological Mineralization and Demineralization

The goal of this Dahlem Workshop is:
to stimulate cooperation between biological and physical sciences in the study of biological minerals in health and disease

Life Sciences Research Reports
Editor: Silke Bernhard

Held and published on behalf of the
Stifterverband für die Deutsche Wissenschaft

Sponsored by:
Deutsche Forschungsgemeinschaft
Stifterverband für die Deutsche Wissenschaft

Biological Mineralization and Demineralization

G. H. Nancollas, Editor

Report of the Dahlem Workshop on
Biological Mineralization and Demineralization
Berlin 1981, October 18 – 23

Rapporteurs:
J. L. Matthews · J. J. Reynolds · W. G. Robertson
K. Simkiss

Program Advisory Committee:
G. H. Nancollas, Chairman · J. Arends · H. Fleisch
A. E. Nielsen · W. G. Robertson · L. H. Smith
E. W. Vahlensieck · R. A. Young

Springer-Verlag Berlin Heidelberg New York 1982

Copy Editors: M. Cervantes-Waldmann, K. McWhirter
Photographs: E. P. Thonke

With 4 photographs, 82 figures, and 16 tables

ISBN-13: 978-3-642-68576-7 e-ISBN-13: 978-3-642-68574-3
DOI: 10.1007/978-3-642-68574-3

CIP-Kurztitelaufnahme der Deutschen Bibliothek:

Biological mineralization and demineralization: report of the Dahlem Workshop on Biolog. Mineralization and Demineralization, Berlin 1981, October 18 – 23 / G. H. Nancollas, ed. Rapporteurs: J. L. Matthews... [Dahlem Konferenzen. Held and publ. on behalf of the Sifterverb. für d. Dt. Wiss. Sponsored by: Dt. Forschungsgemeinschaft; Stifterverb. für d. Dt. Wiss.]. – Berlin; Heidelberg; New York: Springer, 1982.
(Life sciences research report; 23)
(Dahlem workshop reports)

NE: Nancollas, George H. [Hrsg.]; Matthews, James L. [Mitverf.]; Workshop on Biological Mineralization and Demineralization <1981, Berlin, West>; Dahlem Konferenzen; 1. GT

Softcover reprint of the hardcover 1st edition 1982

2131/3014 – 5 4 3 2 1 0

Table of Contents

The Dahlem Konferenzen

DIRECTOR:

Silke Bernhard, M.D.

FOUNDATION:

Dahlem Konferenzen was founded in 1974 and is supported by the Stifterverband für die Deutsche Wissenschaft*, in cooperation with the Deutsche Forschungsgemeinschaft** and the Senat of the City of Berlin.

OBJECTIVES:

The task of Dahlem Konferenzen is:

- to promote the interdisciplinary exchange of scientific information and ideas,
- to stimulate international cooperation in research, and
- to develop and test different models conducive to more effective scientific meetings.

AIM:

Each Dahlem Workshop is designed to provide a survey of the present state of the art of the topic at hand as seen by the various disciplines concerned, to review new concepts and techniques, and to recommend directions for future research.

TOPICS:

The workshop topics (in the Life Sciences and the field of Physicochemistry) should be:

- of contemporary international interest,
- timely,
- interdisciplinary in nature, and
- problem-oriented.

PROCEDURE:

Dahlem Konferenzen approaches internationally recognized scientists to suggest topics fulfilling these criteria and to propose members for a Program Advisory Committee, which is responsible for the workshop's scientific program. Once a year, the topic suggestions are submitted to a scientific board for approval.

---

* The Donors Association for the Promotion of Sciences and Humanities

** German Science Foundation

PARTICIPANTS:

The number of participants is limited to 48 for each workshop. They are selected exclusively by a Program Advisory Committee. Selection is based on international scientific reputation alone and is independent of national considerations, although a balance between Europeans and Americans is desirable. Exception is made for younger German scientists for whom 10% of the places are reserved.

THE DAHLEM WORKSHOP MODEL:

A special workshop model has been developed by Dahlem Konferenzen, the *Dahlem Workshop Model*. The main work of the workshop is done in four small, interdisciplinary discussion groups, each with 12 members. Lectures are not given.

Some participants are asked to write background papers providing a review of the field rather than a report on individual work. These are circulated to all participants 4 weeks before the meeting with the request that the papers be read and questions on them formulated *before* the workshop, thus providing the basis for discussions.

During the workshop, each group prepares a report reflecting the essential points of its discussions, including suggestions for future research needs. These reports are distributed to all participants at the end of the workshop and are discussed in plenum.

PUBLICATION:

The Dahlem Workshop Reports contain:

- the Chairperson's introduction,
- the Background Papers, and
- the Group Reports.

The Dahlem Workshop Reports are available in two series:

1) Life Sciences Research Reports (LS) and

2) Physical and Chemical Sciences Research Reports (PC).

DAHLEM KONFERENZEN, Wallotstrasse 19, D-1000 Berlin 33, Federal Republic of Germany

Biological Mineralization and Demineralization, ed. G.H. Nancollas, pp. 1-4.
Dahlem Konferenzen 1982. Berlin, Heidelberg, New York: Springer-Verlag.

Introduction

G. H. Nancollas
Dept. of Chemistry, State University of New York
Buffalo, NY 14214, USA

This volume represents the fruits of the Dahlem Workshop on "Biological Mineralization and Demineralization" in which some fifty scientists, who share an interest in this field from many points of view, participated to discuss our understanding of biological mineralization and the areas which require further study. Following the unique format established by the Dahlem Workshop, background papers prepared several weeks beforehand formed the basis for the group discussions. For this purpose, the attendees were divided into four study groups:

1. Structure, properties, and functions of mineralized tissue components;
2. Mechanisms of normal mineralization;
3. Mechanisms of pathological mineralization; and
4. Mechanisms of normal and pathological demineralization.

The moderators and rapporteurs of these discussion groups played key roles in preparing the group reports which represented the consensus not only of members of the group, but also of other participants who had an opportunity to participate in the discussions. Emphasized throughout the Workshop, was the need to develop a list of open problems and questions as well as the identification of techniques which might lead to solutions of these problems. By representing the views of all the participants, these proceedings attempt to identify critical problems calling for research.

The mechanism of formation of minerals in biological systems raises a number of questions which await satisfactory answers. In order to prepare the participants for discussions involving such questions, a number of the general papers were written by physical scientists and were concerned with the mechanism of nucleation and crystal growth and the need for calculations of concentrations of species in biological systems. Here, in terms of the physical chemistry of the processes, even defining the degree of supersaturation in biological fluids is a difficult problem because of the uncertainties in the nature and stabilities of metal complexes which can be formed. It is therefore important to take into account as many ion interactions as possible, both in calculating species in vivo and in attempts to prepare, in vitro, solution milieu which will behave similarly to the biological systems. The crystal structures of various minerals of biological importance are also discussed. During the precipitation processes, transformations may take place from one phase to another. Such transformations are very difficult to monitor if the intermediate phases have a very short lifetime or are formed at microscopic concentration levels.

Although it is clear that the formation of the solid phases from supersaturated biological fluids must be interpretable in terms of physical quantities such as supersaturation, specific surface area, ionic strength, pH, and surface energies, the role of the organic constituents in mediating the mineralization reaction is a problem of great importance. Background papers are concerned with those properties of cells which might be important in controlling tissue mineralization as well as the molecular structures of the organic matrix. Both soluble and insoluble components may play a role in controlling the mineralization processes. The Workshop was concerned not only with normal and pathological mineralization in vertebrates but also with the interesting problems involved in the formation of intracellular deposits of calcium oxalate in plants. Here calcium carbonate and silica may also be involved in the mineralization processes. Calcium carbonate is an important component in the formation of mollusc and avian shells. The observation that both calcite and aragonite may be formed in biogenic calcium carbonate raises important questions as to what factors control the formation of the final mineral phase. There is little doubt that thermodynamically less stable phases may be kinetically stabilized for long periods of time by other molecules present in vivo. In normal mineralization, calcium salts may initially be deposited both within the cells and extracellularly. In the latter case, the role of matrix vesicles and the ways in which the matrix components might control mineralization were especially emphasized. There is clearly a need for more structural and functional information involving cells, matrix components, and their associated crystals. The development of further techniques involving mutants, chromofluors, and fixatives for preservation of tissue and the ions present in vivo was recommended.

In dealing with abnormal and pathological mineralization, the Workshop concentrated on urolithiasis, gout and pseudo-gout, and the formation of dental caries. Discussions based on the influence of media of low pH on mineral surfaces highlighted some of the factors of importance in controlling crystal dissolution. Problems involve not only the mechanism of formation of the minerals and the influence of inorganic and organic components in modifying the nature of the mineral phase formed, but also the ways in which the concretions are retained by the tissues. This is of particular importance in the fluid dynamics involved in the kidney. Considerable progress appears to have been made in our knowledge of the mechanism of dental caries formation involving the complexity of calcium phosphate chemistry in which many different phases and defect structures may participate in the mineralization reactions.

This volume does not, of course, serve as a comprehensive treatise on biological mineralization and demineralization. Its aim is to help stimulate cooperative research endeavors between crystal growth physical scientists and biologists interested in understanding and controlling the factors which lead to both normal and pathological mineral formation.

Biological Mineralization and Demineralization, ed. G.H. Nancollas, pp. 5-21.
Dahlem Konferenzen 1982. Berlin, Heidelberg, New York: Springer-Verlag.

The Solubility Concept

W. G. Robertson
M. R. C. Mineral Metabolism Unit, The General Infirmary
Leeds LS1 3EX, England

Abstract. The concept of the solubility of sparingly soluble salts is based on the statistical theory of electrolyte solutions of Debye and Hückel. This requires a knowledge of the relevant free ion activities (or concentrations), ionic strength, activity coefficients, and ion-pairing in the solution concerned. From these the thermodynamic solubility product is developed which describes the absolute solubility of a salt at a given temperature in equilibrium with the pure solid phase of that salt. Aberrant solubility is observed when the surface activity of the salt is altered. This may come about because of the very small crystal size of the salt, because of the presence of a different solid phase at the crystal surface from that in the bulk crystal, or because of the presence of foreign ions on the crystal surface. Examples of these are common in studies on the solubilities of salts in contact with biological fluids. The implications of these effects on the processes of normal and pathological mineralization and demineralization and the complexity of the crystallization mechanisms involved in intracellular formation of mineral are reviewed.

INTRODUCTION

Solubility is considered by many to be one of the most important aspects of biological mineralization and demineralization. This assumption stems from the fact that all hard tissues contain salts which are known to be only sparingly soluble in an aqueous environment. In normal mineralized tissues the main salts are calcium phosphate and calcium carbonate. In terms of pathological mineralization, however, the range of sparingly soluble substances of interest is much broader. For urinary stones, the list includes calcium oxalate, calcium phosphate, magnesium ammonium phosphate, uric acid, cystine, silica, 2,8-dihydroxyadenine, and xanthine; for gallstones it includes cholesterol and the calcium salts of carbonate, phosphate, bilirubinate, and palmitate; for dental and salivary calculi, it involves mainly calcium phosphates; and for gout and pseudogout (or chondrocalcinosis), sodium acid urate, calcium pyrophosphate, and calcium phosphates.

The question arises as to whether the classical solubility theory can account for the formation, growth, and possibly, dissolution of this extensive range of sparingly soluble substances which may be found in contact with a wide variety of complex biological media. In order to appreciate the problems involved in the application of the solubility concept to biological solutions, it is appropriate to review the physical chemistry of ions in aqueous solution (12).

ION ACTIVITY

The first statistical theory of electrolyte solutions, developed by Debye and Hückel in 1923, enabled the successful interpretation of the behavior of ions in dilute solution. According to Debye and Hückel, the effective concentration of an ion (M) in solution (expressed as its chemical activity) differs from its actual concentration by a factor dependent on the electrical field strength (as defined by the ionic strength (I)) of the solution. In simple terms this correction allows for the non-specific effects of other ions in the solution on the true chemical activity of the ion under consideration. The relation between ion activity ({M}) and ion concentration ([M]) is given by the expression:

$$\{M\} = [M]\, y_M, \qquad (1)$$

where y_M is the activity coefficient of ion, M, and is defined as:

$$\log y_M = Az_M^2\ (\sqrt{I}\ /(1 + \sqrt{I})) - 0.3I). \qquad (2)$$

This is the Davies (4) modification of the Debye-Hückel equation. In this expression A is a constant and z_M is the valency of the ion, M. The ionic strength (I) is given by the expression:

$$I = 1/2\Sigma(z_M^2 [M]). \quad (3)$$

The activity corrections are extremely important in physiological solutions. For example, in the case of a binary salt such as calcium oxalate, the activity coefficients for Ca^{2+} and $C_2O_4{}^{2-}$ ions in physiological saline (ionic strength = 0.15 M) are on the order of 0.3. Thus, as a measure of the supersaturation of that solution with respect to calcium oxalate, the activity product is approximately a factor of 11 less than the simple concentration product.

One of the major problems with most biological fluids, however, is that of obtaining an estimate of the ionic strength. Since it is not measurable directly, it has to be determined by combining the individual z_M^2 [M] terms in Eq. 3. This necessitates the determination of the concentrations of all the charged ionic species in the solution concerned, a difficult task indeed. Since the activity coefficients (y_1, y_2, etc.) are dependent on the value of I, any uncertainty in I introduces uncertainty in the values of the ion activities. Fortunately, however, in most extracellular biological fluids $0.12 < I < 0.35$ and within this range, y_1 and y_2 as calculated from Eq. 2 are relatively constant (within 12%), and no appreciable error arises from the omission of small terms in Eq. 3. Within cells, however, the levels of ionic strength are completely unknown.

ION-PAIRING

Another factor of vital importance in the consideration of the solubility of salts in a mixed ionic environment is the possible interaction between specific ions of opposite charge (for example, C^+ and A^-), which results in the formation of a soluble ion-pair or complex (CA) according to the equilibrium:

$$C^+ + A^- \rightleftharpoons CA, \quad (4)$$

where the association constant (or stability constant, K_{CA}) of the equilibrium is given by:

$$K_{CA} = \{CA\}/\{C^+\}\{A^-\} = [CA]y_{C_A}/[C^+][A^-]y_C y_A. \quad (5)$$

This effectively reduces the "free" or "ionized" concentrations of the two ions concerned, although the mass balance of each is preserved. Thus:

$$T_C = [C^+] + [CA], \text{ and} \quad (6)$$

$$T_A = [A^-] + [CA], \quad (7)$$

where T_C and T_A are the total concentrations of ions C and A.

In most physiological solutions, a large number of such interactions is possible, and the resulting "free" ion concentrations (and activities) are usually impossible to measure directly. The exception is ionized calcium which may be measured in a number of biological fluids using colorimetry or ion-selective electrode procedures (19). No electrodes have yet been developed which will satisfactorily measure in biological fluids the free ion concentrations of any of the anions important in the mineralization of the principal salts mentioned in the Introduction, namely, phosphate, oxalate, carbonate, and urate.

To overcome this problem, it has been necessary to devise computer programs to calculate, ab initio, the free ion concentrations of all the relevant ions in a given fluid from a knowledge of the pH, the total concentrations of all the main ions in the fluid, and the stability constants of the mutual complexes which can be formed between these ions. In most instances this covers the ion-pairing of cations with inorganic anions or small organic anions. However, where complexation is significant, the influence of larger organic molecules, such as proteins or glycoproteins, may also be taken into account. A number of such programs have been published which, at least for ionized calcium, give a reasonable approximation to the value measured directly in serum and urine (6,17,19). It has not yet been possible, however, to check the calculated values for the free concentrations of the various anions with the corresponding measured values since, as mentioned above, the techniques for obtaining the latter are not yet available.

The calculation of the free ion concentrations of the various ions also enables the ionic strength (Eq. 3) and the various activity coefficients (Eq. 2) to be estimated. The thermodynamic activities of the ions may then be calculated from Eq. 1. With these techniques it is possible to calculate the activity product of a solution with respect to a given salt by multiplying together the relevant ion activities. Thus, for the salt CA, the ab initio activity product (AP) is given by:

$$AP = \{C^+\}\{A^-\} = [C^+][A^-]\, y_C y_A. \qquad (8)$$

By comparing the value of AP with the thermodynamic solubility product of the salt concerned, an estimate of the degree of supersaturation of the solution may be obtained.

SOLUBILITY PRODUCT

The solubility of a solid is independent of the amount of solid present since, although increasing the surface area of the crystals decreases the time required to reach equilibrium, it does not affect the final saturation level attained at equilibrium (solubility product).

In a solution of our salt, CA, which dissolves according to the equilibrium:

$$(CA)_{solid} \rightleftharpoons (CA)_{solution}, \qquad (9)$$

the equilibrium constant (K) is given by:

$$K = \{CA\}_{solution}/\{CA\}_{solid}. \qquad (10)$$

If $(CA)_{solution}$ further dissociates in solution according to Eq. 4 with a stability constant K_{CA} given by Eq. 5, then combining Eqs. 5 and 10 yields (since $\{CA\}_{solid}$ is arbitrarily given the value 1):

$$K_{sp} = K/K_{CA} = \{C^+\}_e\{A^-\}_e/\{CA\}_{solid} = \{C^+\}_e\{A^-\}_e, \qquad (11)$$

where K_{sp} is the thermodynamic solubility product and $\{C^+\}_e$ and $\{A^-\}_e$ are the activities of C^+ and A^- in equilibrium with pure solid phase $(CA)_{solid}$. It must be emphasized, however, that this K_{sp} is only a constant at a given temperature and if the solution is at equilibrium with pure solid phase. This satisfies the assumption made above that the activity of the solid phase is unity. If the system contains two (or more) separate solid phases coexisting in equilibrium with the aqueous phase, the composition of the solution will be different from that expected at equilibrium with any of the individual solid phases alone. In this situation, however, the intrinsic solubility of each solid phase is unaffected by the presence of the other phases, since the thermodynamic properties of each solid are not altered.

In addition to temperature, there are three factors which may influence the apparent solubility of a solid by altering its surface activity: (a) the size of the crystals, (b) the occurrence of a different crystal phase on the surface from that in the bulk crystal, and (c) the presence of any foreign ions on the crystal surface (10,13). Small particles have an aberrantly high surface energy and solubility, and the presence of a different surface phase or of foreign ions on the surface lattice positions may increase or decrease the solubility of the salt concerned. In the latter case, a state of of "quasi-equilibrium" may be attained. The effect of small particle size (2), the formation of various metastable surface phases (11,21), and the presence of foreign ions on the surface (9) are of particular importance in assessing the solubility of calcium phosphate in various solutions of biological interest. For calcium oxalate, on the other hand, the problem is mainly one of foreign surface ions (6), although phase transformations are also known to occur between the various hydrates and this will influence the solubility. These factors become of vital importance in assessing the true level of relative supersaturation of a given solution with respect to one of these salts.

RELATIVE SUPERSATURATION

In a simple solution of the salt CA, the relative supersaturation can be determined from the ratio of the calculated activity product (AP) to the known thermodynamic solubility (K_{sp}) of the salt. If pure crystals of CA with a single free energy are added to such a solution, then it will be anticipated that they will dissolve at AP/K_{sp} values of less than 1, that they will be at equilibrium with the solution at an AP/K_{sp} value equal to 1, and that they will grow at AP/K_{sp} values greater than 1. In this last situation, the solution is said to be supersaturated. In the absence of seed crystals of the salt concerned and of heterogeneous nucleating material, such solutions may exist for long time periods (several hours or even days) without spontaneous crystallization taking place (metastable supersaturation). There is, however, an ill-defined upper limit of metastability at or above which spontaneous precipitation will take place within short time periods (<5 minutes). This is not a true constant but is dependent, amongst other factors, on the time of incubation and the presence of certain "crystal poisons" which may inhibit the rates of nucleation and/or crystal growth of the salt concerned (7,15,18).

Because of the problems involved in the calculation of activity products in complex biological media, attempts have been made to determine relative supersaturation using a simplified semi-empirical technique in which the activity product of the salt is measured before and after incubation with crystals of the salt concerned (14). After taking into account ionic strength and ion-pairing effects, the activity product ratio (APR) of $(\text{measured AP})_i$ divided by $(\text{measured AP})_f$ should be identical with the ab initio relative saturation value of $(\text{calculated AP})_i$ divided by K_{sp}, where the subscripts i and f refer to the initial and final activity products, i.e., before and after incubation. In practice, the ion-pairing effects are ignored since they cancel out, approximately, in the expression for APR. This introduces a small error of about 2 to 5%, depending on the initial supersaturation level of the prevailing ionic strength (7).

The estimation of relative supersaturation becomes much more complex, however, under nonequilibrium conditions with respect to a pure single solid phase, for example, when there are different surface phases or foreign ions present on the crystal surface. In these situations the semi-empirical determination of APR may differ considerably from the ab initio calculated value. Thus in the determination of the relative supersaturation of calcium oxalate monohydrate in urine, the semi-empirical estimate of relative supersaturation is significantly lower than that derived from ab initio calculations. This is basically due to the fact that, the semi-empirical measurement of the activity product of calcium oxalate at equilibrium (after taking into account ionic strength effects and all the known combinations of ion-pairs) yields an equilibrium activity product apparently higher than the thermodynamic K_{sp} for calcium oxalate monohydrate by a factor of between 1.5 and 15 (14). In this instance, under the conditions employed for the equilibration of urine with crystals, there appears to be an appreciable difference between the (measured AP)$_f$ and the thermodynamic K_{sp}. There are several possible explanations for this discrepancy. The most obvious is that in the calculation of the free ion concentrations of calcium and oxalate, some (as yet identified) factors have been excluded which complex with one or another of these ions. Indeed, there is some evidence that up to 12% of oxalate and 10% of calcium may be bound to urinary macromolecules. However, even these would not account for the large discrepancy described above.

The second possibility emerges from a study which has shown that the equilibrium saturation level for calcium oxalate monohydrate in urine is, in fact, almost identical with the K_{sp} of that salt when a longer equilibration time (7 days) and higher slurry density (1 g of calcium oxalate crystals per ml urine) are employed (Robertson, unpublished results), compared with the corresponding conditions used by Pak (2 days and 0.01 g calcium oxalate per ml urine) (14). This suggests that in Pak's studies "true" equilibrium may not have been reached. One possible reason for this is the presence, in urine, of several known inhibitors of crystal growth such as pyrophosphate, citrate, magnesium, and glycosaminoglycans (8), which may bind to active growth sites on the crystal surface. By initially supplying only a limited crystal surface area, which is reduced even further by the binding of inhibitors, the rate of approach to "true" equilibrium may have been significantly reduced. Ironically, this suggestion is supported by another of Pak's studies (15), which showed that in the presence of ethane-1-hydroxy-1,1-diphosphonate (a well-known inhibitor of calcium oxalate crystal growth), a large excess of crystals is required to enable equilibration to the thermodynamic K_{sp} even in simple inorganic solutions. This suggests that Pak's measured "solubility product" obtained in urine after "equilibration" with calcium oxalate crystals may not be a true constant but dependent on some balance between the remaining level of supersaturation at the time of measurement and the inhibitor/crystal slurry density ratio. The value of this "quasi-equilibrium level" is then a function of the thermodynamic driving force and the various factors affecting the kinetics of the approach to "true" equilibrium. If, however, sufficient crystal surface is introduced to adsorb all the inhibitors present, and leaving sufficient available growth sites, the level of saturation may eventually fall to the true K_{sp}.

Assuming that no new crystals form during the incubation period, then the rate of change of calcium (or oxalate) concentration (-dc/dt) because of seeded crystal growth is given by:

$$-dc/dt = kS' \left((AP)_t^{1/2} - (K_{sp})^{1/2}\right)^n, \quad (12)$$

although it is more common to express this in terms of ion concentrations:

$$-dc/dt = kS' \left((AP/y_2^2)_t^{1/2} - (K_{sp}/y_2^2)^{1/2}\right)^n, \quad (12a)$$

where k is the rate constant, S' is the free surface area available for growth (and is assumed not to change significantly during the growth phase), $(AP)_t$ is the activity product and $(AP/y_2^2)_t$ the concentration product of calcium oxalate at time t, (K_{sp}/y_2^2) the concentration solubility product, and n is the effective order of reaction. For calcium oxalate, n is usually

2 even in the presence of growth inhibitors. S' is a function of the total crystal surface area S (and hence the slurry density), the total concentration of inhibitor(s), and the binding between the inhibitors and the crystal surface as defined by the Langmuir adsorption isotherm:

$$S' = S/(1 + b\,[In]), \qquad (13)$$

where b is a constant and [In] is the concentration of free inhibitor in solution.

Substituting in Eq. 12a gives:

$$-dc/dt = k\,S((AP/y_2^{\,2})_t^{1/2} - (K_{sp}/y_2^2)^{1/2})^2/(1 + b\,[In]), \qquad (14)$$

and it is clear that the crystals only stop growing completely either when S = O, when $(AP)_t = K_{sp}$, or when [In] is very large. However, at low slurry densities and moderate concentrations of inhibitor(s), the value of dc/dt at time t may become very small (approximating zero) at supersaturation levels well within the metastable region. The final approach to "true" equilibrium may then be extremely slow.

If this is the correct explanation for the apparent discrepancy between the semi-empirical and ab initio determinations of the relative supersaturation of urine with calcium oxalate, it raises the following interesting possibility. Although the true relative supersaturation of urine must always be given by AP/K_{sp}, the apparent value AP_i/AP_f may be a better indicator of the overall risk of crystal growth in a given urine, since it is a function of both the prevailing supersaturation and the inhibitory activity of that urine. The main problem is that the apparent relative supersaturation measured in this way is necessarily a function of the arbitrary amount of seed crystal added and the period of incubation chosen. But how should these variables be fixed? For if we choose too small an amount of seed crystals, the amount of inhibitor in the urine will be the dominating factor, and there may even be a risk of a new surface phase forming involving the inhibitor. On the other hand, if we choose too large an amount of crystal, then most, or all, of the inhibitor(s) may be adsorbed, and the net growth rate of the system may be dominated by the supersaturation term. In this situation it is probably important to choose an inhibitor/solid ratio which lies on the rising part of the inhibitor adsorption isotherm.

A third possible reason for the discrepancy between the semi-empirical and ab initio estimates of relative saturation is that there may be so much of one particular inhibitor (say, pyrophosphate) that a new surface phase (in this instance, consisting of calcium pyrophosphate) is formed on all calcium oxalate crystal surfaces. Unless calcium oxalate can grow epitaxially on this new phase, there will be no growth. This seems unlikely, however, on the basis of studies so far which indicate that inhibitors cover only about 70% of the available crystal surface of calcium oxalate (7).

The problems of defining the relative supersaturation in biological fluids become even more complex in relation to calcium phosphate solubility. Not only are there potential non-equilibrium conditions resulting from the adsorption of foreign ions on the crystal surface, but the surface phase itself may be different from that in the bulk crystal ((11,21), and Nancollas, this volume), and for some calcium phosphates the crystal size is sufficiently small to affect the solubility (2). There may even exist intermediate phases in the precipitation of calcium phosphates which are poorly crystalline and which transform to more stable crystalline forms with time ((11,21), and Nancollas, this volume).

A further problem in the case of small crystals of colloidal dimensions is the existence at the crystal-solution interface of an extensive hydration layer. This arises because of the considerable electric field projecting out from the crystal surface (3,13). It is known to contain large amounts of calcium and phosphate ions which are readily exchangeable with ions in the bulk solution (13,20). Thus, depending on the source of apatite and the slurry density used, the amount of calcium in the bulk solution at equilibrium may take any value between 0% and 100% of the total fast exchangeable pool of calcium in the system, the remainder being in the hydration layer around the crystals. When the zeta potential, which reflects the charge on the crystal surface, is changed by adsorption of a polyanion (such as pyrophosphate), then calcium ions move into the hydration layer from the bulk solution and phosphate ions move in the reverse direction (20). This is unlikely to be due to the formation of a calcium pyrophosphate phase on the crystal surface since, even if all the added pyrophosphate had remained in solution, the activity product of calcium pyrophosphate in this study would have still been one tenth of the K_{sp} of that salt. Interestingly, the phenomenon of ion movement into and out of the hydration layer in response to pyrophosphate binding on the crystal surface can be largely reversed by destroying the bound pyrophosphate enzymatically (20).

In spite of these potential problems, the AP of brushite in urine equilibrated with that salt is exactly equal to the K_{sp} for brushite measured under the same conditions of slurry density and time of incubation as used in the corresponding study on calcium oxalate (15). The obvious explanation for this is that the estimate of the AP for brushite in urine is better than that of calcium oxalate because it takes all the important complexes of calcium and phosphate into account in the calculation of ionized calcium and phosphate, whereas some unidentified complex of oxalate is being missed.

There is, however, a second possible explanation for the apparently better results obtained from the equilibration of urine with calcium phosphate. The crystals of most calcium phosphate are relatively small in comparison with those of calcium oxalate. Thus, the total surface area available for adsorption of foreign ions at a given slurry density of calcium phosphate crystals (S in Eq. 15) is much greater than that at an equivalent slurry density of calcium oxalate crystals. Furthermore, in the case of some inhibitors (for example, pyrophosphate), the affinity for calcium phosphate is much higher than for calcium oxalate. Thus, for a given slurry density, calcium phosphate crystals are more likely than calcium oxalate crystals to adsorb all the available inhibitor(s) present and yet leave sufficient growth sites for relief of supersaturation.

The third possibility is that the specific rate constant for calcium phosphate crystal growth may be greater than that for calcium oxalate. In this case, the tendency will be for the supersaturation level to be reduced further in a given calcium phosphate solution within a fixed incubation time than in a calcium oxalate solution with an equivalent initial supersaturation level.

Clearly, then, the definition of solubility in many biological systems is not simple. In solutions whose supersaturation levels are initially higher than the K_{sp} value for the pure salt, for instance, the approach to the "true" solubility product may be so slow that the solution may appear to have a higher solubility value than that anticipated from the K_{sp}. Even when the initial AP is below the K_{sp}, where the rate of dissolution of the crystal phase is usually controlled by the rate of diffusion of solute from the interface into the bulk solution, the approach to "true" equilibrium may be very slow. For example, the rate of dissolution of pyrophosphate-coated crystals of hydroxyapatite is extremely low (5,9), although this may be due to the formation of a surface phase of calcium pyrophosphate or to the blocking of the preferred sites of dissolution by adsorbed pyrophosphate ions. In situations where the surface reaction is rate-controlling, the rate of dissolution may also appear to be reduced to zero at a value different from the thermodynamic K_{sp}.

In summary, it would appear that the empirical determination of solubility of a salt in a biological medium within a restricted time period may yield a wide range of values which are estimates, not of the "true" equilibrium K_{sp} of the salt, but of some "quasi-equilibrium" state. The observed supersaturation value will depend on the initial activity product of the system, the presence of inhibitors of crystal growth, the time of incubation, and the amount of seed crystal added. The question is, then, even if these factors are taken into account, what is the most meaningful definition of solubility from a biological standpoint?

SOLUBILITY WITHIN ORGANIZED MINERALIZED TISSUES

Most of the discussion so far has centered around the solubilities of biologically important minerals found mainly in pathological conditions in which the mineral is in direct contact with or even bathed in the fluid under consideration. What about the situation in normal mineralized tissues such as bone? If such phenomena as described above exist, considering the vast bone mineral/extracellular fluid ratio in the body, it would seem that, within certain limits, ions may be transferred into and out of bone without necessitating the formation or destruction of bone mineral itself, but merely by altering the charge at the mineral surface. In terms of a possible "bone-extracellular fluid equilibrium," this opens up the possibility that this could be controlled by cellular action at the intermediate bone mineral-bone fluid interface and not directly by a simple bone-extracellular fluid solubility equilibrium. The cells could be providing a barrier to direct movement of ions other than from the bone fluid phase to the extracellular fluid phase via some pumping mechanism, or by altering the chemical composition of bone fluid so as to affect the charge at the bone fluid-mineral interface. Few studies have been carried out at this level.

In terms of new mineral formation, the problem becomes even more difficult to study chemically. Recent studies indicate that bone mineral may not be produced extracellularly by precipitation on a matrix or by crystal growth on existing mineral, as once thought, but by an intracellular process involving the sequestration of calcium and phosphate ions into the golgi apparatus of the osteocytes to form calcified particles. These could be enclosed within organic envelopes, which are then actively extruded from the cell to be bound chemotaxially to the collagen matrix (16). The relationships between solubility, saturation, nucleation, and crystal growth must be extremely complex in this situation where the normal chemical processes may be supplemented by energy-producing reactions within the cell which may facilitate the production of vast numbers of organized mineral building blocks per unit time. A good example of this is the formation of coccoliths which are structures of intricate morphology containing calcium carbonate and which are formed within the golgi apparatus of the single cell organisms known as dinoflagellates. They are then extruded to form an intricate pattern covering the entire outer surface of the organism. The amount of calcite produced by this unicellular organism to cover itself in 8 minutes is approximately equivalent to the mass of mineral in a single bone cell "loaded" with calcium phosphate as shown by Aaron and Pautard (1).

Clearly, then, the cell must know all about the physical chemistry of solubility, but what factors are involved in the controlled crystallization of calcium phosphate and other calcium salts at this microenvironmental level will take many years of research to disentangle. Meanwhile, the problems are immense. What do we mean by ion concentrations and activities in the closed microenvironment of intracellular organelle? What is the ionic strength of such a milieu? What are the sources of energy available to assist the processes of crystallization? What are the roles of potential promoters and inhibitors of crystallization? These and many more questions are yet to be answered.

In conclusion, then, whereas the concept of solubility in simple salt solutions may be readily understood, in complex biological media, the picture may be much more complicated. The solubility principle probably only applies, at best, to some types of mineral formation in certain pathological conditions such as urinary, dental, or biliary calculous disease. But in functional mineralized tissues such as bone, teeth, and shell, the formation and re-dissolution of mineral may be more more difficult to interpret in terms of simple solubility theory.

REFERENCES

(1) Aaron, J., and Pautard, F.G.E. 1975. Tetracycline stain of bone in normal and pathological states. In Calcium Metabolism, Bone and Metabolic Bone Diseases, eds. F. Kuhlencordt and H.-P. Kruse. Berlin: Springer-Verlag.

(2) Best, J.B. 1959. Some theoretical considerations concerning crystals with relevance to the physical properties of bone. Biochem. Biophys. Acta 32: 194-202.

(3) Brown, W.E., and Chow, L.C. 1981. Thermodynamics of apatite crystal growth and dissolution. J. Cryst. Growth 53: 33-41.

(4) Davies, C.W. 1962. Ion Association. London: Butterworth.

(5) Evans, J.R.; Robertson, W.G.; Morgan, D.B.; and Fleisch, H. 1980. Effects of pyrophosphate and diphosphonates on the dissolution of hydroxyapatites using a flow system. Calcif. Tissue Int. 31: 153-159.

(6) Finlayson, B. 1977. Calcium stones: some physical and clinical aspects. In Calcium Metabolism in Renal Failure and Nephrolithiasis, ed. D.S. David, pp. 337-382. New York: Wiley.

(7) Finlayson, B. 1978. Physico-chemical aspects of urolithiasis. Kidney Int. 13: 344-360.

(8) Fleisch, H. 1978. Inhibitors and promoters of stone-formation. Kidney Int. 13: 361-371.

(9) Fleisch, H.; Maerki, J.; and Russell, R.G.G. 1966. Effect of pyrophosphate on dissolution of hydroxyapatite and its possible importance in calcium homeostasis. Proc. Soc. Exp. Biol. Med. 122: 317-320.

(10) Lewin, S. 1960. The Solubility Product Principle. London: Pitman.

(11) Meyer, J.L., and Eanes, E.D. 1978. A thermodynamic analysis of the secondary transition in the spontaneous precipitation of calcium phosphate. Calcif. Tissue Res. 25: 209-216.

(12) Nancollas, G.H. 1966. Interactions in Electrolyte Solutions. Amsterdam: Elsevier.

(13) Neuman, W.F., and Neuman, M.W. 1958. The Chemical Dynamics of Bone Mineral. Chicago: University of Chicago Press.

(14) Pak, C.Y.C.; Hayashi, Y.; Finlayson, B.; and Chu, S. 1977. Estimation of the state of saturation of brushite and calcium oxalate in urine: a comparison of three methods. J. Lab. Clin. Med. 89: 891-901.

(15) Pak, C.Y.C.; Ohata, M.; and Holt, K. 1975. Effect of diphosphonate on crystallization of calcium oxalate in vitro. Kidney Int. 7: 154-160.

(16) Pautard, F.G.E. 1978. Phosphorus and bone. In New Trends in Bioinorganic Chemistry, eds. R.J.P. Williams and J.R.R.F. Da Silva. London: Academic Press.

(17) Robertson, W.G. 1969. Measurement of ionized calcium in biological fluids. Clin. Chim. Acta 24: 149-157.

(18) Robertson, W.G. 1973. Factors affecting the precipitation of calcium phosphate in vitro. Calcif. Tissue Res. 11: 311-322.

(19) Robertson, W.G., and Marshall, R.W. 1981. Ionized calcium in body fluids. C.R.C. Crit. Rev. Clin. Lab. Sci. 15: 85-125.

(20) Robertson, W.G., and Morgan, D.B. 1971. Effect of pyrophosphate on the exchangeable calcium pool of hydroxyapatite crystals. Biochem. Biophys. Acta 230: 495-503.

(21) Tomazic, B., and Nancollas, G.H. 1975. The seeded growth of calcium phosphates. Surface characterization and the effect of seed material. J. Colloid Interface Sci. 50: 451-461.

Biological Mineralization and Demineralization, ed. G.H. Nancollas, pp. 23-35.
Dahlem Konferenzen 1982. Berlin, Heidelberg, New York: Springer-Verlag.

Nucleation

J. Garside
Dept. of Chemical Engineering, University of Manchester,
Institute of Science and Technology, Manchester M6O, 1QD, England

Abstract. Different nucleation mechanisms are reviewed and the form of nucleation rate equations discussed. Heterogeneous primary nucleation and secondary nucleation are not well understood although both are likely to be of significant importance in many biological mineralization processes. The measurement of induction times is considered and the difficulties in relating experimental results to fundamental parameters are pointed out.

NUCLEATION PHENOMENA

The first step in a crystallization process is the attainment of a supersaturated state. The condition of supersaturation alone, however, is not sufficient for a system to begin to crystallize. Before crystals can grow, new crystallization centers must exist in the solution. This formation of one phase in another, under conditions where a free energy barrier exists, is the process of nucleation.

If nuclei are formed in perfectly clean solutions in the absence of any foreign particles or surfaces, then the mechanism is one of "homogeneous" nucleation, also sometimes called spontaneous nucleation. Theories of homogeneous nucleation are well-developed and will be outlined below, but in any practical situation the presence of foreign surfaces (in the form of ions, impurity molecules, dust particles, or other surfaces, for example) generally induces "heterogeneous" rather than homogeneous nucleation, the former occurring at lower supersaturations than the latter. Both these nucleation processes are forms of primary nucleation, so called to distinguish them from the second main category, secondary nucleation. The latter occurs only because of the prior presence of crystals of the material being crystallized and has become recognized recently as an important mechanism in its own right, mainly as a result of work on industrial crystallization.

The classification of nucleation phenomena which will be adopted here may therefore be represented as shown in Table 1.

TABLE 1 - Classification of nucleation phenomena.

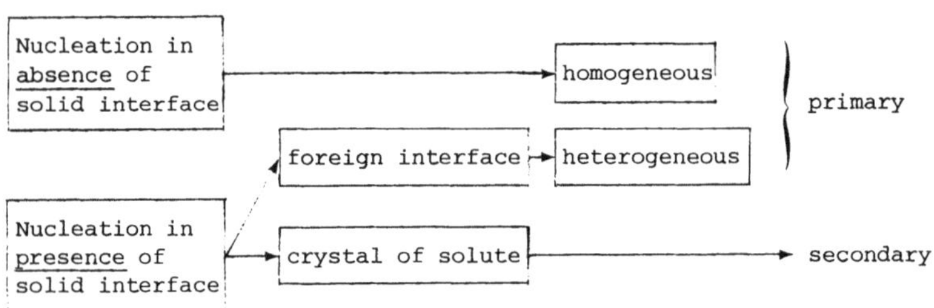

The distinction between these different mechanisms may be illustrated by reference to the ice-water system. Carefully purified water which has been distilled and filtered can be cooled to below $-30^{o}C$ before ice forms, in this case by a primary homogeneous mechanism. In tap water primary heterogeneous nucleation occurs at about $-6^{o}C$. Continuous formation of new ice crystals in a crystallizer already containing a retained bed of ice crystals will, however, take place at -2 or $-3^{o}C$ as a result of secondary nucleation processes.

Note that in all these cases finite undercoolings are necessary before nucleation occurs, indicating the existence of an energy barrier.

PRIMARY NUCLEATION

Homogeneous Nucleation

Homogeneous nucleation is assumed to occur as a result of a series of bimolecular reactions between solute molecules or ions giving rise to ordered aggregates, clusters, or embryos. At steady state a certain size distribution of such clusters will be present in the solution, the distribution depending on, for example, temperature and supersaturation.

The Gibbs free energy of the embryos ΔG_r is a function of their size, and assuming that the embryos are spherical of radius r, it can be written as

$$\Delta G_r = 4\pi r^2\sigma - \frac{4\pi r^3 \Delta G_v}{3V_m} \tag{1}$$

σ is the surface free energy per unit surface area, ΔG_v the free energy change per mole associated with the liquid-solid phase change, and V_m is the molar volume. Inherent in Eq. 1 are two assumptions: that the macroscopic surface free energy can be used for clusters containing a comparatively small number of atoms or molecules, and that the depletion of solute by the formation of nuclei can be neglected (see Simkiss et al., this volume).

The form of Eq. 1 is shown in Fig. 1. The overall free energy goes through a maximum at some critical size r* which by differentiation of Eq. 1 can be shown to be given by

$$r^* = \frac{2\sigma V_m}{\Delta G_v} . \tag{2}$$

The corresponding free energy change associated with the formation of this critical size nucleus is

$$G_r^* = \frac{16\pi\sigma^3 V_m}{3\Delta G_v^2} = \frac{4\pi\sigma r^{*2}}{3} . \tag{3}$$

All the embryos in solution will tend to change their size in such a way as to decrease their overall free energy. For those smaller than the critical size this is achieved by dissolving, whereas clusters larger than r* will continue to grow. The

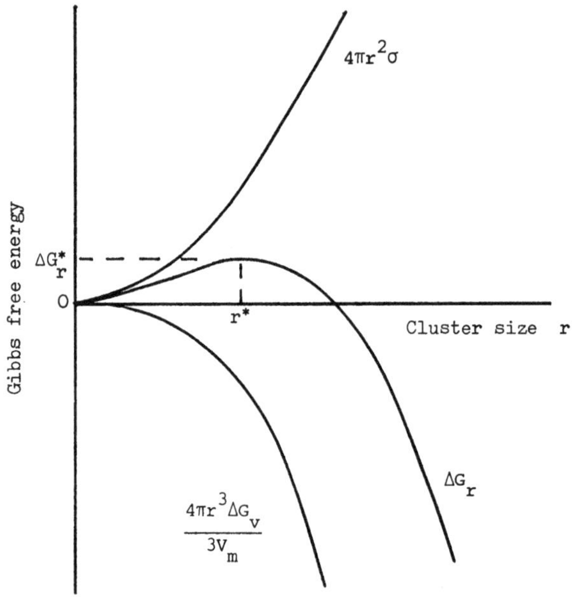

FIG. 1 - Free energy of an atomic cluster as a function of cluster radius.

critical cluster size therefore corresponds to the smallest cluster that can decrease its free energy by growing and so is the minimum size of a stable nucleus.

The Gibbs-Thomson equation describes the relation between the solubility of a crystal and its size. If $a_{\pm r}$ is the mean ionic activity of the solute in equilibrium with a crystal of radius r then

$$\ln \frac{a_{\pm r}}{a_{\pm\infty}} = \frac{2v\sigma}{kTr} , \tag{4}$$

where $a_{\pm\infty}$ is the mean ionic activity of the solute for a large crystal at equilibrium and so represents the normal solubility. v is the molecular volume, k the Boltzmann constant, and T the absolute temperature. If the activity coefficients corresponding to $a_{\pm r}$ and $a_{\pm\infty}$ are assumed to be equal, then Eq. 4 can be written

$$\ln \frac{c_r}{c_\infty} = \frac{2v\sigma}{kTr} , \tag{5}$$

in which c_r and c_∞ are the equilibrium concentrations for crystals of size r and for large crystals respectively. If it is accepted that Eq. 5 can be applied to crystals as small as the critical size, then the value of r* corresponding to a given supersaturation ratio S (=c/c*) can be obtained by rearranging the equation as

$$r^* = \frac{2v\sigma}{kT\ln S} = \frac{2M\sigma}{RT\rho\ln S} \, . \tag{6}$$

M is here the molar mass, R the gas constant, and ρ the crystal density. Furthermore, the critical free energy can be obtained by combining Eqs. 3 and 6:

$$\Delta G_r^* = \frac{16\pi v^2\sigma^3}{3(kT\ln S)^2} \tag{7}$$

Eq. 6 indicates that as the supersaturation of the solution increases, the critical nucleus size, the number of molecules necessary to form a cluster of critical size, and hence the nucleation barrier all decrease. The general form of the relation between r* and S is shown in Fig. 2. For a given supersaturation clusters corresponding to the region below the curve will dissolve whereas those above it will be stable and so will grow. The curve may be thought of as defining the boundary between unstable embryos or clusters and stable nuclei.

The rate at which new nuclei are formed is a problem of kinetics and is determined by the rate at which nuclei surmount the maximum in the free energy curve (Fig. 1), this being the rate limiting step in the nucleation process. Volmer (17) wrote the nucleation rate J (number of nuclei per unit time per unit volume) as proportional to exp $(-\Delta G_r^*/kT)$, i.e.,

$$J = k_n \exp\left[-\frac{16\pi v^2\sigma^3}{3k^3T^3\ln^2 S}\right] \, . \tag{8}$$

The free energy of a critical nucleus plays a similar role to an activation energy in chemical kinetics, but while the latter is effectively constant, ΔG_r^* is extremely sensitive to supersaturation as can be seen from Eq. 7.

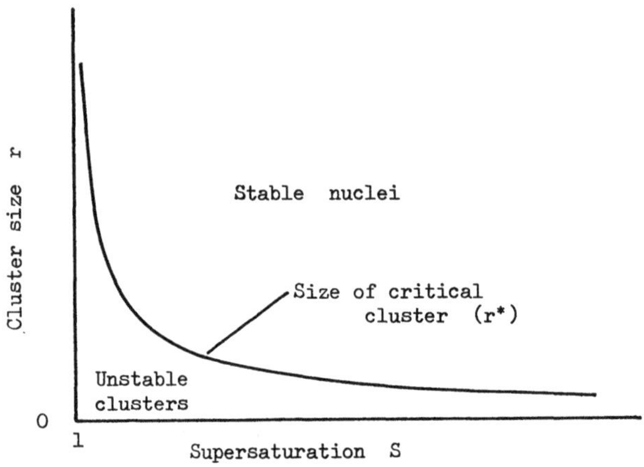

FIG. 2 - Relation between solution supersaturation and the critical nucleus size.

Attempts have been made to put the above theory on a sounder theoretical basis, and a comprehensive review of much of this work is given by Zettlemoyer (20). Most effort has been devoted to making better estimates of the pre-exponential factor k_n. The form of Eq. 8 is, however, dominated by the exponential term, and so the essential features of the equation remain unchanged no matter what approach is adopted. Figure 3 illustrates the relation between nucleation rate and supersaturation, as predicted by Eq. 8. Below a critical value of the supersaturation, S*, the nucleation rate is extremely small; once this supersaturation is reached, however, the nucleation rate increases very rapidly. On a plot such as this the values on the ordinate are not important, as very small changes in S can produce order of magnitude changes in J. The supersaturated region thus is divided essentially into two parts; at low supersaturations the nucleation rate is effectively zero and the solution is "metastable" - crystals will grow but nucleation is negligible. At high supersaturations rapid nucleation takes place and the solution is "labile" (11).

Heterogeneous Nucleation

Most primary nucleation that occurs in practice is likely to be

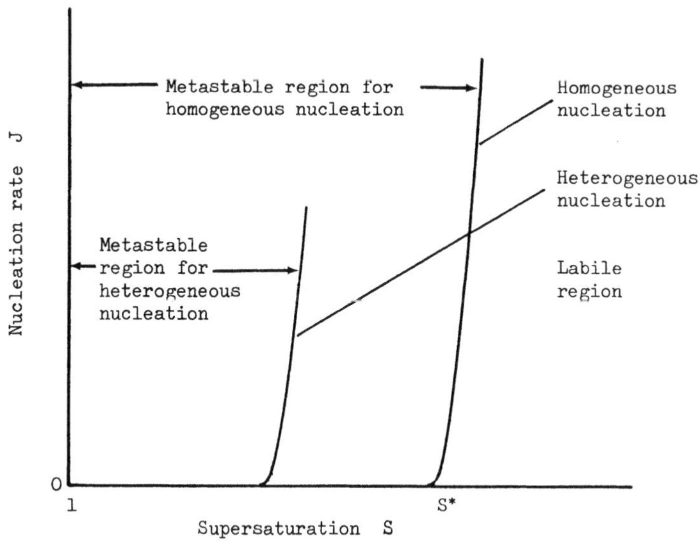

FIG. 3 - Form of the primary nucleation rate equation; homogeneous and heterogeneous nucleation.

heterogeneous, i.e., induced by foreign particles. This requires significantly lower supersaturations than homogeneous nucleation, and so the width of the metastable zone is usually greatly reduced.

Nucleation on a foreign particle will in general require a lower free energy change than that necessary to induce nucleation in the bulk solution. The exact mechanism by which this occurs is unclear, but it is thought to be the result of a local ordering process brought about by interactions across the interface. Consequently, the free energy of formation of a critical nucleus is reduced by a factor ϕ $(0<\phi<1)$ and so $\Delta G^*_{r(het)} = \phi\ \Delta G^*_r$. ϕ has frequently been related to the contact angle, Θ, between the crystal and the foreign substrate as shown in Fig. 4 where a simple force balance shows $\cos\Theta = (\sigma_{SL} - \sigma_{CS})/\sigma_{CL}$.

ϕ is then given by (19).

$$\phi = (2 + \cos\Theta)\ (1 - \cos\Theta)^2/4. \qquad (9)$$

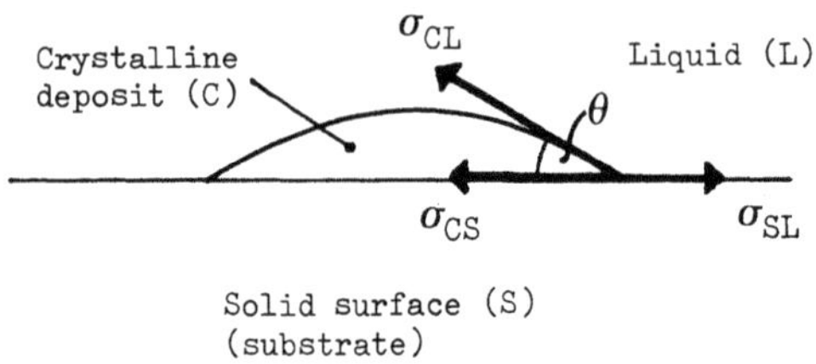

FIG. 4 - Cap-shaped nucleus model of heterogeneous nucleation. The interfacial surface energies and the contact angle Θ are illustrated.

When $\Theta = 180^{\circ}$, $\phi = 1$, there is complete non-affinity between the crystalline solid and the foreign solid surface, and the foreign surface is ineffective in promoting heterogeneous nucleation. On the other hand, complete affinity between crystal and substrate corresponds to $\Theta = 0^{\circ}$ and hence $\phi = 0$. Partial affinity is represented by intermediate cases with $0<\Theta<180^{\circ}$, heterogeneous nucleation then being promoted by the surface.

Turnbull and Vonnegut (16) related the catalytic power of a substrate to the lattice mismatch between it and the growing crystal surface. Experimental evidence in general supports the idea that the power of a substrate to initiate heterogeneous nucleation depends on such parameters (see, for example, Walton (19)), and recent studies with uric acid (1) and calcium phosphates (7) are of particular interest in this context.

The nucleation rate equation for heterogeneous nucleation appears to be of a similar form to that of Eq. 8, but the catalytic effect of heteronuclei reduces the critical supersaturation at which a rapid increase in nucleation occurs and which defines the limit of the metastable zone (see Fig. 3).

There is some evidence that the size of a potential heteronucleus affects its nucleating capabilities, and Walton (18) has suggested that the limited amount of data available indicates the range 0.1 - 1 μm as being the most effective.

SECONDARY NUCLEATION

A number of reviews of secondary nucleation have been published in recent years (2-4,6). These have demonstrated that secondary nucleation can result from several different mechanisms. For example, under certain conditions fluid shear forces are sufficient to produce secondary nuclei from an existing crystal surface (15). Needle breeding sometimes arises at high supersaturations where dendrites may develop or needles grow from existing crystals. If these crystals break, new centers for crystal growth are formed. Contact nucleation, sometimes also referred to as either collision breeding or collision nucleation, arises from contacts between two growing crystals or between a crystal and some other solid surface (6).

The mechanisms by which these various processes take place are not clear. In most cases the hydrodynamic interactions between the solution and the crystal surface are important, as is the nature of the crystal surface itself. Secondary nucleation rates depend on the supersaturation, and Strickland-Constable (14) has suggested that this arises since the initial size distribution of potential secondary nuclei is independent of supersaturation but that the particles are produced in the range of the critical size. The effect of supersaturation is observed since variation of the critical nucleus size with supersaturation allows different proportions of the original nuclei to survive and grow to populate the measured size range. This has since become known as the "survival theory."

Most secondary nucleation studies have been performed in agitated systems such as are usually found in industrial crystallizers. Consequently, their importance and relevance in biological systems is unclear at the present time. It is known, however, that during continuous precipitation of sparingly soluble salts, such as the calcium oxalates, in a crystallizer containing a retained bed of the precipitating phase, nucleation takes place at comparatively low supersaturations (e.g., (8,13)), and secondary nucleation mechanisms may well be important in such cases.

INDUCTION TIMES AND NUCLEATION KINETICS

Perhaps the most common method of investigating nucleation phenomena in solution systems is by the measurement of induction times. A supersaturated solution is prepared, usually by mixing two reactant solutions, and after some time τ crystals are detected. This has in general been known as the induction time, and to some extent, its value depends on the observation and detecting system that is employed. Crystals can usually be seen in the solution before the point at which changes in solution concentration can be measured, these changes resulting from deposition of the solid phase, and Mullin (9) has suggested that this latter time should be called the latent period.

The time that elapses before the detection of crystals depends both on the time taken to form stable nuclei and on the time for their subsequent growth to observable size. It has often been assumed that the induction time can be used as a measure of the nucleation rate, and so, noting that $\tau \propto J^{-1}$, Eq. 8 can be rewritten as

$$\tau \propto \exp\left[\frac{A\,\sigma^3}{T^3 \ln^2 S}\right] . \qquad (10)$$

Therefore, for the case of homogeneous nucleation, a plot of log τ against $(\ln S)^{-2}$ enables the interfacial surface energy σ to be determined.

Nielsen (10) has considered the case when homogeneous nucleation is followed by diffusion controlled growth, and by using order of magnitude estimates of the various parameters, he has again shown how induction times can be used to evaluate σ. Nielsen and Christoffersen (this volume) have also demonstrated that a simple correlation appears to exist between the interfacial energy and the solubility of the corresponding salt in water.

Primary nucleation can often be induced by increasing the degree of agitation in the liquid as well as by other mechanical stimuli (9). The simple relation shown in Eq. 10 must therefore be treated with caution, and doubt must still be cast on much of the literature concerning induction periods. Values of τ are strongly influenced by the presence of impurities and by the crystal growth rate and the relation between homogeneous and heterogeneous mechanisms, induction times, and surface energies remains unclear.

One technique that has recently been used to estimate nucleation rates, especially in systems where secondary nucleation predominates, involves the "mixed suspension mixed product removal" (MSMPR) crystallizer. This is operated in a continuous manner such that solution constantly flows into the crystallizer, where crystals nucleate and grow, while an outlet stream containing these crystals constantly flows from the crystallizer. At steady state conditions it can be shown (12) that the crystal size distribution, both in the crystallizer and in the outlet stream, is given by

$$n = \frac{J}{G} \exp\left(-\frac{L}{G\bar{t}}\right), \qquad (11)$$

where n is the crystal number density, L the crystal size, G the linear growth rate (=dL/dt), and $\bar{t}$ the mean residence time of crystals in the crystallizer. This latter quantity is calculated from

$$\bar{t} = V/Q , \qquad (12)$$

V being the crystallizer volume and Q the volumetric flow rate through the crystallizer. Measurement of the steady state crystal size distribution in such a crystallizer thus allows both the nucleation and growth rates (J and G, respectively) to be determined via Eq. 11. It has been suggested (5) that such a crystallizer is a good model of the urinary tract and the technique has been used to study the crystallization of calcium oxalate from urine-like solutions (8,13).

CONCLUSIONS

Three categories of nucleation have been defined. Homogeneous primary nucleation is the best understood from a theoretical point of view but is unlikely to be encountered in real biological systems. Heterogeneous primary nucleation is probably the most important mechanism in the present context. However, the efficiency of different substrates in catalyzing such nucleation is unclear and the importance of epitaxy is uncertain. Very little is known about secondary nucleation mechanisms and the information that is available is not directly applicable to biological systems. Nevertheless, this process too may be of significance. Finally, it should be noted that experiments designed to elucidate nucleation mechanisms and from which to deduce fundamental parameters are difficult to devise.

REFERENCES

(1) Boistelle, R., and Rinaudo, C. 1981. Phase transition and epitaxies between hydrated orthorhombic and anhydrous monoclinic uric acid crystals. J. Cryst. Growth 53: 1-9.

(2) Botsaris, G.D. 1976. Secondary nucleation: a review. In Industrial Crystallization, ed. J.W. Mullin, pp. 3-22. New York: Plenum Press.

(3) de Jong, E.J. 1979. Nucleation: a review. In Industrial Crystallization 78, eds. E.J. de Jong and S.J. Jancic, pp. 3-17. Amsterdam: North Holland.

(4) Estrin, J. 1976. Secondary nucleation. In Preparation and Properties of Solid State Materials, ed. W.R. Wilcox, vol. 2, pp. 1-42. New York: Marcel Dekker.

(5) Finlayson, B. 1972. The concept of a continuous crystallizer: its theory and application to in vivo and in vitro urinary tract models. Invest. Urol. 9: 258-263.

(6) Garside, J., and Davey, R.J. 1980. Secondary contact nucleation: kinetics, growth and scale-up. Chem. Eng. Commun. 4: 393-424.

(7) Koutsoukos, P.G., and Nancollas, G.H. 1981. Crystal growth of calcium phosphates - epitaxial considerations. J. Cryst. Growth 53: 10-19.

(8) Miller, J.D.; Randolph, A.D.; and Drach, G.W. 1977. Observations upon calcium oxalate crystallization kinetics in simulated urine. J. Urol. 117: 342-345.

(9) Mullin, J.W. 1972. Crystallization, 2nd ed. London: Butterworth.

(10) Nielsen, A.E. 1967. Nucleation in aqueous solutions. In Crystal Growth, ed. H.S. Peister, pp. 419-426. New York: Pergamon.

(11) Ostwald, W. 1897. Studien über die Bildung und Umwandlung fester Körper. Z. Physik. Chem. 22: 289-330.

(12) Randolph, A.D., and Larson, M.A. 1971. Theory of Particulate Processes. New York: Academic Press.

(13) Rodgers, A.L., and Garside, J. 1981. The nucleation and growth kinetics of calcium oxalate in the presence of some synthetic urine constituents. Invest. Urol. 18: 484-488.

(14) Strickland-Constable, R.F. 1972. The breeding of crystal nuclei - a review of the subject. A.I.Ch.E. Symposium Series No. 121. 68: 1-7.

(15) Sung, C.Y.; Estrin, J.; and Youngquist, G.R. 1973. Secondary nucleation of magnesium sulphate by fluid shear. A.I.Ch.E.J. 19: 957-962.

(16) Turnbull, D., and Vonnegut, B. 1952. Nucleation catalysts. Ind. Eng. Chem. 44: 1292-1298.

(17) Volmer, M. 1939. Kinetik der Phasenbildung. Dresden and Leipzig: Steinkopff.

(18) Walton, A.G. 1967. The Formation and Properties of Precipitates. New York: Interscience.

(19) Walton, A.G. 1969. Nucleation in liquids and solutions. In Nucleation, ed. A.C. Zettlemoyer, pp. 225-307. New York: Marcel Dekker.

(20) Zettlemoyer, A.C. 1969. Nucleation. New York: Marcel Dekker.

Biological Mineralization and Demineralization, ed. G.H. Nancollas, pp. 37-77.
Dahlem Konferenzen 1982. Berlin, Heidelberg, New York: Springer-Verlag.

The Mechanisms of Chrystal Growth and Dissolution

A. E. Nielsen and J. Christoffersen
Medicinsk-Kemisk Institut, Panum Instituttet
University of Copenhagen, 2200 Copenhagen N, Denmark

Abstract. Mechanisms of crystal growth and dissolution are generally best studied by means of the kinetic method, that is, by measuring the rate as a function of different parameters (first of all the concentrations) and by comparing the empirical rate laws and the absolute rates measured with theoretical deductions from reasonable hypotheses. In many cases the rates are completely accounted for by transport processes in the solution (diffusion and convection). In other cases processes in the crystal surface are rate-determining - adsorption, surface nucleation, spiral step displacements, and the integration process. The rate of electrolyte crystal growth is often controlled by the partial dehydration of the ions during their integration in the crystal lattice. Also ripening, agglomeration, epitaxy, and inhibition are treated. Future progress will be assisted by measurements of well-defined parameters in simple solutions parallel with experiments in "biological media," and interpretation in terms of precise molecular models.

INTRODUCTION

The use of the word "mineralization" for processes where crystals are formed in a biological system indicates that the crystals are considered as nonbiological elements. This may be reasonable because inside a crystal one will not find the usual biological system with hundreds (or thousands) of different kinds of organic substrate molecules, enzymes, and coenzymes, etc. Within the crystal there are only one or a few kinds of molecules or ions.

A crystal will grow when the constituent material is present at its surface with a certain supersaturation, and it will dissolve when the contacting medium is undersaturated. But unimportant for the crystal growth or dissolution rate is whether the concentration of the constituent material and of the inhibitors is maintained by means of biological processes on, or close to, the crystal surface, or by diffusion or convection. Therefore, biologically important information about the growth and dissolution of crystals can be obtained advantageously with simple nonbiological solutions having only a few, well-known components and can later be applied to in vivo processes.

One of the great advantages of carrying out crystal growth experiments in simple solution with very few components is that information can be obtained concerning the molecular steps in the surface by means of the kinetic method, i.e., by measuring how the rate of the process depends on different parameters, first of all on the concentration. The rate of the elementary growth reaction is proportional to the concentration of the molecular species that reacts. The absolute value of the rate also provides us with important information for identifying the mechanism.

In order to apply the kinetic method successfully, the experimental results should be expressed quantitatively in terms of well-defined parameters which can be treated mathematically according to the different hypotheses that are being tested. The most important step in the interpretation is the mathematical analysis of functional relationships. A certain amount of mathematics is thus needed. In most of the following cases we shall, however, give the necessary equations without proof - readers finding this intellectually dissatisfying are referred to the literature. As this paper deals with methods rather than with facts, we found it relevant to provide the reader with a number of methods problems. The solutions to the problems can be found in the appendix.

CRYSTAL STRUCTURE

An ideal crystal is a piece of matter consisting of identical units arranged in a three-dimensional pattern, which is periodic (repeats itself at a certain displacement) in three directions in space (the three directions are not lying in one plane). The units may be atoms, molecules, ions, or groups of such particles. They are kept in the crystal lattice by covalent, ionic, or hydrogen bonding, or by dispersion forces, etc. In real crystals the units are not always exactly identical, and deviations from strict periodicity called dislocations are common. The periodicity, of course, terminates at the surface of the crystal and is perturbed by the thermal motion (oscillation and possibly rotation of the building units).

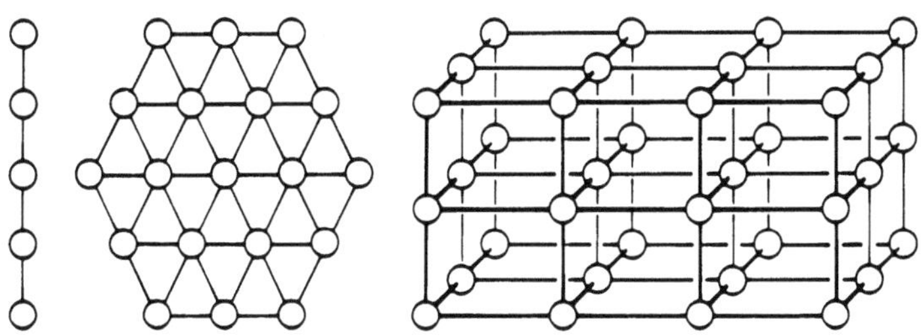

FIG. 1 - One-, two-, and three-dimensional periodic patterns.

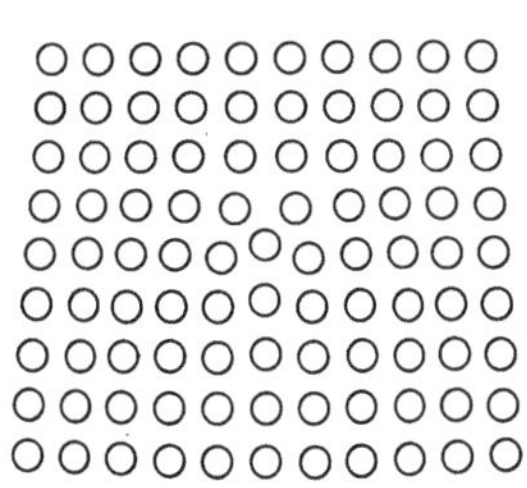

FIG. 2 - Two-dimensional almost periodic pattern with a dislocation.

The (approximate) periodicity makes it possible to determine the inner atomic structure of crystals by X-ray diffraction methods. For many crystals it has been possible to determine the surface structure in vacuum, but the surface structures of crystals in solution are not well-known.

In our examples we shall not strive for a very accurate numerical treatment - in most cases estimates to within a factor of 2 or 5 will be accurate enough to discriminate between the different mechanisms that can be suggested.

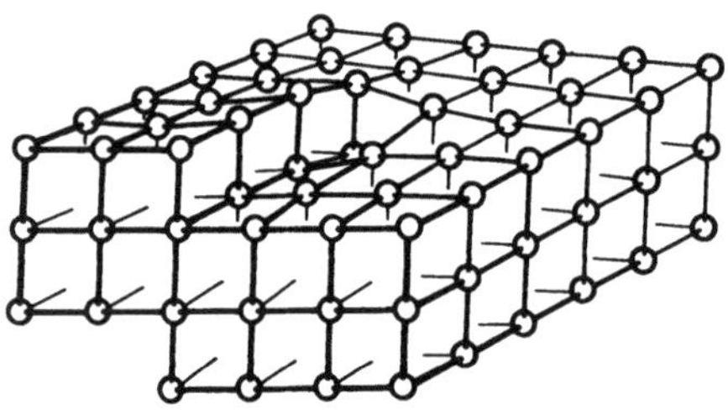

FIG. 3 - A three-dimensional almost periodic pattern with a screw dislocation.

Definition of Growth Rate

The growth rate of a crystal face is defined as the velocity v_g along its normal of that face relative to a fixed point in the crystal. As the growth rate of a crystal we shall use the derivative $\dot{r} = dr/dt$, where r is the radius of a sphere with the same volume as the crystal, and t is the time. In practice we need not distinguish between v_g and $\dot{r}$ unless the crystal is strongly aspheric (a thin plate, a needle, a dendrite, or something similar). For both of v_g and $\dot{r}$ the SI unit is m/s. Growth rates may also be expressed as the amount deposited (or dissolved) per unit surface area per unit time, J (SI unit, mol/m^2s).

Problem 1 A box-shaped crystal grows steadily from the dimension $10.0\times15.0\times20.0\ \mu m^3$ to $11.0\times17.0\times24.0\ \mu m^3$ in 10 seconds. A small bubble in the crystal shows that opposite sides grow equally fast. Find the growth rates v_{g1}, v_{g2}, v_{g3}, and $\dot{r}$. Answer: $v_{g1} = 5.0\times10^{-8}$ m/s; $v_{g2} = 1.0\times10^{-7}$ m/s; $v_{g3} = 2.0\times10^{-7}$ m/s; $\dot{r} = 1.3\times10^{-7}$ m/s.

Problem 2 On a crystal surface, $J = 2\times10^{-3}$ mol per square meter per second. The molar mass is 100 g/mol and the density is 2 g/ml. Find the growth rate v_g. Answer: 10^{-7} m/s.

Problem 3 In a solution of $CaCl_2$ and Na_2CO_3, 0.40 mmol of calcite ($CaCO_3$) per liter has precipitated at a certain time, forming 3100 almost equally large crystals per cm^3. During the next 1.00 minute period the total calcium concentration of the solution decreases by 0.008mM due to the consumption by the growing calcite crystals (ρ = 2.7 g/mol). Find $\dot{r}$. Answer: $\dot{r} = 1.2\times10^{-9}$ m/s.

Problem 4 A paper states that the rate of a precipitation process is $-dC/dt = 0.02$ mM/s when 300 ml reaction mixture contains 10 ml of a slurry of seed crystals, 50 mg/ml with the specific area 10 m^2/g dry substance. The molar mass is 125 g/mol and the density 2.5 g/ml. Find the growth rates v_g and J of a face, and $\dot{r}$ of the crystal. (Assume that the particles are equally large and of cubical shape.) Answer: $J = 1.2\times10^{-6}$ mol/m^2s; $v_g = 6.0\times10^{-11}$ m/s; $\dot{r} = 7.4\times10^{-11}$ m/s.

Crystal Growth

The high degree of structural order at the molecular level in a crystal is thermodynamically explained by being the most stable at the given temperature, pressure, and concentration in the solution. "Historically" it is established by the addition of growth units at growth sites on the crystal surface. Growth units are molecules or ions (or segments of a flexible linear polymer). The growth units may be very large (for instance, protein molecules or even virus particles), but we shall only treat low-molecular substances, primarily electrolytes. (For more comprehensive general treatments, see (1, 2, 8, 9, 10, 11, 12, 13, 14, 16, 19).)

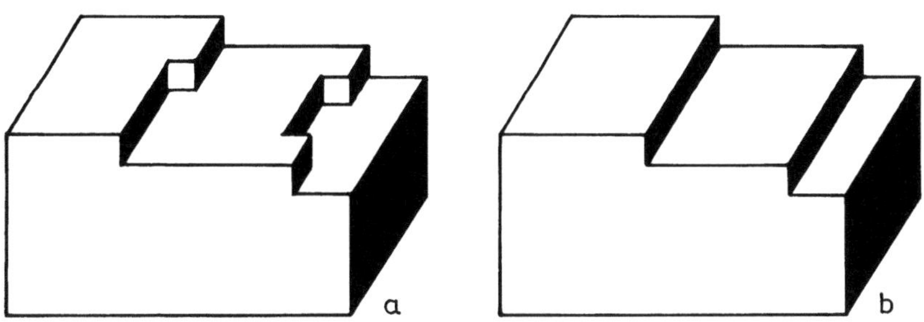

FIG. 4 - a) Steps with kinks (growth sites); b) Growth sites disappear when the kinks are filled up.

The growth sites can structurally be described as kinks in steps on the crystal surface (see Fig. 4a). When the kink reaches the end of the step it disappears (Fig. 4b). However, new kinks will appear spontaneously. If a growth unit is added at a step in a non-kink site (a on Fig. 5), the surface area of the crystal is increased by $2a^2$, where a is the length of the edge of the cube describing the geometry of the growth unit (accurately enough for our purpose), and two kinks are created.

The energy e (work) needed to creat a kink (in a saturated solution) is thus $a^2\sigma$, where σ is the surface tension (assumed uniform on all sides of the growth unit in our primitive model).

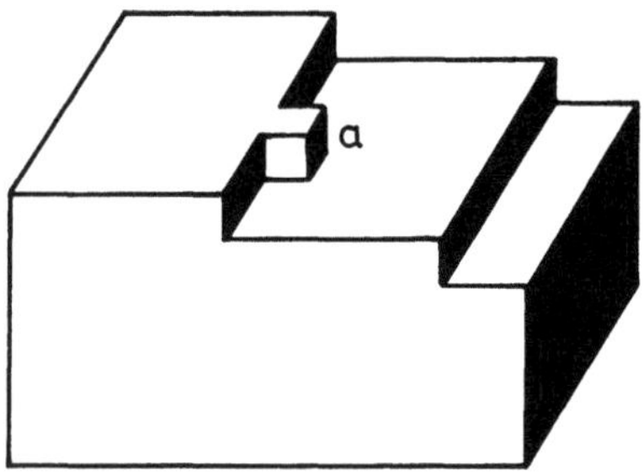

FIG. 5 - A growth unit adsorbed at a step in a non-kink site.

For low-molecular, sparingly soluble electrolytes, $a^2\sigma$ is on the order of 10^{-20} to 2×10^{-20} J, and as the thermal energy kT is 4×10^{-21} J, the ratio e/kT is between 2.5 and 5, the Boltzmann factor exp(-e/kT) is between exp(-2.5) = 0.08 and exp(-5) = 0.007. Thus the probability of having a kink in any chosen position along a step is between 8% and 0.7%, which is far from negligible. Consequently, there will always be an active population of kinks in the steps, and the steps will continue to grow to the edge where they disappear. Will new steps similarily form by adsorption of growth units on a "perfect" surface as shown in Fig. 6? Adsorption of a single growth unit increases the energy by $4a^2\sigma$ corresponding to a Boltzmann probability of between $e^{-10} = 4\times10^{-5}$ and $e^{-20} = 2\times10^{-9}$. For answering the problem the following reasoning is necessary:

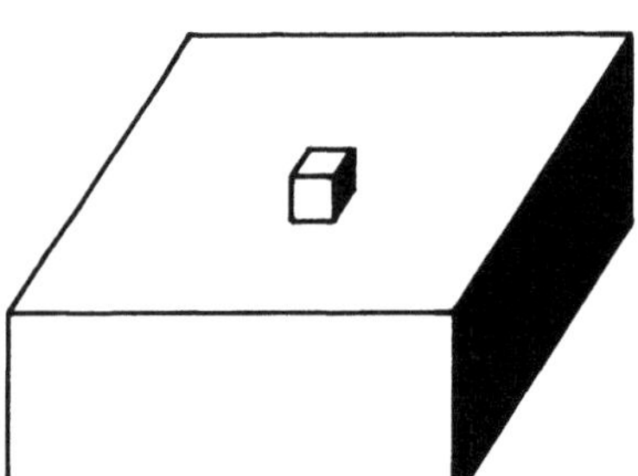

FIG. 6 - A growth unit adsorbed on a perfect surface.

For adsorbed groups of n growth units ("surface nuclei"), the energy is proportional to the edge length and thus to $n^{\frac{1}{2}}$. But ΔG for the formation of a surface nucleus also depends on the supersaturation which we shall express as the <u>saturation ratio</u> $S = c/c_s \equiv$ (concentration)/(solubility). $S>1$, $S=1$, and $S<0$ for supersaturated, saturated, and undersaturated solutions, respectively. For a square surface nucleus

$$\Delta G = 4n^{\frac{1}{2}}a^2\sigma - nkT\ln S. \qquad (1)$$

For $S>1$, ΔG has a maximum at $n = n^*$,

$$n^* = [2a^2\sigma/kT\ln S]^2. \qquad (2)$$

For surface nuclei with $n>n^*$, $d\Delta G/dn < 0$ and thus growth is a natural process. For $n<n^*$ the opposite is true. The maximal, or critical, value of ΔG is

$$\Delta G^* = 4(n^*)^{\frac{1}{2}}a^2\sigma - n^*kT\ln S = 4a^4\sigma^2/kT\ln S. \qquad (3)$$

The probability of having a critical surface nucleus in a specified place on the surface is

$$P = \exp(-\Delta G^*/kT) = \exp[-(2a^2\sigma/kT)^2/\ln S]. \qquad (4)$$

If we let $a^2\sigma/kT = 5$, $S = 1.1$, we find the probability

$$P = \exp(-100/0.1) = \exp(-10^3) = 10^{-(10^3/\ln 10)} = 10^{-434}.$$

which is of course vanishingly small, meaning that the crystals do not grow (by surface nucleation) at 10% supersaturation. For what values of S will the growth rate become measurable? Let us say that we can measure rates of growth corresponding to one monolayer per second on a crystal with a 1 μm^2 large face. The growth units are assumed to be 0.3 nm cubes. The fastest rate of a molecular process (corresponding to activation energy zero) is about 10^{13} s^{-1}. The number of growth units in the crystal face layer is 1 $\mu m^2/0.09$ $nm^2 \approx 1\times10^7$. If one monolayer is added per second, one surface nucleus must form per second. The probability of getting a surface nucleus on a certain site within a period of 10^{-13} second is thus $P = 10^{-13}\times10^{-7} = 10^{-20}$. Solving $\exp(-100/\ln S) = 10^{-20}$, one finds $S = 8.8$.

Problem 5 (a) Which value of the surface tension σ did we use in the calculation of S? (b) How fast is the linear growth rate at S = 8.8 if each surface nucleus gives rise to a complete monolayer before the next surface nucleus is formed? Answer: (a) $\sigma = 0.228\ J/m^2$; (b) $v_g = 3\times10^{-10}$ m/s.

Problem 6 How many growth units (n*) does the critical surface nucleus contain at S = 8.8? Answer: n* = 21.

Spiral Steps

In practise it turns out that most crystals grow at much lower supersaturations than those calculated above for a surface with steps originating from surface nucleation. This indicates that some other, more efficient, step source must be operating. Such a step source is a screw dislocation (2) which emerges on the surface where a step has one of its ends. The other end may be at another screw dislocation of the opposite twist, or at the edge of the face (see Fig. 3). While the crystal grows, the step will wind itself into a spiral with the center fixed at the dislocation. The step does not disappear in growth, and this is why the crystal may grow continuously at very low supersaturation, without the need for surface nucleation.

Homogeneous Nucleation

In a moderately supersaturated homogeneous solution, crystals cannot form because very small crystals are unstable - this is analogous to the phenomena discussed in connection with surface nucleation (see Garside, this volume). For a three-dimensional cubic critical nucleus at the supersaturation S, we have (by analogy to Eq. 2)

$$n^* = (4\sigma a^2/kT\ln S)^3. \tag{5}$$

Some authors approximate the shape with a sphere instead of a cube. In that case

$$n^* = \frac{32\pi}{3}(\sigma a^2/kT\ln S)^3, \tag{6}$$

(still assuming that the volume of the crystal growth unit is $a^3 \equiv V_m/\nu L$; ν is the number of growth units per formula unit, and L is the Avogadro constant). The radius of the critical nucleus is then

$$r^* = 2\sigma a^3/kT\ln S, \quad \text{(three-dimensional critical nucleus)}, \tag{7}$$

and the Gibbs activation energy of homogeneous nucleation is

$$\Delta G^* = \frac{16\pi}{3}\,\sigma^3 a^6/(kT\ln S)^2. \tag{8}$$

The corresponding Boltzmann factor $\exp(-\Delta G^*/kT)$ is vanishingly small for moderate S-values, so that homogeneous nucleation is negligible up to a certain relatively high critical value, and increases rapidly above this value.

The rate of homogeneous nucleation may be estimated as the product of the particle concentration N^* of critical nuclei and a reaction frequency for combining with one more growth unit. N^* may be estimated as

$$N^* = a^{-3}\exp(-\Delta G^*/kT), \tag{9}$$

and the reaction frequency as xD/a^2, where a is a molecular scale length (diameter or jump distance), x the mole fraction of growth units, and D the diffusion coefficient. Thus

$$J = (xD/a^5)\,\exp(-\Delta G^*/kT). \tag{10}$$

The preexponential factor is on the order of $10^{36\pm3}\ m^{-3}s^{-1}$, and measurable nucleation rates are between $10^{6\pm3}\ m^{-3}s^{-1}$. Thus $\exp(\Delta G^*/kT) \approx 10^{30\pm5} = \exp(70\pm10)$. Inserting $\Delta G^*/kT = 75$, and kT in Eq. 8 we obtain

$$a^6\sigma^3/(\ln S)^2 = \frac{3(\Delta G^*/kT)(kT)^3}{16\pi} = 3\times10^{-61}\ J^3. \tag{11}$$

It is normally impossible to measure the surface tension of a solid by macroscopic methods. But from precipitation experiments the critical supersaturation for homogeneous nucleation can be found for many crystals. Therefore, this is a valuable method for determining σ through Eq. 11 - or more sophisticated expressions developed from the same basic assumptions. The σ values obtained are correlated with the solubilities (14),

$$a^2\sigma = [11.6 - 1.12\ \ln(c_s/\text{mM})]\times10^{-21}\ J. \tag{12}$$

Heterogeneous Nucleation

When a series of solutions are prepared with increasing supersaturations, one usually finds that below a certain value of $S(1<S<S_1)$, the solutions are stable for an appreciable period of time (hours or days). More concentrated solutions $(S_1<S<S_2)$ develop crystals, although the number of crystals is irreproducible and depends more on the cleaning of the vessel and of the solutions than on the concentrations. But when S exceeds S_2, the number of crystals formed increases very strongly with

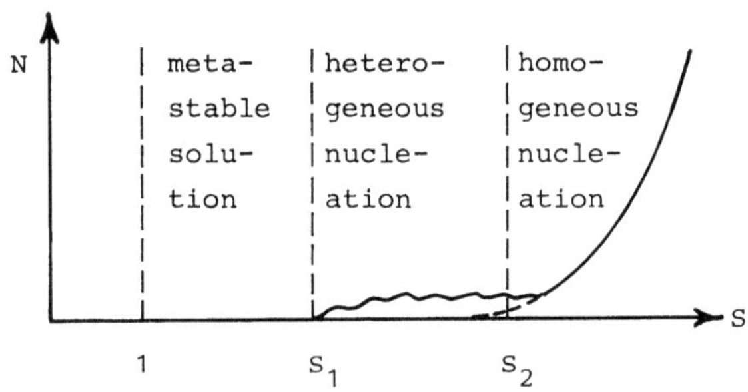

FIG. 7 - Increasing supersaturations lead through intervals of metastability, heterogeneous nucleation, and homogeneous nucleation.

S - typically by the factor 10^3 to 10^6 within less than one decade of S values. S_1 is called the metastability limit, and S_2 the limit for homogeneous nucleation. The crystals appearing between S_1 and S_2 were nucleated by means of heterogeneous impurities (dust particles) which are normally present in all liquids not particularly clean (10) (see also the section on "Epitaxy").

Growth Controlling Mechanisms and Kinetics

The growth rate may be determined by processes in the solution (chemical reactions and transport from the bulk solution to the crystal) and by processes on the crystal surface. With bulk solution concentration = c, concentration in the solution at the crystal surface (but outside any adsorption layer, etc.) = c', and the solubility = c_s, we have the following three possibilities (see also Fig. 8):

a) Transport control, $c' = c_s$;

b) General case, $c_s < c' < c$; and

c) Surface control, $c' = c$

Transport Control

In many cases - especially with large crystals and high concentrations - the growth rate is determined by the transport (by convection or diffusion, or both) through the solution surrounding the crystal. When this is the case, measurements of the growth rate cannot give any information about the processes on the surface - except that they are relatively fast.

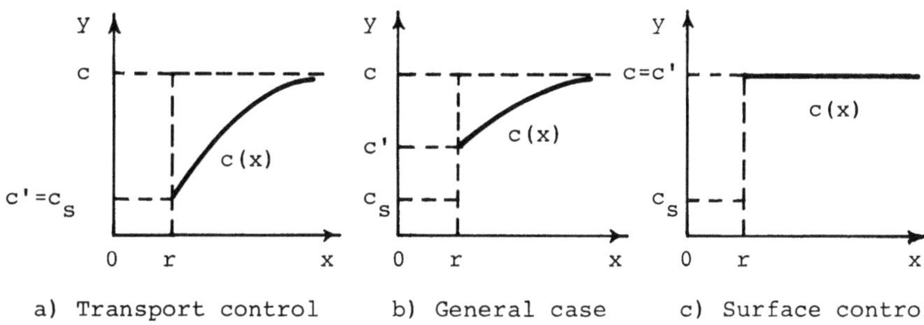

FIG 8 - x = distance from crystal center (assuming spherical shape). y = concentration.

In the absence of liquid flow, transport is by diffusion only, and the growth rate of a (spherical) crystal containing only one kind of growth unit is (10,13)

$$\dot{r} = \frac{DV_m \Delta c}{r}, \tag{13}$$

with D = diffusion coefficient, V_m = molar volume, and $\Delta c = c - c_s$ = excess concentration, relative to solubility. The same equation is valid for diffusion controlled dissolution ($\dot{r} < 0$, $c < c_s$). For binary, ternary, etc., compounds (such as electrolytes) the equation becomes more complicated.

For a binary electrolyte AB with the ionic concentration a and b, the solubility product K_s, and the ionic diffusion coefficients D_A and D_B,

$$\dot{r} = (V_m/2r)[aD_A + bD_B - \sqrt{(aD_A - bD_B)^2 + 4D_AD_BK_s}]$$

$$\approx (DV_m/2r)[a + b - \sqrt{(a - b)^2 + 4K_s}], \quad (14)$$

where the approximation assumes $D_A = D_B$ ($\equiv$ D) (13). For all low-molecular ions (except H^+ and OH^-) in aqueous solution at 25 ^oC, $D_A \approx D_B \approx 10^{-9}$ m^2/s. As we can always calculate $\dot{r}$ for a diffusion controlled crystal growth process from easily accessible data (such as r and Δc, or r, a, b, and K), there should never be any doubt whether or not a measured rate is diffusion controlled when convection influence can be excluded, i.e., when the system is mechanically stagnant (e.g., a gel) or the crystals are smaller than about 10 µm.

Problem 7 Show that Eq. 14 transforms to Eq. 13 for a = b, if $D_A = D_B = D$.

Problem 8 At $[Ba^{2+}] = [SO_4^{2-}] > 3\times10^{-4}$ M, barium sulfate crystals grow approximately by a diffusion controlled rate. Find $\dot{r}$ for $[Ba^{2+}] = [SO_4^{2-}] = 10^{-3}$ M, r = 2 µm. (The molar mass of $BaSO_4$ is 0.233 kg/mol, the density is 4.5 kg/dm^3, and the solubility product 1×10^{-10} mol^2/l^2).
Answer: $\dot{r} = 3\times10^{-8}$ m/s.

Problem 9 At $[Ca^{2+}] = [CO_3^{2-}] = 3\times10^{-4}$, calcite crystals with r = 4 µm grow by the rate, 0.01 nm/s. Is this rate determined by convection (by sedimentation), diffusion, or by a surface process? ($V_m = 3.7\times10^{-5}$ m^3/mol; $K = 10^{-8}$ mol^2/l^2). Answer: The rate is controlled by a surface process.

Equation 13 shows that in the same solution the smaller particles grow faster than the larger ones when the growth rate is diffusion controlled. We also know that small particles sediment more slowly than large particles. It follows that particles below a certain size (actually 5-10 µm, depending on the density and shape) will grow by diffusion control even at moderate stirring as if they did not move in the solution. But when a particle is larger than 10-20 µm, the flow of solution relative to the particle will bring solution of the full bulk concentration close to the particle, and thus make it grow faster.
For particles from r = 20 µm to a few millimeters, the rate may be calculated as if the liquid keeps the concentration equal to the bulk concentration right down to a distance, δ, from the surface, on the order of 10-100 µm, depending on the "stirring rate" - the liquid remaining unstirred within the

layer of thickness δ. This model is a simplification (as all models are).

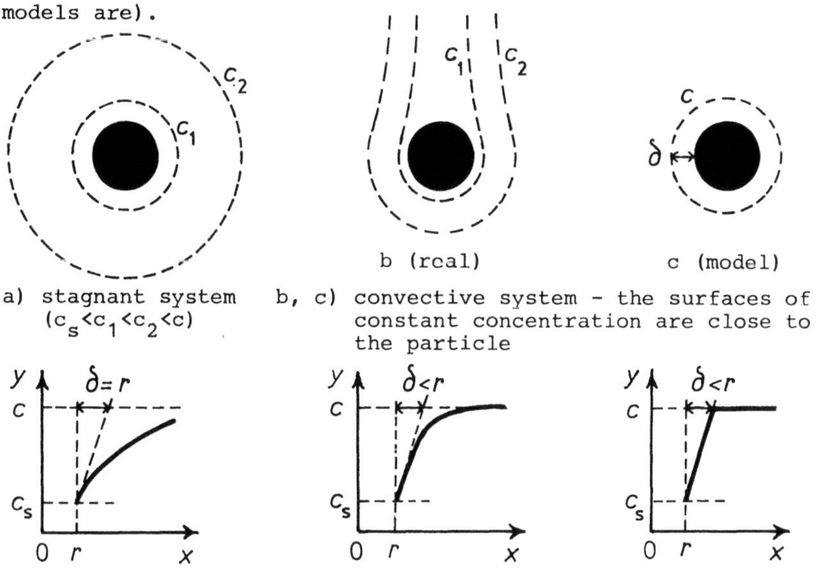

a) stagnant system ($c_s < c_1 < c_2 < c$)

b, c) convective system - the surfaces of constant concentration are close to the particle

d) stagnant diffusion

e) convective diffusion; realistic picture

f) convective diffusion; model for calculation of $\dot{r}$

FIG. 9 - Concentration profiles close to a falling particle (transport control).

The liquid will of course move relative to the particle also within this layer. However, δ is a convenient fitting parameter, and the model provides us with the simple equation

$$\dot{r} = \frac{DV_m \Delta c}{\delta} \quad . \tag{15}$$

For small spheres δ may be calculated quite accurately (12,13).

Problem 10 According to Fick's first law, the current of matter (SI unit, mol/s) through a surface of area, A, is DAdc/dx, where dc/dx is the concentration gradient normal to the surface. For stagnant diffusion controlled growth of a sphere $dc/dx = (c-c_s)/r$. Derive Eq. 13.

Problem 11 Derive Eq. 15 for the "unstirred layer model."

Problem 12 Which of the Eqs. 13 or 15 would you use to calculate $\dot{r}$ for a particle growing by a transport controlled rate if a) r = 1 μm? b) r = 1 mm? c) In which of the two cases would you expect the stirring intensity to have influence on the crystal growth rate?

Surface Control

Although the solute must always be transported from the bulk solution to the crystal surface, this is not always the rate-determining step. At the surface the growth units may (a) get adsorbed, (b) move along the surface, or (c) integrate into the crystal lattice at a growth site, i.e., a kink in a surface step. Each of these elementary processes may be rate-determining. For electrolyte crystals growing from aqueous solution, most evidence has been found for rate control by (a) adsorption and (c) integration; (b) surface migration may be more or less concealed by the other mechanisms. Growth units (ions) constantly bombard the crystal surface and are attracted by the electric charges of the ions on the surface. Therefore an adsorption layer is being formed and constantly renewed. The ions are hydrated in the solution and must give off some of their hydration water molecules in order to get close to the ions already integrated on the surface. As the water molecule leaves the ion, it passes through an activated state, and the needed activation (Gibbs, or free) energy is several times the thermal energy (kT). The activation Gibbs energy ($G_w^{\ddagger}$) for giving off a water molecule from the hydration shell is normally higher for cations than for anions. Furthermore, the ion must penetrate the hydration layer of the crystal which is stabilized by similar forces (ion-dipole interaction) as the hydration shell of the single ions, and also possibly by water structural forces (hydrogen bonding, etc.).

Therefore, the adsorption rate may be controlled by the partial dehydration of the cations or by the penetration through the hydration layer of the crystal. In Table 1 we have collected some activation energies for the dehydration of cations. These data were obtained by means of high frequency sound adsorption (17). The Gibbs activation energies were calculated from the relaxation time τ or from $\nu = \tau^{-1}$ by means of the Eyring equation (with $\nu = k_w$)

$$G^{\ddagger} = kT\ln(kT/h\nu) \qquad (16)$$

($kT = 4.12\times10^{-21}$ J; $kT/h = 6.21\times10^{12}$ s^{-1} at 25 °C).

TABLE 1 - Frequencies (k_w or ν) and Gibbs activation energies for release of a water molecule from the hydration shell.

Ion	Li^+	Na^+	K^+	Be^{2+}	Mg^{2+}	Ca^{2+}	Sr^{2+}	Ba^{2+}
$\log(k_w/s^{-1})$	8.2	8.6	9.0	2.3	4.7	8.2	8.7	9.2
$G_w^{\ddagger}/10^{-21}$ J	43	40	36	99	77	43	39	34
Ion	Al^{3+}	La^{3+}	Zn^{2+}	Ni^{2+}	Co^{2+}	Fe^{2+}	Mn^{2+}	Cu^{2+}
$\log(k_w/s^{-1})$	0.4	7.8	6.8	4.3	5.1	6.2	6.7	8.6
$G_w^{\ddagger}/10^{-21}$ J	118	47	57	81	73	62	58	40

When the rate of crystal growth is determined by the adsorption of the cations, the anions following without delay in stoichiometric ratio, we may calculate the number of ions entering the adsorption layer per ionic site (not growth site - any site on the surface) as the product of the probability that an ion is present in the liquid just above the site and the frequency of reaction. The first factor is the mole fraction $c_A V_w$ and the second factor the rate constant for adsorption k_{ad}. The current of ions entering per site is then

$$j_+ = k_{ad} c_A V_w. \tag{17}$$

Adsorbed ions also have a chance of escaping into the solution which gives a reverse current j_-. This is independent of c_A. The adsorption layer is in equilibrium with the crystal in the discussed case because the adsorption is assumed to be the only rate limiting elementary reaction step. At solubility equilibrium $j_+ = j_-$ which gives $j_- = k_{ad} c_{A,eq} V_w$, and thus the net current is

$$j \equiv j_+ - j_- = k_{ad}(c_A - c_{A,eq}) V_w. \tag{18}$$

Each time an ion is adsorbed per ionic site, the surface is displaced by one ionic diameter. Consequently, the linear growth rate is

$$v_g = aj = k_{ad} a (c_A - c_{A,eq}) V_w. \tag{19}$$

For a stoichiometric solution of the electrolyte $A_\alpha B_\beta$, we have $c_A = \alpha c$, and using $c = Sc_s$ we find

$$v_g = k_{ad} a\alpha c_s V_w (S-1) = k_1 (S-1). \quad (20)$$

Equation 20 is a linear rate law, since a plot of v_g as a function S (or c) is a straight line. By inserting empirical values of v_g, a, α, c_s, V_w, and S, one may find k_{ad} and may calculate $G^{\ddagger}_{ad}$ by inserting $\nu = k_{ad}$ in Eq. 16. The very few data treated in this way so far (unpublished) are presented in Table 2. (The actual rates measured were typically on the order of 1/10 of the rates calculated for transport control).

TABLE 2 - Activation Gibbs energies of adsorption.

Crystal:	Ag_2MoO_4	$NH_4H_2PO_4$	$MgSO_4 \cdot 7H_2O$	$Ni(NH_4)_2(SO_4)_2 \cdot 6H_2O$	$Na_2S_2O_3 \cdot 5H_2O$
$G^{\ddagger}_{ad}/10^{-21}$ J	72	74	68	76	74

These values are generally similar to, or slightly larger than, the values for the removal of a water molecule from the hydration shell of the cation (Table 1). But they are much closer to each other than to the different values of Table 1. Maybe this is because the rate-determining mechanism is the penetration of the hydration layer of the crystal rather than the removal of a water molecule from the ion. We have no basis today for proceeding further with this discussion.

Problem 13 For crystallization of $MgSO_4 \cdot 7H_2O$ at 30 °C, Mullin (7) reports that $v = 8.2\times10^{-8}$ m/s at S = 1.01 and $v = 1.5\times10^{-7}$ m/s at S = 1.02. Verify that this substance obeys the linear rate law and find $G^{\ddagger}_{ad}$. (Molar mass = 246 g/mol; density = 1.68 g/cm^3; solubility = 2.4 M; V_w = 18 ml/mol).

In many cases the growth rate follows a parabolic rate law

$$v_g = k_2 (S-1)^2, \quad (21)$$

or an "exponential rate law" of the form,

$$v_g = A\exp(-B/\ln S). \quad (22)$$

(Theoretically A should be proportional to $(S-1)^{5/6}$, see Eq. 29).

Both these rate laws may be explained by assuming that it is the integration of the growth units into kinks or "adsorption" at steps (with following integration at kinks) that is rate-determining. In the parabolic case the step source is screw dislocations, giving rise to spiral steps on the surface. The distance y_o between two consecutive steps (see Fig. 10) can be shown to be very close to 19 times the radius of a critical circular surface nucleus. A critical square surface nucleus containing n* growth units has the edge length $a\sqrt{n^*} = 2a^3\sigma/kT\ln S$, taking n* from Eq. 2.

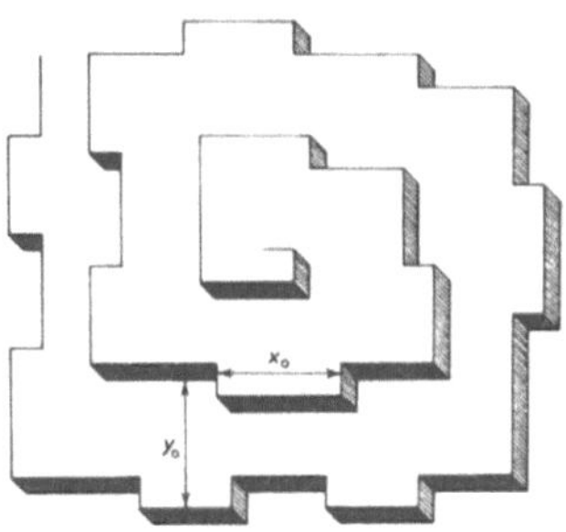

FIG. 10 - Growth spiral step with kinks.

A circular critical nucleus can be shown to be just the inscribed circle in a square surface nucleus. It follows that its radius will be

$$r^* = a^3\sigma/kT\ln S, \text{ (two-dimensional critical nucleus)}, \quad (23)$$

and thus the step distance of a growth spiral is

$$y_o = 19a^3\sigma/kT\ln S. \quad (24)$$

The density of the steps y_o^{-1} is thus proportional to lnS, and if the steps move laterally with a rate proportional to S-1 (due to "adsorption" of growth units along the edge), the total rate of growth will be proportional to (S-1)lnS. For S-values close to unity we have the approximation $\ln S \approx S-1$, and thereby a <u>parabolic rate law</u>. In some cases we have found the parabolic law obeyed far outside the range where the approximation is valid (1<S<ca. 2). This is in agreement with the assumption

that it is not the adsorption at steps but the integration at kinks that is rate-determining. The distance between kinks along a step is

$$x_o = a\ S^{-\frac{1}{2}}\exp(a^2\sigma/kT). \tag{25}$$

Consequently, the density (or surface concentration) of kinks on a surface is proportional to $S^{\frac{1}{2}}\ln S$, and the rate will be proportional to $(S-1)S^{\frac{1}{2}}\ln S$. This mathematical function happens to be almost proportional to $(S-1)^2$ in the interval $1<S<30$, which explains the validity of the "parabolic" rate law even for large supersaturations. In order to derive a theoretical expression for the factor k_2 in Eq. 21, and thereby for the absolute rates in this group of crystal growth processes, several estimates of questionable accuracy have to be made. Probably the best estimate (to within a factor of 10 in v_g) is (14)

$$k_2 = \frac{k_i c_s kTK_I M(a,b)}{a\sigma\exp(a^2\sigma/kT)}, \tag{26}$$

where k_i is the integration rate constant (s^{-1}), K_I the ion pair equilibrium constant (m^3/mol), and $M(a,b)$ a dimensionless factor depending on the ionic type of the electrolyte ($A_\alpha B_\beta \rightleftharpoons \alpha A^{a+} + \beta B^{b-}$,) $M(1,1) = 4\times10^{-3}$; $M(1,2) = M(2,1) = 5\times10^{-4}$; $M(2,2) = 2\times10^{-4}$. Ion pair equilibrium constants have been measured in many cases (18). For ion pairs where K_I has not been measured, one may estimate K_I (to within an order of magnitude) for ions that do not form covalent bonds (i.e., excluding silver ions): $K_I(1,1) = 1.5\times10^{-3}$ m^3/mol, $K_I(1,2) = K_I(2,1) = 0.06$ m^3/mol, and $K_I(2,2) = 1$ m^3/mol. If these values are inserted in Eq. 26, one obtains

$$k_2 = \frac{k_i c_s kTN(a,b)}{a\sigma\exp(a^2\sigma/kT)}, \tag{27}$$

with $N(1,1) = 6\times10^{-6}$ m^3/mol; $N(1,2) = N(2,1) = 3\times10^{-5}$ m^3/mol; and $N(2,2) = 2\times10^{-4}$ m^3/mol. When empirical values of k_2 are inserted and k_i calculated from Eq. 26 (or 27), and the corresponding activation Gibbs energy is determined by Eq. 16, the resulting $G_i^{\ddagger}$ values normally show a good correlation with the $G_w^{\ddagger}$ values of the actual cation (Table 1), which may be taken as an indication that it is the dehydration of the cation that is determining in the rate of integration at the growth site. The ratio $G_i^‡/G_w^‡$ is usually about 1.5, meaning that 50% more than the energy needed for the removal of a water molecule from an ion is required for integration of the ion. Considering the sterics of the integration situation, it is easy to understand that a rather serious distortion of the hydration shell must take place.

Problem 14 For K_2SO_4 at 20 °C, Mullin (8) reports that v_g is 2.8×10^{-8} m/s at S = 1.09 and 1.4×10^{-7} at S = 1.18. a) Verify that potassium sulphate crystal growth follows a parabolic rate law, and b) find k_2, k_i, $G_i^‡$, and $G_i^‡/G_w^‡$. (a = 3.3Å, c_s = 0.65 M, $\sigma \approx 40$ mJ/m^2, K_I = 0.010 m^3/mol). Answer: a) (80% agreement in $v/\Delta S^2$); b) $k_2 = 3.9\times10^{-6}$ m/s; $k_i = 1.1\times10^7$ s^{-1}; $G_i^‡ = 5.5\times10^{-20}$ J; $G_i^‡/G_w^‡ = 1.53$.

Problem 15 Find n^*, r^*/a, y_o/a, and x_o/a for K_2SO_4 at 20°C, S = 1.09. Answer: $n^* = 600$; $r^*/a = 12$; $y_o/a = 230$; $x_o/a = 1.1$.

In some cases, the growth rate varies more strongly with S than $(S-1)^2$, and fitting a power law $(S-1)^n$ leads to a value of n varying (e.g, through 3-5) with S. These cases are explained by assuming that the step source is surface nucleation. Very small crystals (< 100 nm) without dislocations will grow very slowly at moderate supersaturation because the probability per unit of time that a surface nucleus shall appear is proportional to the surface area - and any time a surface nucleus is formed it will give rise to only one monolayer ("Mononuclear growth"). With increasing size (or supersaturation), the nucleation frequency increases, and the growth rate accelerates. When new nuclei are formed, before all the older ones have reached the edge, several nuclei intergrow in each layer, and the growth rate is no longer size-dependent. This is called polynuclear (10), birth-and-spread (16), or nuclei-above-nuclei growth.

The rate of surface nucleation (nuclei formed per unit area in unit time; see also Eq. 4) can be expressed as

$$J = (\Gamma L v_+/a)\exp(-\Delta G^*/kT), \tag{28}$$

where Γ is the surface concentration (mol/m^2) of growth units,

L the Avogadro constant (ΓL = the number of adsorbed growth units per unit area), and v_+ the lateral velocity of a surface step corresponding to the forward elementary reaction only. [The net lateral velocity is $v_\infty = v_+ - v_- \approx v_-(S-1) = v_+(1-S^{-1})$.] $\Delta G^* = \pi a^4\sigma^2/kT\ln S$ (Eq. 3, adapted to a circular surface nucleus). For polynuclear growth

$$v_g = (a^2\Gamma L)^{1/3} v_+ (S-1)^{5/6} \exp(-\Delta G^*/3kT). \tag{29}$$

$a^2\Gamma L$ may be interpreted as the mole fraction of growth sites within the adsorption layer. This is always larger than in the solution, and the exponent 1/3 brings $(a^2\Gamma L)^{1/3}$ close to unity.

If the concentration in the adsorption layer approaches that in the crystal (a^{-3} growth units per volume unit), the fastest lateral rate v_+ will not exceed aj_+ = (a)(area)(diffusion coefficient)(concentration gradient) = $(a)(a^2)(D)(a^{-3}/a)$ = $D/a \approx (10^{-9}m^2/s)/(3\times10^{-10}m) \approx 3$ m/s. The slowest rate v_+ (when inhibition by non-constituent ions is absent) will be obtained if the integration at kink sites is rate-determining. The density of kinks along a step is $a/x_o = S^{\frac{1}{2}}\exp(-a^2\sigma/kT)$, Eq. 25. The integration rate (growth units per unit time per kink) may be expressed as $j = \nu_o \exp(-G_i^{\ddagger}/kT)x_{ad}$, $\nu_o = kT/h = 6.21\times10^{12}s^{-1}$ at 25 °C. With $x_{ad} = x_{ad,s}S$ we find

$$v_+ = a(a/x_o)j = a\nu_o x_{ad,s} S^{3/2} \exp(-a^2\sigma/kT)\exp(-G_i^{\ddagger}/kT) \tag{30}$$

$[a\nu_o \approx (3\times10^{-10}m)(6.2\times10^{12}s^{-1}) = 2000 \text{ m/s}]$

Approximating $a^2\Gamma L \approx x_{ad,s} \approx 1, S \approx 1$, we find, respectively,

with surface diffusion control (fastest):

$$v_g = (D/a)(S-1)^{5/6}\exp[-\pi a^4\sigma^2/3(kT)^2\ln S] \tag{31}$$

with integration rate control (slowest):

$$v_g = a\nu_o(S-1)^{5/6}\exp[-(a^2\sigma + G_i^{\ddagger} + \pi a^4\sigma^2/3kT\ln S)/kT] \tag{32}$$

Problem 16 Let $a^2\sigma = 3kT$, $G_i^{\ddagger} = 10$ kT and $v_g = 10$ nm/s. Find the corresponding value of S from Eq. 31 and Eq. 32. Answer: S = 1.6 and 2.0, respectively.

Equations 31 and 32 also apply when a dissolution process is controlled by surface nucleation. The surface nuclei are in this case holes in the top layer of the crystal. S is less than unity, and $(S-1)^{5/6}$ should be replaced by $-(1-S)^{5/6}$. In the exponential part lnS should be substituted by $-\ln S = |\ln S|$.

For analyzing a series of (S, v_g) data, one may plot $\ln[v_g/(S-1)^{5/6}]$ (or simply $\ln v_g$) as a function of $1/\ln S$ and see if a straight line is obtained. If so, σ may be obtained from the slope. An analysis of the preexponential factor will only be meaningful with very accurate rate data.

The Interplay Between Different Growth Processes

The different mechanisms suggested for rate control of crystal growth all have as an essential element the molecular jump of a growth unit - either in the solution (diffusion), at the surface (adsorption), at the step (surface diffusion), or at a kink (integration). The activation energy ranges from that of diffusion (about 6kT) to about 1.5 times the activation energy of ion dehydration (about 15 to 25 kT). The low-energy mechanisms (diffusion) take place in a relatively dilute solution, whereas the high-energy mechanisms occur in the crowded surface layer. Therefore, the rates of the different mechanisms will often be comparable, and the rate control may go from one mechanism to another upon change of concentration or particle size. In order to understand such changes, one must remember

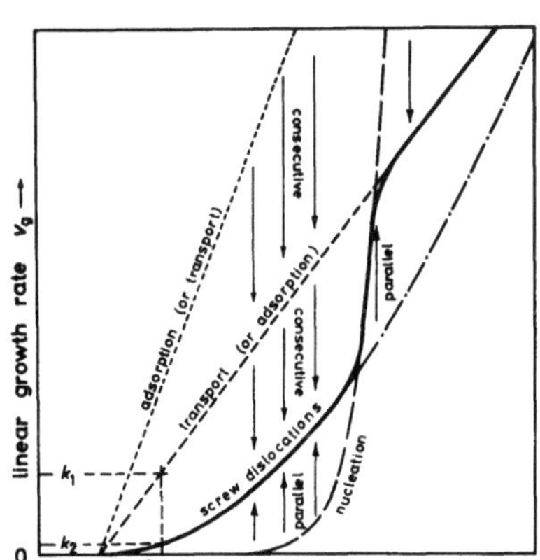

FIG. 11 - The competition for rate control. The winning mechanism is the slower (↓) or faster (↑) one for consecutive and parallel pairs of mechanisms, respectively.

that when two consecutive mechanisms "compete," the slower one (at equal driving force) "wins," and with two parallel mechanisms the faster one wins. The relationships are:

Consecutive: Transport → surface adsorption → integration

Parallel: Screw dislocation steps/surface nucleated steps

What has been said here can probably also be adapted to dissolution processes. However, one should be aware that many crystals that grow by a surface controlled process dissolve by a transport controlled process because steps are formed at the edges of the crystal. However, some crystals do dissolve by a surface controlled process.

Problem 17 Draw and discuss a diagram of the type in Fig. 11, where the point of intersection between the parabolic and the exponential curve is situated above the linear curves.

Problem 18 The following growth rates were measured for calcium fluoride in aqueous solution at 25 °C (interpolated) (unpublished):

$c \equiv [Ca^{2+}]/mM = \frac{1}{2}[F^-]/mM$ =	0.48	0.70	1.10	1.58	2.95	6.31
$\dot{r}/ms^{-1}$ =	10^{-12}	10^{-11}	10^{-10}	10^{-9}	10^{-8}	10^{-7}

The rate was found to be independent of the particle size. The solubility of CaF_2 is 0.25 mM and $a = 2.4\times10^{-10}$ m. Analyze the data.

Integration of Rate Laws

If an equation for the size r as a function of the time t is wanted, the rate law $\dot{r} \equiv dr/dt = f(r,c)$ must be integrated. When $\dot{r}$ is independent of the size and the concentration is constant,

$$r-r_o = \int_o^t f(c)dt = f(c)\cdot t. \tag{33}$$

If $\dot{r} = f_1(r)f_2(c)$, we may separate the variables and integrate

$$\int_{r_o}^{r} \frac{dr}{f_1(r)} = \int_o^t f_2(c)dt. \tag{34}$$

For diffusion controlled growth $\dot{r} = DV_m\Delta c/r$, and we find for growth at constant c (using $\int rdr = \frac{1}{2}r^2$)

$$r - r_o = \sqrt{2DV_m(\Delta c)t}. \tag{35}$$

Similar equations are valid for dissolution. If the concentration is changing during the process - which is normal in most laboratory experiments where the reacting solutions are depleted through the consumption of the solute by the growing crystals - the mathematical integration becomes more complicated (10).

Problem 19 A spherical particle of radius 1 mm, molar volume 100 cm^3/mol, and solubility 1 mol/l is placed in pure water and dissolves by a transport controlled rate. How long will it take to dissolve the particle if a) the liquid is stagnant (the particle lies on the bottom, or (better) the water is immobized by 1% gelatine), and b) the water is gently stirred? (Use $\delta \cong 10$ µm for $r > 10$ µm). Answer: a) 5000 s; b) 100 s.

Nonstoichiometric Solutions - Isotachs

When the concentration of the ions of an electrolyte deviate from stoichiometry, the "solubility" cannot be expressed in terms of a single concentration of an actually present species. The condition for saturation for a binary electrolyte of ionic type $A_\alpha B_\beta \rightleftharpoons \alpha A^{a+} + \beta B^{b-}$ is of course well-known - the activity ionic product should be equal to the solubility product

$$\Pi_a \equiv (A^{a+})^\alpha (B^{b-})^\beta = K_s. \quad (36)$$

Whereas the solubility c_s of a nonelectrolyte at a given temperature, pressure, etc., may be indicated graphically by a point on an axis, the "solubility" of a binary electrolyte corresponds to a line in a diagram with two axes, such as the activities (A^{a+}) and (B^{b-}) or the concentrations $[A^{a+}]$ and $[B^{b-}]$. The curve

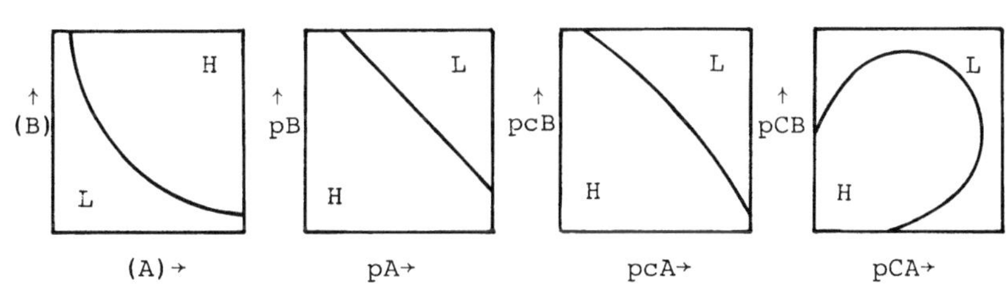

FIG. 12 - The solubility curve of an electrolyte in four different A,B-diagrams. L and H indicate the corners with the lowest and the highest concentrations, respectively.

in the first diagram will be an exact hyperbola (if a = b), and with [A] and [B] it will resemble a hyperbola. In the corresponding logarithmic diagrams with the axes pA, pB or pcA, pcB (pA ≡ $-\log\{(A^{a+})/\text{mol l}^{-1}\}$, pcA ≡ $-\log\{[A^{a+}]/\text{mol l}^{-1}\}$), the curves are straight lines with the slope $-\alpha/\beta$ and weakly bent curves, respectively. It is difficult to plot experimental data in these two kinds of diagram because activity coefficients, ion pair concentrations, and complex ion concentrations must be known. In practice we shall therefore prefer to plot curves in a pCA, pCB diagram, pCA = $-\log\{C_A/\text{mol l}^{-1}\}$, where $C_A \equiv [A]_{st}$ is the stoichiometric (or "total") A concentration. Dropping for simplicity the charge symbols (a+, b-, etc.) we have

$$C_A \equiv [A] + [AB] + 2[A_2B] + [AB_2] + [AX] + 2[A_2Y] + \dots, \tag{37}$$

where AB, A_2B, ... AX, etc., are ion pairs and complexes with constituent and non-constituent ions.

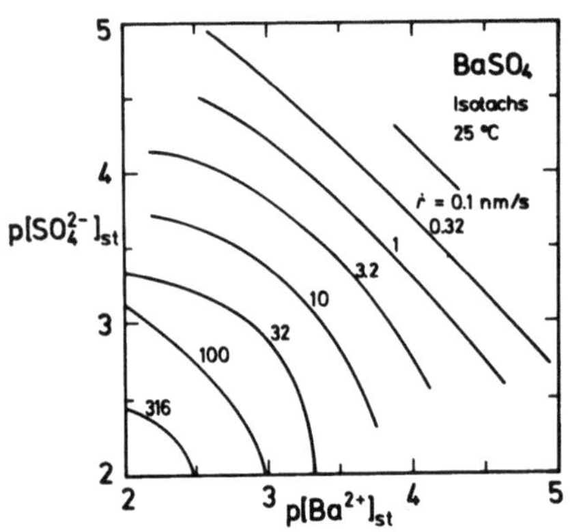

FIG. 13 - Barium sulphate isotachs (curves for constant growth rate) in a logarithmic diagram. The predominant slope is -1 corresponding to constant values of the ionic product, $[Ba^{2+}][SO_4^{2-}]$.

For an electrolyte the supersaturation is defined

$$S \equiv (\Pi_a/K_s)^{1/\nu}, \tag{38}$$

where Π_a is the activity ionic product, e.g., $\Pi_a = (A)^{\alpha}(B)^{\beta}$ for $A_\alpha B_\beta$, and ν is the number of ions per formula unit, $\nu = \alpha + \beta$. Each value of S corresponds to a curve in each of the diagrams of Fig. 12. As the affinity of the crystallization process ($-\Delta G$ per mole of growth units) equals RTlnS, the curves for constant S-values are <u>isergons</u>. The solubility curve is a special case of isergon corresponding to $S = 1$ or $-\Delta G = 0$.

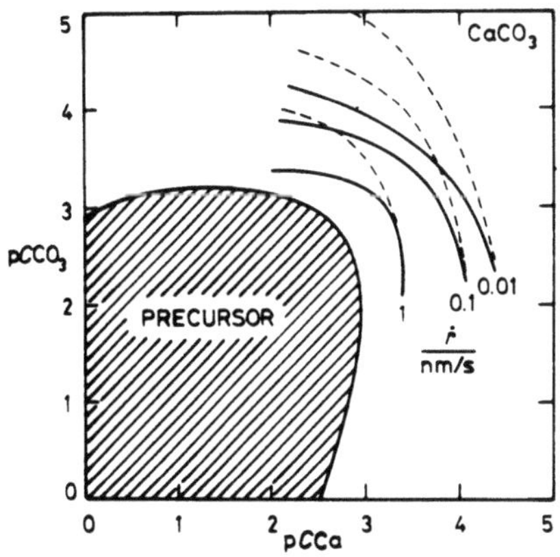

FIG. 14 - Isotachs for calcite crystals. In the hatched area a precursor (unstable hydrated calcium carbonate) is formed prior to calcite. (Broken lines: the ordinate does not include HCO_3^- and CO_2).

Empirical data on electrolyte growth rates may be presented in the form of an A,B diagram, if possible as <u>isotachs</u> - curves for constant growth rates (v_g, $\dot{r}$ or J). For surface controlled rates where $\dot{r}$ is a function of C_A and C_B only a vast amount of

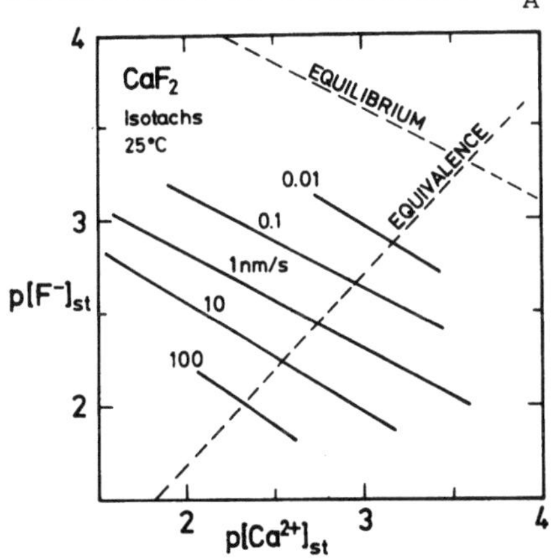

FIG. 15 - Calcium fluoride isotachs. The slope, $-\frac{1}{2}$, corresponds to constant values of the ionic product, $[Ca^{2+}][F^-]^2$, or constant affinity for crystallization.

experimental material may be condensed and surveyed in this way. With transport controlled rates, $\dot{r}$ also depends on r. For particles smaller than r = 10 μm, curves for constant $r\dot{r}$ may be convenient - but if r does not vary much, simple isotachs may be applicable (see Figs. 13-16).

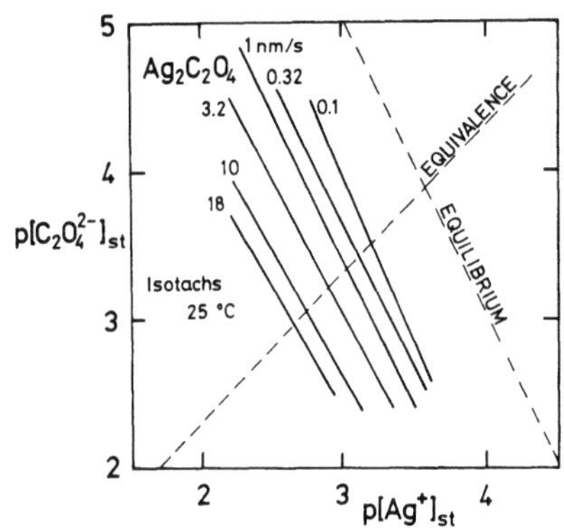

FIG. 16 - Silver oxalate isotachs. The figure illustrates that also for a 1,2-electrolyte the isotachs follow the isergons.

The isotachs for diffusion controlled growth of spherical particles of any fixed size (or curves for constant values of the product, $r\dot{r}$) can be calculated theoretically. As can be seen from Eq. 14, these isotachs do not coincide with isergons. When the ratio [A]/[B] is far from unity, the smaller of [A] and [B] dominates the rate. This may also be expressed: The transport controlled rate is not a function of the supersaturation S defined by Eq. 28. However, when the growth rate is surface controlled, $\dot{r}$ is a function of S. This seems to be the case for all the three basic kinetic types - linear, parabolic, and exponential growth. This is of course an important observation and supports the mechanisms described above (14).

Problem 20 Estimate the rate $\dot{r}$ in a solution with $[Ca^{2+}] = 10^{-4}$ M, $[CO_3^{2-}] = 10^{-3}$ M by means of Fig. 14. Answer: $10^{-11} < \dot{r} < 10^{-10}$ m/s.

Problem 21 For $BaSO_4$ the solubility product is 1.0×10^{-10} $mol^2 l^{-2}$. Draw curves for S = 1, a) in an [A],[B], and b) in a pA,pB diagram.

Problem 22 For $CaCO_3$ the ion pair constant is $[CaCO_3]/[Ca^{2+}][CO_3^{2-}] = K_I = 1.6\times10^3 l\ mol^{-1}$, and the solubility product is 8×10^{-9} mol^2/l^{-2}. Calculate points of and draw the curve corresponding to S = 10 in: a) a pA,pB, and b) a pCA,pCB diagram. c) Calculate the affinity ($-\Delta G$) per mole of growth units and per mole of $CaCO_3$. Answer: 5700 J/mol, and 11400 J/mol, respectively.

Problem 23 Discuss (assisted by the results of Problem 22) the condition for obtaining a precursor when calcium carbonate is precipitated by mixing $CaCl_2$ and Na_2CO_3 solutions.

Ripening

A small spherical particle with radius r will be in equilibrium with a solution of supersaturation S, when

$$r = r^* \equiv 2a^3\sigma/kT\ln S = 2(V_m/\nu)\sigma/RT\ln S. \tag{39}$$

This is the same equation as Eq. 7, valid for homogeneous nucleation. For a particle with $r < r^*$ the solution is undersaturated, and for particles with $r > r^*$ it is supersaturated. If Eq. 39 is valid for a single particle in a very large volume, the equilibrium will be unstable, but in a small volume, partial dissolution will increase S so as to stabilize the equilibrium. However, two particles, 1 and 2, are not stable together even if they are equally large. If a little substance is transferred from 1 to 2 by a fluctuation, making $r_1 < r_2$ the corresponding equilibrium supersaturations will be related as $S_1 > S_2$, and the transfer of substance will continue until 2 has consumed all the substance of 1. This process is called ripening (or Ostwald ripening). Such a transfer of material will also take place for any larger number of particles, and the average size will increase with time. If the rate is determined only by the diffusion through the solution, the average size $\bar{r}$ will be a function of time, t, as given (10) by

$$\bar{r} = (\sigma V^2 Dc_s t/\nu RT)^{1/3}. \tag{40}$$

For growth and dissolution processes so close to equilibrium, diffusion control is not very likely, and Eq. 40 will only give an estimate of the upper limit of the average size after ripening for a period, t.

Problem 24 A binary electrolyte, AB, $\nu = 2$ has the solubility $c_s = 1.234$ mol/m^3, $\sigma = 50$ mJ/m^2, and $V_m = 50$ ml per formula unit. Find the equilibrium concentration c at the surface of a spherical particle with r = 0.1 μm. Answer: 1.246 mol/m^3.

Problem 25 Calculate with $\sigma = 0.1\ J/m^2$, $D = 10^{-9}\ m^2/s$, $V_m = 5\times10^{-5}\ m^3/mol$, $\nu = 2$, $c_s = 2\times10^{-8}\ mol/m^3$, the average radius after diffusion controlled ripening of an initially very fine precipitate for 1 second, 1 minute, 1 day (24 h), and 1 year. Answer: $\bar{r}/nm$ = 10, 39, 440, and 6800.

Agglomeration

When a solid is disintegrated (e.g., in a liquid), the surface (or interface) is increased, and as the solid is stable in contact with the liquid, the interfacial tension must be positive, i.e., work has to be done in order to disintegrate the solid. Therefore, the reverse process - the union of the particles into a homogeneous body - is a natural process ($\Delta G < 0$). The ripening described above is one such process. A more direct one would be the agglomeration by collision between the particles. However, dispersions of very small particles are often rather stable, caused first of all by electric repulsion between the particles. A particle of the electrolyte $A_\alpha B_\beta$ will adsorb an excess of A ions when [A] >> [B] and thus become positively charged. Similarly, it becomes negative for [A] << [B]. Somewhere in the vicinity of equivalence, $[A]/[B] = \alpha/\beta$, there will be an isoelectric point (which may be determined by electrophoresis) with no repulsion, and the dispersion force will then exert attraction between the particles. Therefore, many electrolyte suspensions will coagulate in a narrow zone along the equivalence line in a pA,pB diagram (see Fig. 12). At high ionic strength the electric fields are shielded, which makes the suspension less stable. Specifically adsorbed ions may provoke coagulation, and surfactants and macromolecules will often be able to hinder contact and thus stabilize suspensions. They may even redisperse a coagulated sediment. Agglomeration may speed up the ripening process because of the shorter distances and because it allows surface migration to contribute to the transport. A microcrystalline or apparently amorphous precipitate may thus change into clear crystals.

The maximum rate of aggregation calculated (assuming that the particles move by Brownian motion, do not repel or attract each other at distance, but stick together whenever two particles touch) can be expressed as follows (for details and references, see (10)): At time zero, we have a monodisperse suspension with the particle concentration c_1^o, and at time t we have a suspension with the particle concentration c_1 monomers, c_2 dimers, c_3 trimers, etc. For rapid coagulation (or Smoluchowski coagulation)

$$c_1 = c_1^o/(1 + t/\tau)^2; \quad (41)$$

$$c_2 = c_1^o(t/\tau)/(1 + t/\tau)^3; \quad (42)$$

......

$$c_i = c_1^o(t/\tau)^{i-1}/(1 + t/\tau)^{i+1}; \quad (43)$$

$$\text{and } c_{total} \equiv \sum_1^{i_{max}} c_i = c_1^o/(1 + t/\tau), \quad (44)$$

where $\tau \equiv 3\eta/4kTc_1^o$, and η = viscosity of the liquid (see Fig. 17).

Problem 26 A monodisperse suspension contains initially 2.7×10^{10} particles per cm^3. The temperature is 25°C, and the solution has the viscosity (in SI units) $\eta = 9\times10^{-4}$ Pa s. Find the number concentration of monomers, dimers, and all particles after one minute of rapid coagulation.
Answer: $c_1 = 2.2\times10^{14}\ m^{-3}$; $c_2 = 2.0\times10^{14}\ m^{-3}$; $c_{total} = 2.5\times10^{15}\ m^{-3}$.

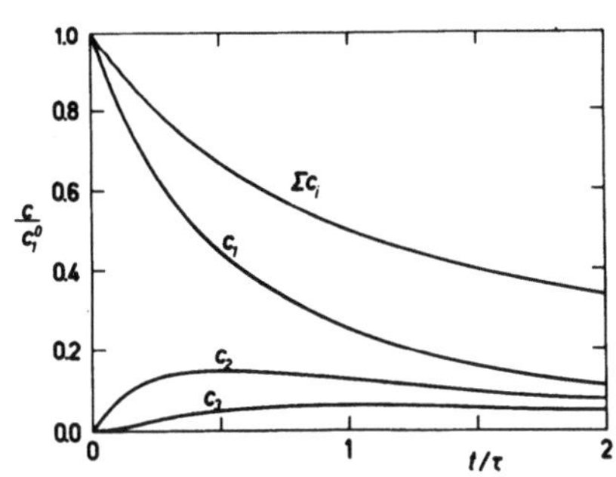

FIG. 17 - The variation of the concentrations of monomers, dimers, trimers, and all particles together during rapid (Smoluchowski) coagulation.

Epitaxy

When crystals are formed at moderate or low supersaturation, their nucleation cannot be homogeneous because the work of forming the nuclei is too large. Any solid surface may act as a heterogeneous catalyst (or heteronucleus) if the surface tension of the contact interface is sufficiently small compared to the sum of the surface tensions of the two solids. When two crystals made up of ions or oriented dipoles are in close contact, the following electric interaction patterns are possible: (a) no correlation between the lattices (see Fig. 18), electric attractive and repulsive forces cancel; (b) local areas of good correlation with strong attraction and evenly scattered dislocations at the contact plane; (c) good correlation almost everywhere in the contact plane - obtained through elastic deformation (and a low concentration of dislocations in the interior of the crystals).

As a rule of thumb, the different cases correspond to different ranges of the lattice mismatch, δ, defined as $\delta = |a-b|/a$ where a and b are lattice parameters of the unstrained crystals (see Fig. 18). The influence of δ (i.e., the similarity in lattice parameters) is much more important than any chemical similarity. When $\delta <$ ca. 0.02, one of the crystals may grow to macroscopic (millimeter-centimeter) size in close contact with the other one in a simple relative crystallographic orientation (epitaxy) (see also Koutsoukos and Nancollas(7)).

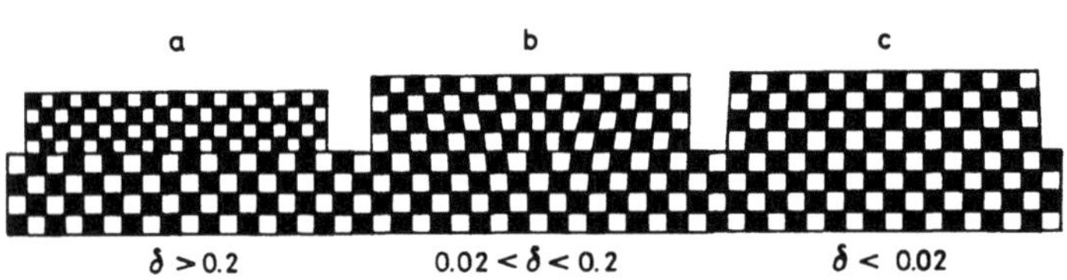

FIG. 18 - Schematic illustration of the degrees of correlation between two crystal lattices at different values of the so-called mismatch, δ.

Inhibition

Whenever crystals grow or dissolve, the linear rate of growth or dissolution may be strongly affected by the presence of non-constituent ions or molecules. If the rate is reduced, the components responsible for the reduction are called inhibitors.

The presence of inhibitors may explain why, e.g., calcium phosphates and calcium oxalates are not spontaneously precipitated in some biological systems, despite the solutions being supersaturated with respect to them. The formation of subsurface lesions in connection with caries processes can probably only be explained as being due to a combination of acid attack and inhibitor protection of the enamel crystallites.

In a review article, Fleisch (6) has recently discussed inhibition of calcium phosphate precipitation and the role of inhibitors in biological mineralization. The existing information concerning inhibition of crystal growth and dissolution is mainly of a qualitative nature. It is of importance to develop methods for explaining the effect of inhibitors in a more quantitative way. The effects of inhibitors on screw dislocation growth was discussed by Cabrera and Vermilyea (3) as early as 1958.

When studying inhibition of crystal growth and dissolution in biological systems, some fundamental problems have to be considered.

1) How should the effect of inhibition of biological mineralization and demineralization processes be detected and described?
2) How can the effect of inhibitors be explained?
3) How can in vivo inhibitors be isolated and identified?

The rate of crystal growth and dissolution depends not only on the concentrations of inhibitors present, but also on many other reaction parameters, the most important of which are: the amount of solid material present, the specific surface area (size distribution of particles), the surface structure, e.g., concentration of dislocations in the original seed material, etc., and the solution composition.

Inhibitors cannot influence the diffusion in the solution. Therefore, the reduced growth rate due to an inhibitor must be surface controlled. In the following we shall only discuss cases where a surface process is rate controlling, and we shall use dissolution of calcium hydroxyapatite, HAP, microcrystals in aqueous solution as an example of a surface controlled process. The rate of dissolution of HAP is controlled by a polynuclear dissolution mechanism (4). Dissolution nuclei (holes) are formed in the crystal surface and grow laterally, whereby they intergrow. A nucleus of the critical size has the same chances for growing larger as for being annihilated. The radius of the critical surface nucleus depends on the saturation ratio $S = c/c_s$ ($S < 1$ for dissolution conditions), according to Eq. 23 (counting r as negative for holes).

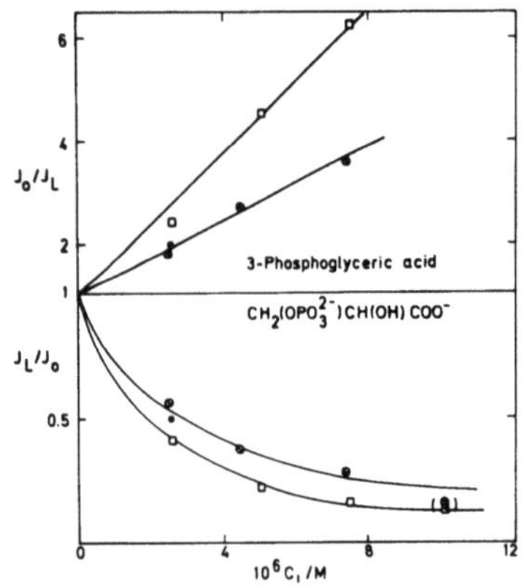

FIG. 19 - The effect of 3-phosphoglyceric acid on the rate of dissolution of HAP, $C/C_s = 0.25$. The symbols used in Figs. 19-21 are defined as follows: Circles: pH = 6.76±0.02; squares: pH = 7.16±0.02. Open symbols: $m/m_o>0.75$; lines through symbols: $0.75>m/m_o>0.65$; solid symbols: $0.65>m/m_o>0.55$.

A simple method for describing the effect of an inhibitor, L, the concentration of which is denoted C_L, is to plot J_L/J_o against C_L. Index L denotes an inhibitor is present, index o indicates that no inhibitor is present. J_L and J_o are the growth rates determined with and without the inhibitor present, keeping all other parameters constant as far as possible. The ratio J_L/J_o is thus equal to the ratio of the corresponding linear rates. Typical plots showing the effects of 3-phosphoglyceric acid and of 3-amino-1,1,3-propanetricarboxylic acid (γ-carboxyglutamic acid) on the rate of dissolution of HAP are given in the lower plots in Figs. 19 and 20.

The adsorption of inhibitors on a crystal surface can often be described by a Langmuir adsorption isotherm

$$K_L = \frac{X_L}{C_L(1 - X_L)} , \tag{45}$$

where X_L is the mole fraction of adsorption sites occupied by the inhibitor. The rate of growth or dissolution may be practically zero for a certain value of the mole fraction $X_L = X^*$. In the case of dissolution of HAP, the rate is practically zero, when the concentration of adsorbed inhibitor is so high that critical dissolution nuclei (holes) cannot be formed without inhibitor units being removed from the crystal surface.

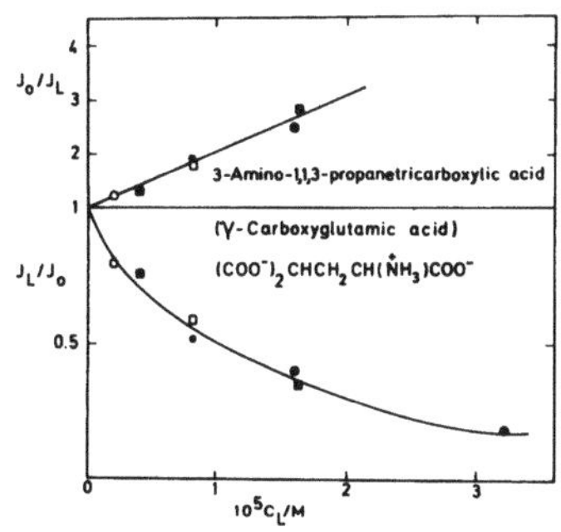

FIG. 20 - The effect of 3-amino-1,1,3-propanetricarboxylic acid on the rate of dissolution of HAP. $C/C_s = 0.25$.

From a kinetic viewpoint, how much inhibitor can be adsorbed on the crystal surface is not so important; what is important for effectively blocking the reaction is to obtain $X_L > X^*$. Christoffersen and Christoffersen (4) assumed that for inhibitors for which the adsorption can be described by Eq. 45, J_L/J_o can be approximated by $1 - X_L$, which results in $J_o/J_L = 1/(1 - X_L) = 1 + K_LC_L$. A plot of J_o/J_L against C_L gives a straight line, the slope of which is K_L (see Fig. 19). The values of K_L determined in this way depend in general on pH and on C/C_s, where C and C_s are the concentration of HAP in solution and the saturation concentration, respectively. A better approximation is to assume $J_L/J_o = 1 - X_L/X_L^*$, which leads to $J_o/J_L = 1 + K^*C_L$, with

$$K^* = K \frac{(1 - X_L)}{X^*(1 - X_L/X^*)}. \tag{46}$$

For low values of X_L/X^*, Eq. 46 reduces to

$$K^* = K/X^* .$$

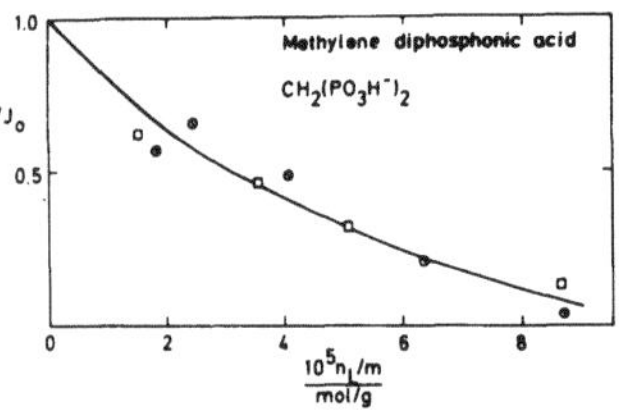

FIG. 21 - The effect of methylene diphosphonic acid (MDP) on the rate of dissolution of HAP. $C/C_s = 0.25$.

In cases where the concentration of inhibitor in the aqueous solution is so low that the concentration can hardly be measured accurately, a useful device may be a plot of J_L/J_o against n_L/m, where n_L is the total amount of inhibitor added and m is the mass of crystals. For adsorption of methylene diphosphonic acid (MDP) on HAP, a plot of this type is given in Fig. 21. Christoffersen and Christoffersen (4) show that the amount of adsorbed MDP, which for $C/C_s \approx 0.3$ and pH ≈ 7 can reduce the rate of dissolution of HAP by about 90%, corresponds to about 30% of the amount of phosphate groups in the crystal surface. Similar results were obtained for adsorption of diphosphoric acid on HAP.

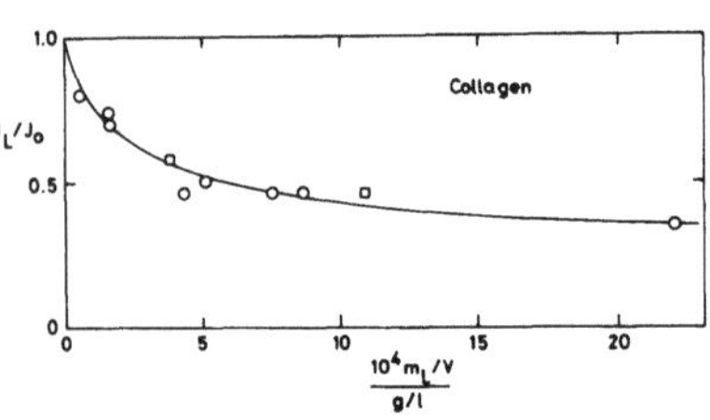

FIG. 22 - The effect of a partly peptized collagen, "Polypep," M ≈ 15 kg/mol, on the rate of dissolution of HAP. Circles: pH = 6.76; squares: pH = 7.16. $C/C_s = 0.25$.

For bone formation and resorption it is of interest to know how collagen, diphosphate (pyrophosphate, PP), adenosine triphosphate (ATP), adenosine diphosphate (ADP), and adenosine monophosphate (AMP) adsorb on HAP and thereby affect the exchange rate of Ca^{2+} and PO_4^{3-} between the crystalline phase and the solution. Figure 22 shows how varying amounts of partly peptized collagen affects the rate of dissolution of HAP. The collagen used was Sigma "Polypep" - a partially peptized, slightly soluble form of collagen with a molar mass of 15 kg/mol. Adsorption of both collagen and MDP on HAP showed that the effects of collagen and of MDP are not simply additive. Tracer experiments showed normal adsorption of MDP despite the addition of collagen either before or after the addition of MDP. From the kinetic experiments one can conclude that the presence of MDP reduces the inhibitory effect of collagen. The effects on HAP dissolution due to adsorption of PP, ATP, ADP, and AMP are shown in Fig. 23. The lower curve gives the effect of PP (4); the points close to this curve give the effect of ATP.

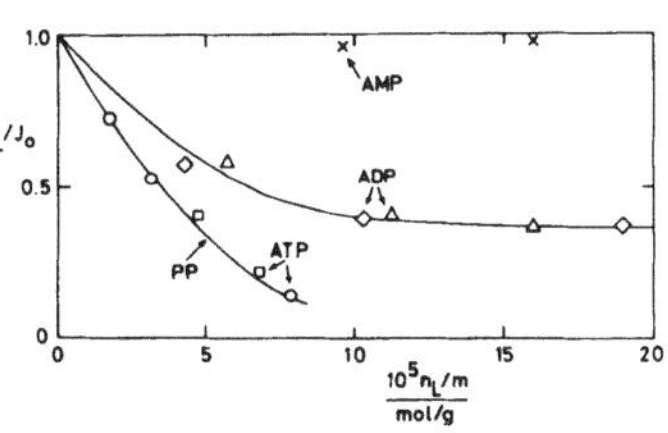

FIG. 23 - The effects of pyrophosphoric acid, PP (lower curve), ATP (circles: pH = 6.76; squares: pH = 7.16), ADP (triangles: pH = 6.76; diamonds: pH = 7.16), and AMP (crosses: pH = 6.76) on the rate of dissolution of HAP. C/C_s = 0.25.

The upper curve shows the effect of ADP. The two crosses show the effect of AMP. From this figure we may conclude that PP and ATP have the same effect on the rate of dissolution of HAP. The effect of ADP is lower and the effect of AMP is negligible compared to ATP and ADP.

The effect of an inhibitor should in general be explained on the basis of knowledge of the rate controlling mechanism for the growth or dissolution process studied. For HAP dissolution the rate controlling mechanism has been shown (4,5) to be a polynuclear dissolution mechanism. If the concentration of inhibitors adsorbed is so high that new dissolution nuclei (holes) can hardly be formed without inhibitor units being removed, the rate of dissolution will be severely reduced.

In cases of abnormal formation of crystals in biological systems, one may naturally look for elevated concentrations of the ions or molecules forming the crystals. If these concentrations are not elevated, one has to look for a lack of inhibitors which when present in normal concentrations are able to suppress crystal growth sufficiently. In order to test for inhibitors, one can perform reference (or "blind") experiments with seeded growth of the crystals in a reproducible medium. Addition of biological fluids will cause the rate of growth to be retarded if the biological fluid contains inhibitors for growth of the crystals. If the normal biological fluid contains inhibitors it is important to identify these inhibitors, or at least to determine whether such inhibitors are organic molecules or organic or inorganic ions. In order to make sure that the chemical treatment of the biological fluid has not changed the inhibitors, one should check that the total inhibitory effect be maintained when the various fractions into which the original fluid is separated are added together.

<u>Problem 27</u> The rate of dissolution of calcium hydroxyapatite (HAP) crystals was found to be reduced in the ratio $J_L/J_o = 0.10$ when 8.1×10^{-5} mole of diphosphate (pyrophosphate) was adsorbed per gram of HAP, having the specific area 28.6 m^2/g. From the structure of HAP one knows that in the crystal face from which the dissolution primarily takes place there are 3.1 phosphate groups per nm^2. The dissolution process was found to be polynuclear, and the radius of a critical surface nucleus (Eq. 23) was found to be $r^* = 0.30$ nm for typical conditions.

a) Find the surface concentration Γ_{PP} of adsorbed diphosphate groups (PP).
b) Find the number of PP groups adsorbed per nm^2.
c) Each of the adsorbed PP groups are assumed to substitute one phosphate group (P) in the top layer of the crystal face. What percentage of the P group is substituted by PP groups when the dissolution rate is reduced by 90%?
d) Assume for simplicity that the adsorbed PP groups form a simple square planar lattice, each occupying a circular area with radius $r_o = 0.25$ nm. Calculate the radius r_{max} of the greatest circle that may be placed between the PP groups.

e) Discuss why the dissolution rate is almost reduced to zero at the inhibitor surface concentration mentioned.

Answers: a) $\Gamma_{PP} = 2.83\times10^{-6}$ mol/m^2; b) $N_{PP} = 1.71$ nm^{-2}; c) 55%; d) 0.29 nm = 2.9×10^{-10} m.

CONCLUSION

The authors have been asked to suggest future needs. Being physical chemists, we find it reasonable to suggest that the emphasis should be laid on measuring well-defined physico-chemical parameters that can be compared with deductions from simple molecular models. We feel that this advice may be particularly relevant to biologists accustomed to thinking in terms of cells, metabolites, enzymes, genetic control, etc. - realizing that crystals belong to a different and simpler world. As stated in the introduction, the number of chemical components in a crystal is usually one or at most a one-hand number. Crystal growth, therefore, ought to be described in terms of simple mechanisms and ought to have simple kinetics when described in terms of the right parameters.

Acknowledgement. We thank M. Christoffersen and J. Toft for permission to use unpublished data. One of us (J.C.) acknowledges the support from The Danish Medical Research Council, grants 12-0323 and 12-2215.

Solutions to the problems

1. $v_{g1} = \Delta l/2\Delta t = (11.0-10.0)/20 = 0.05$ µm/s; v_{g2} and v_{g3} analogously. Volume $V_o = 10\times15\times20 = 3000$ µm^3; $V_{10} = 11\times17\times24 = 4488$ µm^3; $V = 4\pi r^3/3$; $r_o = (3V_o/4\pi)^{1/3} = 8.947$; $r_{10} = 10.232$; $dr/dt \approx \Delta r/\Delta t = (r_{10}-r_o)/10$ s $= 0.1285$ m/s $\approx 1.3\times10^{-7}$ m/s.

2. $V_m = (100/2)10^{-6} = 5\times10^{-5}$ m^3/mol; 2×10^{-3} mol $= 2\times10^{-3}\times5\times10^{-5} = 10^{-7}$ m^3 $= 10^{-7}$ m$\times$1 m^2; $v_g = 10^{-7}$ m/s.

3. $V_m = 100.1/2.71 = 36.9$ ml/mol; the volume per particle is (4×10^{-4} mol/l)$\times$(36.9 cm^3/mol)/(3.1×10^6 l^{-1}) $= 4.76\times10^{-9}$ cm$^3 = 4\pi r^3/3$; $r = 1.044\times10^{-3}$ cm; $\Delta V/V = 0.008/0.40 = 0.02 \Rightarrow \Delta r/r = 0.02/3$; $\dot{r} = \Delta r/\Delta t = 1.044\times10^{-3}\times0.02/3\times60 = 1.16\times10^{-7}$ cm/s $= 1.16\times10^{-9}$ m/s.

4. In 0.3 l, 0.5 g seeds with 5 m^2 surface area consume $0.3\times2\times10^{-5}$ mol/s $= 6\times10^{-6}$ mol/s, or $J = 6\times10^{-6}/5 = 1.2\times10^{-6}$ mol/m^{2}s. The molar volume is $125/2.5 = 50$ cm^3/mol $= 5\times10^{-5}$ m^3/mol; $v_g = JV_m = 6\times10^{-11}$ m/s; $L \equiv$ cube length, $L^3 = 4\pi r^3/3$; $dL/dt = 2v_g$; $\dot{r} \equiv dr/dt = (3/4\pi)^{1/3}\, dL/dt = (3/4\pi)^{1/3}\cdot2\cdot v_g = 1.24v_g = 7.4\times10^{-11}$ m/s.

5. a) $a^2\sigma/kT = 5$; $\sigma = 5kT/a^2 = 5(4.12\times10^{-21}J)/(3\times10^{-10}m)^2 = 0.228\ J/m^2$.
b) $v_g = 0.3$ nm/s $= 3\times10^{-10}$ m/s.

6. $n_c = [2a^2\sigma/kT\ln S]^2 = [2\times5/\ln 8.8]^2 = 21$.

7. $\dot{r} = (V_m/2r)[2Dc - \sqrt{(4D^2c_s^2)}] = (DV_m/r)(c-c_s)$ q.e.d.

8. $c_s = K^{\frac{1}{2}} = 1\times10^{-5}M << c = 10^{-3}M = 1\ mol/m^3 \approx \Delta c$
$\dot{r} = DV_m\Delta c/r = 10^{-9}\times0.233\times1/(4500\times2\times10^{-6}) = 2.6\times10^{-8}$ m/s.

9. The diffusion controlled rate would be $\dot{r}_D = DV_m\Delta c/r = 10^{-9}\times3.7\times10^{-5}\times0.2/4\times10^{-6} \approx 2\times10^{-9}$ m/s $<< r_{exp} = 10^{-11}$m/s $\Rightarrow$ the rate is surface controlled.

10. $\dot{r} = V_m j = V_m A^{-1}\ dn/dt = V_m A^{-1} DA\ dc/dt = V_m D(c-c_s)/r$ q.e.d.

11. $\dot{r} = V_m j = V_m D$ grad $c = V_m D(c-c_s)/\delta$ q.e.d.

12. a) $r << \delta \approx 10\ \mu m$, Eq. 13; b) $r >> \delta$, Eq. 15; c) When convection is rate determining, i.e., case b.

13. $a = (M/\nu\rho L)^{1/3} = (246/2\times1.68\times6\times10^{23})^{1/3} = 5\times10^{-8}$ cm $= 5\times10^{-10}$ m;
$\alpha = 1$; $c_sV_w = 2.4\times0.018 = 0.0432$; $k_{ad} = \dot{r}/a\alpha c_sV_w(S-1) \approx 8\times10^{-6}/5\times10^{-10}\times 0.0432 = 3.7\times10^5$; $G^{\ddagger}_{ad} = kT\ln[(kT/h)/k_{ad}] = 4.12\times10^{-21}\ln(6.21\times10^{12}/3.7\times10^5) = 6.9\times10^{-20}J$.

14. a) $(v_1/\Delta S_1^2)/(v_2/\Delta S_2^2) = 3.5\times10^{-6}/4.3\times10^{-6} = 0.80 \approx 1$
b) $k_2 \approx \frac{1}{2}(3.5+4.3)\times10^{-6}$m/s $= 3.9\times10^{-6}$m/s

$$k_i = \frac{k_2 a\sigma\exp(a^2\sigma/kT)}{c_s kTK_I M(1,2)}$$

$$= \frac{3.9\times10^{-6}\times3.3\times10^{-10}\times0.040\times\exp[(3.3\times10^{-10})^2\times0.040/4.12\times10^{-21}]}{650\times4.12\times10^{-21}\times0.01\times5\times10^{-4}}$$

$= 1.1\times10^7\ s^{-1}$ (neglecting for convenience the difference between 20 and 25 °C)
$G^{\ddagger}_i = kT\ln(kT/hk_i) = 4.12\times10^{-21}\ln(6.21\times10^{12}/1.1\times10^7) = 5.5\times10^{-20}$ J
$G^{\ddagger}_i/G^{\ddagger}_w = 5.5\times10^{-20}/3.6\times10^{-20} = 1.53$
(with Eq. 22 and $N(1,2) = 3\times10^{-5}$ one finds the less certain values $k_i = 1.8\times10^{-6}\ s^{-1}$, $G^{\ddagger}_i = 6.2\times10^{-20}$ J, $G^{\ddagger}_i/G^{\ddagger}_w = 1.72$).

15. $n^{\square*} = [2a^2\sigma/kT\ln S]^2 = [2\times(3.3\times10^{-10})^2\times0.040/4.12\times10^{-21}\ \ln 1.09]^2 = 600$
$n^{\circ*} = (\pi/4)600 = 470$; $r^*/a = a^2\sigma/kT\ln S = (3.3\times10^{-10})^2\times0.040/4.12\times 10^{-21}\ \ln 1.09 = 12$; $y_o/a = 19r^*/a = 230$; $x_o/a = S^{-\frac{1}{2}}\exp(a^2\sigma/kT) = 1.09^{-\frac{1}{2}}\exp[3.3\times10^{-10})^2\ 0.04/4.12\times10^{-21}] = 1.1$(i.e., the step is "rough" - any site at the step is capable of acting as growth site).

16. Eq. 31 with $\pi/3 \approx 1$:v_g/ms$^{-1} = 10^{-8} = 3(S-1)^{5/6}\exp(-9/\ln S) \approx 3\exp(-9/\ln S)$;
$9/\ln S = \ln 3\times10^8$; $\ln S = 0.46$; $S = 1.59$; $(S-1)^{5/6} = 0.59^{5/6} = 0.64$;
$\ln S = 9/\ln(3\times0.64\times10^8) = 0.47$; $S = 1.60$; Eq. 32: $10^{-8} \approx 2000\exp(-3-10-9/\ln S)$;
$9/\ln S = \ln 2\times10^{11}-13 = 13$; $\ln S = 0.697$; $S = 2.00$.

17.

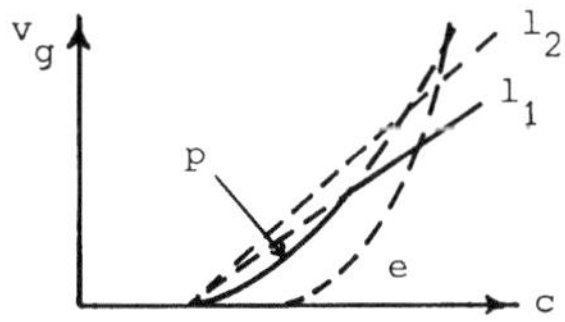

The rate is determined by the process with a parabolic law (screw dislocation) at small supersaturations, and by the process with a linear law, and having the smallest rate, (transport or adsorption) at large supersaturation.

18. Plots of $\dot{r}$ and $\dot{r}^{\frac{1}{2}}$ against c are bent curves. A plot of $y = \ln[(S-1)^{-5/6}\dot{r}]$ against $x = (\ln S)^{-1}$ is linear except for small S-values, indicating the logarithmic rate law being followed for large S and probably transition to parabolic or linear at low S values. From the slope $\alpha \equiv dy/dx = 14$, one finds $\sigma = (kT/a^2)(3\alpha/\pi)^{\frac{1}{2}} = 0.26$ J/m^2 in agreement with literature values based on nucleation experiments (0.26-0.33 J/m^2)(14).

19. a) $dr/dt = DV_m(c-c_s)/r \approx -DV_m c_s/r$; $r_0^2 = 2DV_m c_s t$; $t = r^2/2DV_m c_s = 10^{-6}/2\times10^{-9}(0.1\times1) = 5000$ s. b) 1 mm > r > 10 μm: $dr/dt = -DV_m c_s/\delta = -10^{-10}/10^{-5} = -10^{-5}$m/s; $\Delta t_1 = (\Delta r_1)/(-10^{-5}m/s) = 9.9\times10^{-4+5} = 99$ s;
10 μm > r > 0: $\Delta t_2 = r_0^2/2DV_m c_s = 10^{-10}/2\times10^{-10} = 0.5$ s;
$t = \Delta t_1 + \Delta t_2 = 99.5$ s ≈ 100 s.

20. The rate will be between 10^{-11} and 10^{-10} m/s.

21.

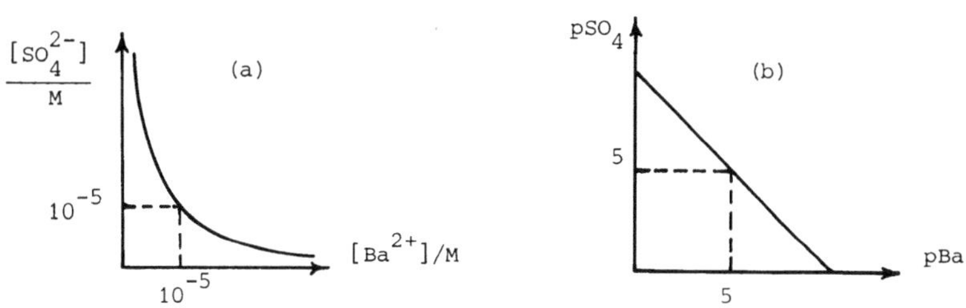

22. a) $[Ca^{2+}] = a$; $[CO_3^{2-}] = b$; $[CaCO_3] = c$; $C_A = a+c$; $C_B = b+c$; $S = (ab/K_s)^{\frac{1}{2}} = 10$; $ab = S^2K_s = 8\times10^{-7}M^2$; $pA+pB = -\log\times8\times10^{-7} = 6.1$.
b) $c = a\cdot b\cdot K_I = S^2K_sK_I = 1.28\times10^{-3}$M; $(C_A-c)(C_B-c) = ab = S^2K_s$;

$$C_B = \frac{8\times10^{-7}}{C_A-1.28\times10^{-3}} + 1.28\ 10^{-3} ;$$

$C_A = C_B$: $(C_A-c)^2 = 8\times10^{-7} \Rightarrow C_A = 9.0\times10^{-4} + 1.28\times10^{-3} = 2.18\times10^{-3}$ M.

C_A/mM	C_B/mM	pCA	pCB
2.18	2.18	2.66	2.66
3.00	1.75	2.52	2.76
5.00	1.50	2.30	2.82
10.00	1.37	2.00	2.86
100.00	1.29	1.00	2.89

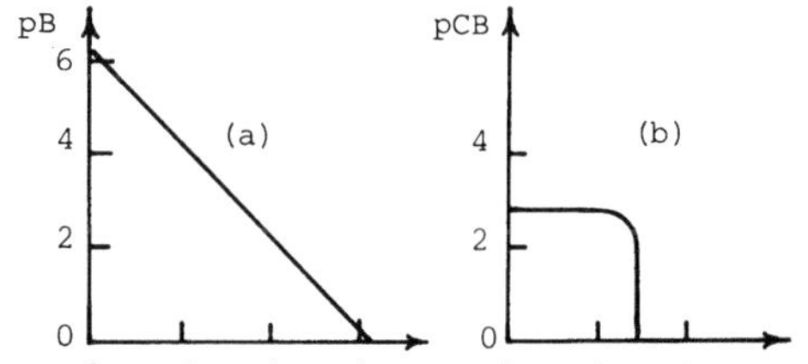

c) $-\Delta G = RT\ln S = 8.31\times298\times\ln10 = 5700$ J/mole of growth units = 11400 J/mole $CaCO_3$.

23. The limit of the precursor area in Fig. 14 seems to follow a curve for constant S (an isergon). So the condition for having a precursor is $S > S_{cr}$, S_{cr} being a definite, critical value. From the diagram we read off $[Ca^{2+}] \approx [CO_3^{2-}] \approx 10^{-2.7} = 2\times10^{-3}$ M and find $S_{cr} \approx 2\times10^{-3}/(8\times10^{-9})^{\frac{1}{2}} = 22 \approx 20$. [$K_S$(precursor) = $(2\times10^{-3}M)^2 = 4\times10^{-6}M^2$ (?)]

24. $\ln S_r = 2\sigma V_m/\nu rRT = 2\times0.050\times5\times10^{-5}/(2\times10^{-7}\times2500) = 0.010$; $S_r = 1.01$; $c = 1.01$; $c_s = 1.246\ mol/m^3$.

25. $\bar{r} = (\sigma V_m^2 D c_s/\nu RT)^{1/3} t^{1/3} \simeq (10^{-8} m)(t/s)^{1/3} = (10\ nm)(t/s)^{1/3}$

t =	1 s	1 min	1 d	1 y
t/s =	1	60	86400	3.16×10^7
$(t/s)^{1/3}$	1	3.9	44.2	681
$\bar{r}$/nm	10	39	442	6810

26. $\tau = 3\eta/4kTc_1^o = 3\times9\times10^{-4}/(4\times4.12\times10^{-21}\times2.7\times10^{16}) = 6.0$ s; $t/\tau = 60/6.0 = 10$; $c_1 = 2.7\times10^{16}/11^2 = 2.2\times10^{14}m^{-3}$; $c_2 = 2.7\times10^{16}\times10/11^3 = 2.0\times10^{14}m^{-3}$; $c_{total} = 2.7\times10^{16}/11 = 2.5\times10^{15}m^{-3}$.

27. a) $\Gamma_{PP} = 8.1\times10^{-5}/28.5 = 2.83\times10^{-6}\ mol/m^2$; b) $N_{PP} = 2.83\times10^{-6}\times6.02\times10^{23} = 1.71\times10^{18}\ m^{-2} = 1.71\ nm^{-2}$; c) $1.71\times100/3.1 = 55\%$; d) Area per "unit cell" square = $1\ nm^2/1.71 = 0.585\ nm^2$; length of diagonal = $(2\times0.585)^{\frac{1}{2}}$ = 1.08 nm = diameter of circle + diameter of PP unit ⇒ $r_{max} = (1.08-0.50)/2 = 0.29$ nm; e) $r_c \approx r_{max}$; $v_{max} \approx 0$.

REFERENCES*

(1) Bennema, P., and Gilmer, G.H. 1973. Kinetics of crystal growth. In Crystal Growth, ed. P. Hartman, pp. 263-327. Amsterdam: North-Holland.

(2) Burton, W.K., Cabrera, N., and Frank, F.C. 1951. The growth of crystals and the equilibrium structure of their surfaces. Phil. Trans. Roy. Soc. A 243: 299-358.

(3) Cabrera, N., and Vermilyea, D.A. 1958. The growth of crystals from solution. In Growth and Perfection of Crystals, eds. R.H. Doremus, B.W. Roberts, and D. Turnbull, pp. 393-410. New York: Wiley; London: Chapman & Hall.

(4) Christoffersen, J., and Christoffersen, M.R. 1981. Kinetics of dissolution of calcium hydroxyapatite. J. Cryst. Growth 53: 42-54.

(5) Christoffersen, J. 1981. Dissolution of calcium hydroxyapatite. Calc. Tiss. Int. 34, in press.

(6) Fleisch, H. 1981. Inhibitors of calcium phosphate precipitation and their role in biological mineralization. J. Cryst. Growth 53: 120-134.

* The references are mostly to review articles rather than to the original papers.

(7) Koutsoukos, P.G., and Nancollas, G.H. 1981. Crystal growth of calcium phosphates - epitaxial conditions. J. Cryst. Growth 53: 10-19.

(8) Mullin, J.W. 1972. Crystallisation. London: Butterworths.

(9) Nancollas, G.H. 1979. The growth of crystals in solution. Adv. Colloid Interface Sci. 51: 215-252.

(10) Nielsen, A.E. 1964. Kinetics of Precipitation. Oxford: Pergamon Press.

(11) Nielsen, A.E. 1972. Nucleation and Growth. In Problems in Materials Science 1, ed. H.D. Merchant, pp. 145-167. New York: Gordon and Breach.

(12) Nielsen, A.E. 1979. Kinetics of crystal growth during precipitation of a binary electrolyte. In Industrial Crystallization 78, eds. E.J. de Jong and S.J. Jančić, pp. 159-168. Amsterdam: North-Holland.

(13) Nielsen, A.E. 1980. Transport control in crystal growth from solution. Croatica Chem. Acta 53: 255-279.

(14) Nielsen, A.E. 1981. Theory of electrolyte crystal growth. The parabolic rate law. Pure Appl. Chem. 53: 2025-2039.

(15) Nielsen, A.E., and Söhnel, O. 1971. Interfacial tension electrolyte crystal-aqueous solution, from nucleation data. J. Cryst. Growth 11: 233-242.

(16) Ohara, M., and Reid, R.C. 1973. Modelling Crystal Growth Rates from Solution. New Jersey: Prentice Hall.

(17) Petrucci, S. 1971. Kinetic approach to the study of ionic association and complexation: Relaxation kinetics. In Ionic Interactions, II. Kinetics and Structure, ed. S. Petrucci, pp. 39-124. New York: Academic Press.

(18) Sillen, L.G., and Martell, A.E. 1964. Stability constants of Metal Ion Complexes. London: The Chemical Society.

(19) Walton, A.G. 1967. The Formation and Properties of Precipitates. New York: Interscience.

Biological Mineralization and Demineralization, ed. G.H. Nancollas, pp. 79-99.
Dahlem Konferenzen 1982. Berlin, Heidelberg, New York: Springer-Verlag.

Phase Tranformation During Precipitation of Calcium Salts

G. H. Nancollas
Dept. of Chemistry, State University of New York
Buffalo, NY 14214, USA

Abstract. A number of minerals of biological importance can precipitate as different phases and polymorphs under conditions typical of those in vivo. The calcium phosphates which form in solution are markedly dependent on pH, supersaturation, ionic strength, temperature, and the nature of the solid phases already present. Although thermodynamic considerations yield the driving force for precipitation, the course of the reaction may be mediated much more by kinetic factors. Metastable phases may persist for considerable periods of time in the supersaturated solutions. The same factors are important in the mineralization of the calcium oxalate hydrates. The thermodynamically most stable monohydrate may be formed from the di- and tri-hydrates by nucleation and growth processes, the rates of which are sensitive to temperature, solution composition, and the presence of growth inhibitors typical of those in vivo.

INTRODUCTION

In biological systems, the precipitation of calcium salts is involved in vital problems dealing with beneficial mineralization such as the formation of bone and tooth and pathological processes such as bone disease, carious lesions, and renal stone formation. The mechanism of the precipitation process still awaits satisfactory elucidation. Even defining the degree of supersaturation in biological systems is a difficult problem because of uncertainties in the nature and stabilities of the possible metal complexes which can be formed. The number of mineral phases which is involved is relatively small, being limited, for the most part, to solids such as the calcium phosphates, carbonates, oxalates, and some urate and other alkaline metal phosphate salts. However, many of these salts can precipitate in a number of different phases and polymorphs which makes the elucidation of the mechanism of their precipitation even more difficult.

In many cases, mixed phases may form by the growth of one phase upon another in metastable supersaturated solutions. Thus anhydrous and hydrated uric acids are often found as constituents of urinary calculi, and since the hydrated form is nearly always associated with the anhydrous phase, it has been suggested that an epitaxial formation may take place (3). When crystals of one modification are present in solution, they may act as nucleation substrates for crystals of the other hydrate. Similar epitaxial relationships have been proposed for the growth of calcium phosphate phases and the formation of phosphate/oxalate calcium mixed deposits often observed in renal calculi (25). It should be noted, however, that nucleation of one phase by another does not necessarily imply a true epitaxial growth.

Although it is not possible at the present time to confirm unequivocally the true epitaxial growth of these mixed phases, the presence of the substrate allows nucleation and growth of the new phase at supersaturations considerable lower than those required for homogeneous precipitation. In the formation of these solid phases in supersaturated biological fluids such as serum, saliva, and urine, the process must be mediated not only by biological restraints but also by factors, familiar to the crystal growth scientist, which control the formation of nuclei and their subsequent crystallization. The role of the organic matrix in the nucleation and precipitation process is also an important problem which will be considered elsewhere in this volume. The present paper is concerned with the nature of the phases which form in two calcium salt precipitation processes, namely, the phosphates and the oxalate hydrates.

CALCIUM PHOSPHATES

Biological apatites are the main constituents of the bones and teeth of vertebrates as well as almost all hard tissues of humans. Pathological mineral deposits such as those which occur in dental calculus, bursitis, arthritis, and various other forms of ectopic calcification also involve apatite-like materials. Many of the kinetic studies which have been made have been concerned with mineralizing solutions considerably more supersaturated than those typical of in vivo conditions. These solutions have been supersaturated with respect to at least five calcium phosphate phases, in order of increasing solubility: hydroxyapatite ($Ca_5(PO_4)_3OH$,HA), tricalcium phosphate ($Ca_3(PO_4)_2$,TCP), octacalcium phosphate ($Ca_8H_2(PO_4)_6.5H_2O$,OCP), dicalcium phosphate anhydrous ($CaHPO_4$,DCPA), and dicalcium phosphate dihydrate ($CaHPO_4.2H_2O$,DCPD).

In order to calculate the free ionic concentrations in the solutions, it is necessary to take into account ion-pair and complex formation. Typical thermodynamic data are given in Table 1 for the protonation of phosphoric acid, calcium phosphate ion-pair formation, and the solubilities of the calcium phosphate phases at 25°C. Solubility isotherms are plotted as a function of pH in Fig. 1 and are expressed as $\log(T_{Ca}T_p)$,

TABLE 1 - Thermodynamic Equilibrium Data at 25°C.

Equilibrium	Equilibrium Constant
$Ca^{2+} + H_2PO_4^- \rightleftharpoons CaH_2PO_4^+$	31.9 ℓ mol^{-1}
$Ca^{2+} + HPO_4^{2-} \rightleftharpoons CaHPO_4$	681 ℓ mol^{-1}
$Ca^{2+} + PO_4^{3-} \rightleftharpoons CaPO_4^-$	2.9 x 10^6 ℓ mol^{-1}
$H^+ + H_2PO_4^- \rightleftharpoons H_3PO_4$	164.1 ℓ mol^{-1}
$H^+ + HPO_4^{2-} \rightleftharpoons H_2PO_4^-$	1.58 x 10^7 ℓ mol^{-1}
$H^+ + PO_4^{3-} \rightleftharpoons HPO_4^{2-}$	2.33 x 10^{12} ℓ mol^{-1}
$Ca^{2+} + OH^- \rightleftharpoons CaOH^+$	32.4 ℓ mol^{-1}
HAP solubility product	4.7 x $10^{-59}(mol\ \ell^{-1})^9$
TCP solubility product	1.20 x $10^{-29}(mol\ \ell^{-1})^5$
OCP solubility product	1.25 x $10^{-47}(mol\ \ell^{-1})^8$
DCPA solubility product	1.26 x $10^{-7}(mol\ \ell^{-1})^2$
DCPD solubility product	2.49 x $10^{-7}(mol\ \ell^{-1})^2$

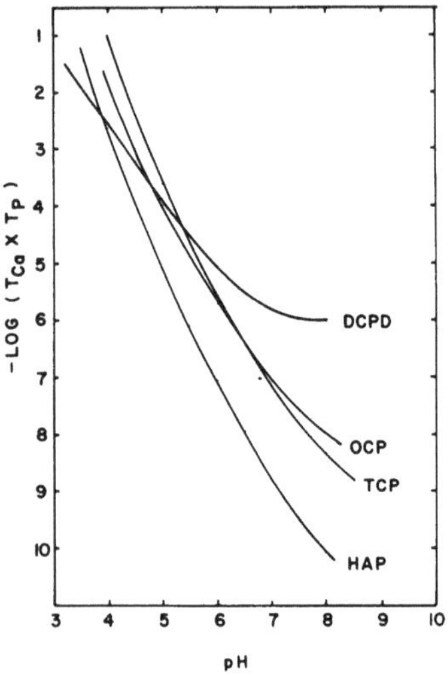

FIG. 1 - Log of the product of calcium and phosphate concentration plotted against pH values of solutions at 37°C, saturated with respect to each phase in the ternary system $Ca(OH)_2 - H_3PO_4 - H_2O$ (Shyu, Perez, Zawacki, and Nancollas, to be published).

where T_{Ca} is the total molar calcium concentration and T_P is the total molar phosphate concentration. Figure 1 has been constructed using the data in Table 1 assuming that the solutions contain calcium chloride and potassium phosphate. The calculations of the free ionic species for the construction of Fig. 1 can be made using mass balance, electroneutrality, and the appropriate equilibrium constants in Table 1 by successive approximations for the ionic strength (30). Activity coefficients of the ionic species may be calculated at ionic strengths up to about 0.2M, to within ± 1%, by using an extended form of the Debye-Hückel equation such as that proposed by Davies (12). It is important to note that in Fig. 1, the positions of the curves and the singular points will be dependent upon the ionic strength of the background electrolyte. Each computed figure is valid only when the ions which are not constituents of the ternary system are kept at constant concentration.

Although HAP is the most stable phase under many conditions, it becomes less stable than DCPD if the solution is sufficiently acid. The calcium phosphate phase which is exposed to a solution more acidic than the corresponding singular point may be expected to become covered by a surface coating of a more acid calcium phosphate. The apparent solubility behavior may therefore be quite different from that of the phase itself and this has probably been the source of the widely discrepant data in the literature for the solubilities of calcium phosphate phases. This was emphasized by Brown in 1973 (9) who suggested that this uncertainty may have led many workers to conclude that the properties of calcium phosphates in general and HAP in particular are rather variable (26). In considering diagrams such as that in Fig. 1, it is essential to establish that the systems are at equilibrium in order to be able to use thermodynamic solubility products which refer to the free energies of single solid phases. It is now quite well established that kinetic factors may be considerable more important in determining the nature of the solid phase present than are considerations based solely upon equilibrium data. Thus different calcium phosphate phases may be stabilized or destabilized by the presence of various cations and anions which may not be significantly incorporated into the calcium phosphate crystal lattice.

Much work has been done in attempts to identify the phases which form during the spontaneous precipitation of calcium phosphate from solutions supersaturated with respect to all five phases. The stoichiometric molar ratio of calcium to phosphate calculated from changes in concentration of the solution species during the precipitation is frequently in the range 1.45 ± 0.05 which is considerably lower than the value, 1.67, required for the thermodynamically favored HAP. In order to reach approximate HAP composition, extended solid/solution equilibration is required. It is usually assumed that the first formed phase during the precipitation process is an unstable amorphous phase, ACP, consisting of a highly hydrated cryptocrystalline calcium phosphate which appears as a precursor in the precipitation of OCP and HAP from neutral and slightly basic solutions (5,13,15,16,36,37). ACP usually appears in the form of chain-like aggregates of sphereolytic rather than monodispersed particles, and it is still not clear whether this phase is a single stoichiometric unit. Phases with Ca/P ratios ranging from 1.3 to 1.67 could be prepared depending upon the solution conditions. Thus in the physiological region of pH the stoichiometry (Ca/P approximately 1.5) is close to TCP, but it is clear that the solid phase also contains a certain amount of hydrogen phosphate, HPO_4^{2-} ions (5,36). X-ray radial distribution measurements made by Betts and Posner (2) on well washed ACP with Ca/P ratio of 1.5 suggested a basic structural unit consisting of a spherical cluster of about 9.5 Å in diameter with a stoichiometry of $Ca_9(PO_4)_6$. The spherical particles seen in the electron microscope were thought to consist of clusters of these units with water molecules located in interstices. Adsorbed phosphate ions or occluded calcium ions were suggested as accounting for the lower and higher Ca/P ratios, respectively, without changing the basic ACP structural unit. The lifetime of this precursor phase was also extremely sensitive to experimental conditions used for the preparation (37). Transformation into crystalline calcium phosphate phases could proceed by secondary precipitation as indicated by marked decreases in pH (5) and calcium and phosphate concentrations (13). It has been proposed that the conversion of the precursor ACP to the thermodynamically stable HAP proceeds by an autocatalytic mechanism (15).

An alternative theory to the clusters of $Ca_9(PO_4)_6$ is that ACP contains OCP-like structures. This is based on the idea that the ribbons first formed in the enamel matrix, which are best described as single unit cell thicknesses of OCP, are a prototype for the simplest particles in ACP. For reasons described in the paper by Young and Brown (this volume), these single unit-cell thicknesses of OCP would be stable against hydrolysis. Addition of another unit cell thickness, however, would permit hydrolysis of one unit cell thickness into a pair of unit cells of hydroxyapatite to form a unit that has been termed "sesquiapatite." The addition of other layers of OCP or apatite to these units would provide crystallites of various thicknesses and Ca/P ratios. This unit has advantages over the one proposed by Posner since no new structural entities need be defined, and the particles have properties that can be rationally derived from structural considerations.

Recently there has been an increased interest in attempts to follow concentration changes during the spontaneous precipitation process at a physiological pH of 7.4 (16) in which the ripening of the crystals appears to be accompanied by inflections in the time profiles of calcium and phosphate concentrations in the solution. Calculation of the decrease of supersaturation with respect to the various calcium phosphate phases indicates that these inflections occur in regions corresponding to the solubility of OCP. This supported the original suggestion by Brown (7,10), who presented considerable X-ray and other evidence for the existence of OCP in the physiological range of pH. Other workers subsequently have also provided evidence for OCP as an intermediate phase in the spontaneous precipitation of calcium phosphate (13,19).

Studies of the conversion process from ACP to more thermodynamically stable calcium phosphate phases have also been made in the presence of various stabilizing agents (7,20,21). Thus, adenosine triphosphate (ATP), some diphosphonates, pyrophosphate, and proteoglycans are thought to delay the transformation by inhibiting the growth of HAP embryos. Stabilization by added metal ions such as magnesium, however, is considered to involve interference with the formation of HAP critical nuclei by entering the pre-nuclei structures and creating a structural mismatch. A molar Mg/Ca ratio greater than 0.2 effectively prevents the transformation of ACP and will also delay the formation of apatite from solutions of low supersaturation (2). Magnesium also plays an important role in moderating the formation of OCP. In the presence of 1 mM Mg, the hydrolysis of OCP is dramatically inhibited in boiling water. The relatively large concentration of magnesium ion in blood serum is thought to be the principal component which prevents the precipitation of HAP in vivo.

Instead of invoking the formation of ACP and other defect apatites during the spontaneous precipitation of calcium phosphate, other workers have argued in terms of the Ostwald-Lussac Law of stages, according to which the phase with the highest solubility forms preferentially during a sequential precipitation. It is interesting to note that the Ostwald-Lussac sequence of stages is also the sequence of decreasing hydration for the calcium phosphates. As suggested by Young and Brown (this volume), the formation of the more highly hydrated crystals first may relate to (a) a low surface energy of hydrated crystals (especially during the nucleation stage, when a low interfacial energy would be important) or (b) the decreased dehydration of the ions required in the formation of the more highly hydrated crystals. DCPD was proposed as a precursor by Francis and Webb (22) and the formation of OCP, as mentioned above, was suggested by Brown and his co-workers (6,10) who showed that OCP and HAP can occur as intracrystalline interlayered mixtures since OCP contains a layer in which the structure closely resembles that of HAP. This factor gives a possible insight into the reason for the observed Ca/P ratios of the precipitated calcium phosphates. Scanning electron microscopic investigations of the crystals which frequently form plate-like morphologies which are not consistent with the hexagonal symmetry of HAP but may relate to the formation of OCP as a precursor phase during nucleation and growth (7,24). The formation of pyrophosphate upon heating the solid phases also suggests OCP as a precursor phase (42).

The difficulty of identifying the precursor phase by measuring the changes in stoichiometry of calcium phosphate solutions is principally that of analyzing for small differences in concentration during the reactions. In addition, the decrease in lattice ion concentration during the precipitation process results in supersaturated solutions metastable with respect to different calcium phosphate phases which can form and subsequently redissolve during the precipitation process as indicated by the solubility isotherms in Fig. 1. In the precipitation of calcium phosphate, these factors probably account for the variable stoichiometry and complex kinetics which have been observed experimentally. At physiological pH, the precipitating media contain HPO_4^{2-} ions in large excess over PO_4^{3-}, and there is little doubt that an appreciable protonation of the apatite surfaces can take place (11). In addition, cations such as sodium may be incorporated to a significant extent and other ions such as Cl^- may change the morphology of the developing crystals. Clearly, the phases which form depend upon the initial precipitating solution supersaturation. Thus at point A in Fig. 1, it has been shown that precipitation of HAP takes place directly without the participation of precursor phases (4) as is to be expected from thermodynamic considerations. However, it is at just such conditions at very low concentration that difficulties arise in attempts to calculate stoichiometric molar Ca/P ratios from changes in solution concentrations.

In the light of the above, we can conclude that the almost ubiquitous Ca/P ratio of 1.45 $\pm$ 0.05 during the major part of the precipitation reactions may be interpreted either in terms of "defect apatites" or mixtures of discrete calcium phosphate phases. There is a continuum of materials between hydroxyapatite and octacalcium phosphate which could be referred to as "defect OCP." A variety of hypotheses would account for the compositions of these materials as well as their structures (defect apatite, interlayered OCP, DCPA + HAP, sesquiapatite, or "collapsed" OCP), and quite possibly each has its own conditions of existence. Both the structures of these materials and the ways that two or more of them can be associated in the same "crystallite" deserve serious consideration. In terms of the calcium deficient arguments, when these phases are exposed to solutions containing calcium ions, they are taken up in an amount inversely proportional to the Ca/P ratio (34). The solids subsequently become less calcium deficient with Ca/P values approaching 1.67 for stoichiometric HAP. Unlike the discrete calcium phosphate phases, it is not possible to define solubilities, in the thermodynamic sense, for such calcium deficient phases, even though this has been attempted by many workers. Solubility products should be used only for solid phases which have a single free energy value, but attempts continue to be made to express "ACP" precipitation reaction in terms of thermodynamic driving forces. For systems in such a state of rapid kinetic change, the use of the solubility concept, which refers to true thermodynamic equilibrium, is questionable. As mentioned above (32), kinetic factors may be considerably more important in determining the nature of the phase which precipitates in calcium phosphate solution than are considerations based purely upon equilibrium thermodynamic factors.

In addition to the spontaneous precipitation studies, a number of seeded growth measurements have been made, since these offer experimental techniques which yield more reproducible results and, in addition, may be much better models for the precipitation in biological systems which invariably takes place in the presence of existing organic and/or mineral surfaces. The application of this method has resulted in some progress in the elucidation of the mechanism of calcium phosphate growth from studies both at variable and at constant pH (31,35). The results suggest that the overall precipitation reaction involves not only the formation of different calcium phosphate phases but also the concomitant dissolution of thermodynamically unstable OCP formed rapidly in the initial stage of the reaction. One method which was used in an attempt to identify the first-formed crystalline phase was the selective dissolution of the grown solids in carefully controlled kinetic experiments. However, such measurements made at constant pH controlled by the addition of alkali solutions also suffered from the disadvantage of marked reductions in calcium and phosphate concentrations, making it difficult to characterize the solid phases formed. Decreases in concentrations from the very high supersaturations (e.g., point B, Fig. 1) to the low values, indicated by point A in Fig. 1, resulted in marked changes in the thermodynamic driving forces for the precipitation of the intermediate phases.

The problems outlined above, associated with the substantial decreases in calcium and phosphate concentrations during the precipitation reactions, have been overcome by using a technique in which the precipitation is studied under conditions of constant T_{Ca}, T_p, pH, and ionic strengths typical of those in vivo. Following the addition of seed crystals to metastable supersaturated solutions of calcium phosphate, the activities of the crystal lattice ions are maintained constant by the simultaneous addition of titrant solutions from mechanically coupled burets, controlled by specific ion electrodes (glass and calcium electrodes for calcium phosphates and oxalates, respectively) (24). The application of this method in OCP (24) and HAP (1) crystallization has confirmed, kinetically, the formation of these phases during the precipitation reactions. These studies have provided direct evidence for OCP as a precursor to HAP formation, while studies at very low supersaturation have shown that macroscopic quantities of stoichiometric, highly crystalline, HAP can be formed. This is an important consideration because it confirms the observation that conditions where OCP cannot participate in the formation of apatite (high pH, high temperature, high fluoride) always appear to lead to the formation of an apatite that is nearly stoichiometric. This is a strong indication that the participation of OCP as a precursor is the cause for most, if not all, non-stoichiometric apatites. The hydrolysis of OCP in the presence of carbonate results in incorporation of CO_3^{2-} into the apatitic product, suggesting that this is the mechanism by which impurities are often incorporated into apatite. The constant composition method has enabled investigations of the influence of pH, supersaturation, seed-type and concentration, and traces of added substances, with a precision hitherto unattainable. The method also enables the morphology of the HAP crystallites to be investigated as a function of the extent of crystallization. Ions such as lithium and chloride, normally regarded as "inert," markedly affect the crystal shapes.

Unless special precautions are taken to exclude carbon dioxide during the precipitation process, the formation of calcium phosphates is always accompanied by the incorporation of carbonate ion. The presence of this ion is easily detected by characteristic peaks in the infrared spectra. In samples prepared at both high and low temperatures, the CO_3^{2-} ions are thought to be in specific sites within the apatite structure. The optical properties and the content of phosphate and various other ions appear to vary monotonically with carbonate content. In the presence of this ion, the radial distribution functions decrease in amplitude at larger interatomic spacings with increasing carbonate content as a consequence of the reduced regularity of atomic positions resulting from the strain induced by the substitution of PO_4^{3-} ions for CO_3^{2-} ions (2). It is generally held that in low temperature apatites, the CO_3^{2-} occupies predominantly the sites of the PO_4^{3-} ions, whereas in apatites prepared at high temperature, the CO_3^{2-} ion is in the vicinity of the sixfold axis, where it replaces two OH^- ions (8). It is clear that appreciable amounts of CO_3^{2-} ions can be incorporated into the structures of carbonate - HAP with a linear change of the unit cell dimension with CO_3^{2-} content. Synthetic apatites containing CO_3^{2-} usually show reduced crystallinity, whereas the biological apatites become more crystalline as the carbonate content increases. In general, the low Ca/P ratios in many precipitated calcium phosphates is thought to result from the presence of calcium vacancies with charge compensation achieved by incorporation of interstitial protons. Although many of the proposed lattice substitutions are speculative, it is quite possible that in the highly irreversible conditions of precipitation, appreciable concentrations of vacancies, interstitial ions, and substituents may become frozen into the structures (11).

The conversion of OCP to HAP, readily interpretable in terms of the similar crystallographic interlayered structures, is accelerated in the presence of fluoride ions. Thus Newesely (33) reported that OCP crystals are converted into apatite by a heterogeneous reaction in the presence of low fluoride concentration (approx. 10^{-6}M). This conversion was also verified by Brown (6), when OCP crystals were treated with 0.1M sodium fluoride solution. The importance of fluoride ion in the apparent reduction of dental caries may also be related to this factor, and there is little doubt that there is a positive correlation between fluoride levels in drinking water (usually approximately 1 ppm) and a decrease in dental caries. Seeded growth experiments of HAP in the presence of fluoride ion have shown that this anion reduces the rate of reaction at low concentration ($<10^{-5}$M) but enhances growth at higher concentrations ($>10^{-5}$M) (27). These results have also been confirmed, recently, in spontaneous precipitation experiments (41). Studies of the ionic molar ratios in the precipitating phase as calculated from initial rates of crystallization suggest that this phase is not fluorapatite (FAP) but rather a fluoridated hydroxyapatite in which the degree of substitution of fluoride ion for hydroxide ion in the lattice varies with the initial fluoride concentration in the solution (29). It has also been shown that fluoride retards the formation of ACP which precedes the hydrolysis of OCP to HAP and inhibits or even eliminates the formation of OCP-like precursor phases. This is in contrast to the influence of magnesium during the precipitation of calcium phosphate which apparently stabilizes TCP-like precursor phases (33). One of the problems in these studies, however, is that small errors in the determination of calcium and phosphate can lead to large errors in the calculated molar ratios of the precipitating phases. X-ray powder diffraction evidence pointing to mixed phases containing HAP and TCP may not be convincing, however, because of the difficulty involved in distinguishing between these phases from the diffractograms. Crystallization of HAP at very low constant supersaturation and without the involvement of precursor phases indicate that this phase was formed in the presence of magnesium ion at levels up to 1mM, although the rate of formation was markedly reduced in the presence of the added ion. The retarding effect could be interpreted in terms of a Langmuir adsorption isotherm representing the effective blocking of growth sites on the HAP seed crystal surface by the presence of magnesium ion (Koutsoukos and Nancollas, unpublished results).

CALCIUM OXALATE HYDRATES

Although calcium oxalate monohydrate, $CaC_2O_4.H_2O$ (COM) is the thermodynamically most stable calcium oxalate hydrate and one of the most important constituents of urinary calculi, many renal stones consist of an inner core of monohydrate with an outer surface layer of the thermodynamically less stable dihydrate, $CaC_2O_4.2H_2O$, (COD) (17). Another calcium oxalate phase, the trihydrate, $CaC_2O_4.3H_2O$, (COT), has been identified and the solubility isotherms of the three phases are shown in Fig. 2. In a calcium oxalate precipitation process, the Oswald-Lussac empirical rule would predict preferential formation of the kinetically favored least stable phase. During the formation of renal calculi of COM, therefore, precursor phases consisting of the higher hydrates should be taken into account since solutions saturated with respect to COT and COD are supersaturated in COM. Indeed, COD has frequently been reported as a stone constituent although its role in the overall process is not yet established (18).

Spontaneous calcium oxalate precipitation studies have shown that an initially formed COT readily undergoes transformation to COM (23). The process is markedly influenced by changes in

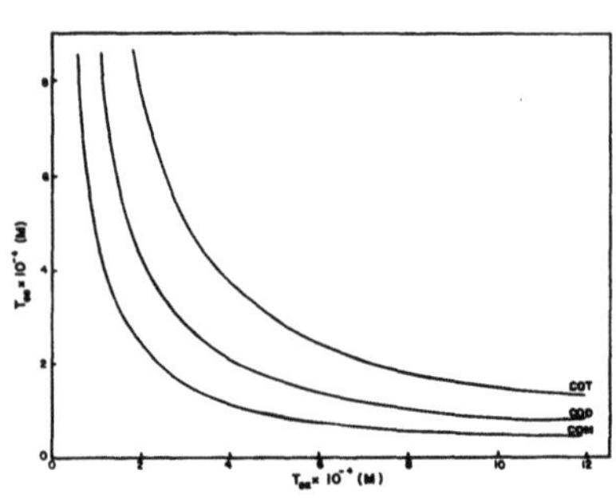

FIG. 2 - Solubility of calcium oxalate hydrates.

temperature and by the presence of potential mineralization inhibitors such as the pyrophosphate ion and organic phosphonates (23). The high apparent activation energy of the transformation process, 188.1 kJ mol^{-1}, was interpreted in terms of structural changes of the COT lattice, nucleation of COM, and solid/solid and solid/liquid interface reorganization. Such spontaneous precipitation studies are difficult to reproduce, inasmuch as the size and size distribution of the solid particles change continually during the course of the reaction. In vivo, the process of formation and dissolution of kidney stones is complicated by the presence of uncharacterized natural inhibitors and promoters which may markedly influence the nature of the hydrate phase which precipitates. Thus it has been shown that natural urine constituents selectively affect the rate of nucleation and crystallization of the calcium oxalates (14). Moreover, the method of pretreatment of the urine may influence the rate of the transformation. Although crystal growth inhibition studies have been made in urines appreciably diluted with water, the resulting changes in the ionic species distribution may have pronounced effects upon the experimental results (28).

Highly reproducible kinetic studies have been made of the dissolution and crystallization of COM, COD, and the least thermodynamically stable COT at temperatures from 15-50°C and a range of ionic media up to 0.3 M in sodium and potassium chlorides (39). The rate of dissolution is proportional to the relative undersaturation, and typical first-order rate constants, k_d, are shown in Table 2 from which it can be seen that the values fall in the sequence k_d (COT)>k_d(COD)>k_d(COM). The dissolution appears to be diffusion controlled with an activation energy of 12-17 k J mol^{-1}.

Both COD and COT readily undergo transformation to the least soluble COM and the process proceeds directly as shown by X-ray and scanning electron microscopic evidence. In contrast to the observed extensive nucleation of COM on COT, the transformation of COD to COM appears to take place through bulk

TABLE 2 - Dissolution, Growth, and Adsorption Reactions.

Phase	k_d	k_g	10^7x Γ(polyphos)	k_g(in $5x10^{-7}$M polyphos)
	$min^{-1}m^{-2}$	$\ell mol^{-1}min^{-1}m^{-2}$	$mol.m^{-2}$	$\ell mol^{-1}min^{-1}m^{-2}$
COM	1.24	$5.87x10^3$	2.71	~0
COD	1.61	$2.77x10^3$	0.44	5.5
COT	2.28	$2.47x10^4$	1.23	20.0

nucleation and growth of COM. In Fig. 3, the percentage transformation is plotted as a function of time, and it can be seen that the process is considerably more rapid for COT than for COD. It is also sensitive to temperature, ionic strength, and ionic medium composition. The transformation appears to proceed through dissolution and subsequent surface or secondary nucleation of the new phase and is also influenced by the presence of traces of crystal growth inhibitors such as polyphosphase, phytate, pyrophosphate, and whole urine (38).

It is clear from these studies that the nature of the calcium oxalate hydrate phase which forms may be markedly influenced by the presence of traces of foreign substances and ions in solution which may affect the precipitation rate of one phase more than another. In order to test this hypothesis, the extent of adsorption of linear polyphosphate (molecular weight approx. 4500) was investigated. It was shown that the three hydrates differ markedly in their surface adsorbability, Γ, falling in the sequence Γ(COM)>Γ(COT)>Γ(COD) (Table 2). This

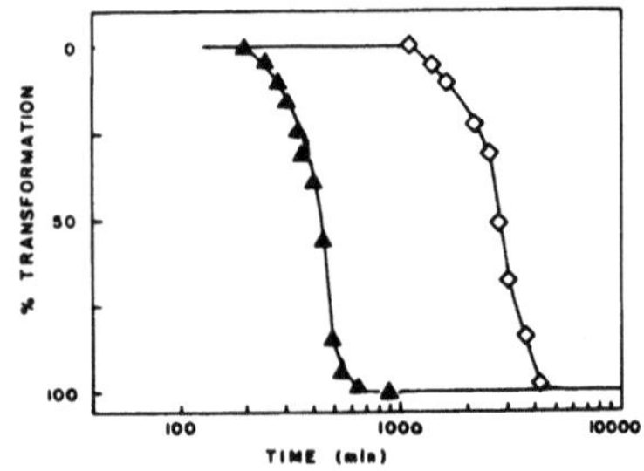

FIG. 3 - Transformation of hydrates into COM, 37°C. ▲COT; ◇COD.

marked difference in the adsorbability of anionic components may explain, in part, the stabilization of metastable hydrates, since the formation of stable COM is readily inhibited by the greater surface adsorption of impurity molecules. It is interesting to note that in the presence of additives such as polyphosphate, the mechanism of dissolution of all three hydrates appears to become surface controlled with a rate proportional to the square of the relative undersaturation (39).

In the corresponding seeded crystallization reactions from metastable supersaturated solutions of calcium oxalate, the rate is proportional to the square of the relative supersaturation provided that, in the case of COD and COT, concomitant transformation to COM is discouraged either by using a high solid/solution ratio or by adding traces of polyphosphate (40). The second-order growth rate constants given in Table 2 fall in the order $k_g(COT) > k_g(COM) > k_g(COD)$. The evidence points to a surface controlled crystallization, and the relatively low rates of COD growth probably account for the fact that this phase does not precipitate directly from pure calcium oxalate solutions. The adsorption of polyphosphate by the calcium oxalate hydrates results in a marked reduction in their rates of crystallization. The rate constants are shown in Table 2. The observed differences in the inhibitor adsorbability, resulting in the complete inhibition of formation of COM and stabilization of the thermodynamically less stable COT and COD, may play an important role in some in vivo remineralization processes. However, the overall process of formation of calcium oxalate calculi is the result of competitive kinetic processes such as nucleation, crystal growth, dissolution, and transformation, all of which may be moderated by the presence or absence of natural urine inhibitors.

Acknowledgment. I thank the National Institutes of Health for grants DE03223 and AM-19048 in partial support of this work.

REFERENCES

(1) Amjad, Z.; Koutsoukos, P.G.; Tomson, M.B.; and Nancollas, G.H. 1978. The growth of hydroxyapatite from solution. A new constant composition method. J. Dent. Res. 57: 909-910.

(2) Betts, F.; Blumenthal, N.C.; and Posner, A.S. 1981. Bone mineralization. J. Cryst. Growth 53: 63-73.

(3) Boistelle, R., and Rinaudo, C. 1981. Phase transition and epitaxies between hydrated orthorhombic and anhydrous monoclinic uric acid crystals. J. Cryst. Growth 53: 1-9.

(4) Boskey, A.L., and Posner, A.S. 1976. Formation of hydroxyapatite at low supersaturation. J. Phys. Chem. 80: 40-45.

(5) Brecevic, Lj., and Füredi-Milhofer, H. 1972. Precipitation of calcium phosphates from electrolyte solutions II. The formation and transformation of the precipitates. Calc. Tiss. Res. 10: 82-90.

(6) Brown, W.E. 1962. Crystal structure of octacalcium phosphate. Nature 196: 1048-1050.

(7) Brown, W.E. 1966. Crystal growth of bone mineral. Clin. Orthop. 44: 205-220.

(8) Brown, W.E. 1971. Progress report on apatites. Nat. Bur. Standards, Report Number 10 455: 1-25.

(9) Brown, W.E. 1973. Solubilities of phosphate and other sparingly soluble compounds. In Environmental Phosphorus Handbook, eds. E.J. Griffith et al., p. 203. New York: Wiley & Sons.

(10) Brown, W.E.; Schroeder, L.W.; and Ferris, J.S. 1979. Interlayering of crystalline OCP and HAP. J. Phys. Chem. 83: 1385-1388.

(11) Christoffersen, J. 1981. Dissolution of calcium hydroxyapatite. Calc. Tiss. Int. 34, in press.

(12) Davies, C.W. 1962. Ion Association. London: Butterworths.

(13) Despotovic, R.; Filipovic-Vincekovič, N.; and Füredi-Milhofer, H. 1975. Precipitation of calcium phosphates from electrolyte solutions III. Radiometric studies of the kinetics of precipitation and aging of calcium phosphates. Calc. Tiss. Res. 18: 13-26.

(14) Drach, G.W.; Thorson, S.; and Randolph, A.D. 1980. Effects of urinary organic macromolecules on crystallization of calcium oxalate. Enhancement of nucleation. J. Urol. 123: 519-523.

(15) Eanes, E.D.; Gilessen, I.H.; and Posner, A.S. 1966. Mechanism of conversion of noncrystalline calcium phosphate to crystalline hydroxyapatite. In Crystal Growth, ed. H.S. Peiser, pp. 373-376. Oxford: Pergamon Press.

(16) Eanes, E.D., and Meyer, J.L. 1977. The maturation of crystalline calcium phosphates in aqueous suspensions at physiologic pH. Calc. Tiss. Res. 23: 259-269.

(17) Elliott, J.S. 1973. Structure and composition of urinary calculi. J. Urol. 109: 82-83.

(18) Elliott, J.S., and Rabinowitz, I.N. 1976. Crystal habit, structure and incidence in the urine of hospital population. In Urolithiasis Research (Proc. Int. Symp. on Urolithiasis Research, Davos, Switzerland), eds. H. Fleisch, W.G. Robertson, L.H. Smith, and W. Vahlensieck, p. 257. New York: Plenum Press.

(19) Feenstra, T.P., and DeBruyn, P.L. 1979. Formation of calcium phosphates in moderately supersaturated solutions. J. Phys. Chem. 83: 475-479.

(20) Fleisch, H.; Russell, R.G.G.; Bisaz, S.; Termine, J.D.; and Posner, A.S. 1968. Influence of pyrophosphate on the transformation of amorphous to crystalline calcium phosphate. Calc. Tiss. Res. 2: 49-59.

(21) Francis, M.D. 1969. Inhibition of calcium hydroxyapatite crystal growth by polyphosphonates and polyphosphates. Calc. Tiss. Res. 3: 151-162.

(22) Francis, M.D., and Webb, N.C. 1971. Hydroxyapatite formation from hydrated calcium monohydrogen phosphate precursor. Calc. Tiss. Res. 6: 335-342.

(23) Gardner, G. 1976. Kinetics of the dehydration of calcium oxalate trihydrate crystals in aqueous solutions. J. Coll. Int. Sci. 54: 298-310.

(24) Koutsoukos, P.; Amjad, Z.; Tomson, M.B.; and Nancollas, G.H. 1980. Crystallization of calcium phosphates. A constant composition study. J. Am. Chem. Soc. 102: 1553-1557.

(25) Koutsoukos, P.G., and Nancollas, G.H. 1981. Crystal growth of calcium phosphates - epitaxial considerations. J. Cryst. Growth 53: 10-19.

(26) LaMer, V.K. 1962. The solubility behavior of hydroxyapatite. J. Phys. Chem. 66: 973-978.

(27) Meyer, J.L., and Nancollas, G.H. 1972. Effect of stannous and fluoride ions on the rate of crystal growth of hydroxyapatite. J. Dent. Res. 51: 1443-1450.

(28) Meyer, J.L., and Smith, L.H. 1975. Growth of calcium oxalate crystals. II. Inhibition by natural urinary crystal growth inhibitors. Invest. Urol. 13: 36-39.

(29) Moreno, E.C., and Varughese, K. 1981. Crystal growth of calcium apatites from dilute solution. J. Cryst. Growth 53: 20-30.

(30) Nancollas, G.H. 1965. Interactions in Electrolyte Solutions. Amsterdam: Elsevier.

(31) Nancollas, G.H., and Mohan, M.S. 1970. The growth of hydroxyapatite crystals. Arch. Oral Biol. 15: 731-745.

(32) Nancollas, G.H., and Tomažić, B. 1974. Growth of calcium phosphate on hydroxyapatite crystals. Effect of supersaturation and ionic medium. J. Phys. Chem. 78: 2218-2225.

(33) Newesley, H. 1961. Changes in crystal types of low solubility calcium phosphates in the presence of accompanying ions. Arch. Oral Biol. 6: 174-180.

(34) Pak, C.Y.C., and Skinner, H.C.W. 1968. Ionic interaction with bone mineral. IV. Varying affinity of synthetic calcium phosphates for Ca^{2+}. Biochim. Biophys. Acta. 165: 274-282.

(35) Strates, B.S.; Neuman, W.F.; and Levinskas, G.I. 1957. The solubility of bone mineral II. Precipitation of near neutral solutions of calcium and phosphate. J. Am. Chem. Soc. 61: 279-282.

(36) Termine, J.D., and Eanes, E.D. 1972. Comparative chemistry of amorphous and apatitic calcium phosphate preparations. Calc. Tiss. Res. 10: 171-197.

(37) Termine, J.D.; Peckauskas, R.A.; and Posner, A.S. 1970. Calcium phosphate formation in vitro. II. Effects of environment on amorphous-crystalline transformation. Arch. Biochem. Biophys. 140: 318-325.

(38) Tomažić, B., and Nancollas, G.H. 1979. A study of the phase transformation of calcium oxalate trihydrate-monohydrate. Invest. Urol. 16: 329-335.

(39) Tomažić, B., and Nancollas, G.H. 1979. The kinetics of dissolution of calcium oxalate hydrates. J. Cryst. Growth 46: 355-361.

(40) Tomažić, B., and Nancollas, G.H. 1980. Crystal growth of calcium oxalate hydrates. A comparative kinetics study. J. Coll. Int. Sci. 75: 149-160.

(41) Van den Hoek, W.G.M.; Feenstra, T.P.; and DeBruyn, P.L. 1980. Influence of fluoride on the formation of calcium phosphates in moderately supersaturated solutions. J. Phys. Chem. 84: 3312-3317.

(42) Winand, L.; Dallemagne, M.J.; and Duyckaerts, G. 1961. Hydrogen bonding in apatitic and calcium phosphates. Nature 190: 164-165.

Biological Mineralization and Demineralization, ed. G.H. Nancollas, pp. 101-141.
Dahlem Konferenzen 1982. Berlin, Heidelberg, New York: Springer-Verlag.

Structures of Biological Minerals

R. A. Young* and W. E. Brown**
*School of Physics and Engineering Experiment Station,
Georgia Institute of Technology, Atlanta, GA 30332
**American Dental Association Health Foundation Research Unit
National Bureau of Standards, Washington, DC 20234, USA

Abstract. The basic chemical and physical properties of ionic crystals reside in their structures. Structures of the following compounds of biological interest are described: calcium phosphates, a urate, calcium carbonates, and calcium oxalates. To a considerable degree their properties are affected by the presence of impurities and lattice defects in the crystals. In keeping with this, the kinds and locations of impurities and defects in hydroxyapatite are discussed, with special reference to how these affect the transport of ions along the hexagonal channels of hydroxyapatite. The crystal structure of octacalcium phosphate, which has a special relationship to that of hydroxyapatite, is described. The ways in which octacalcium phosphate may affect the growth mechanisms, impurity-defect content, morphology, stoichiometry, and chemical properties of hydroxyapatite are described. It is shown that much of the variability in the chemistry of the β-$Ca_3(PO_4)_2$-whitlockite series of compounds resides in one of the 21 calciums in a unit cell. It is suggested that the hydrated compounds in all these series of biominerals may have special roles during the nucleation stage, and subsequently because they usually grow more rapidly than the anhydrous forms. The structure of $CaCO_3 \cdot 6H_2O$ contains $CaCO_3$ ion pairs entirely isolated from each other by surrounding water molecules; this structural unit may participate in various important biochemical processes.

INTRODUCTION

It is generally known that the physicochemical properties of a material depend on the kinds and relative amounts of atoms present. But the properties also depend strongly on just how the constituent atoms are located relative to one another, i.e., on the details of the atomic scale architecture. Maintaining the same atoms but shifting them around can drastically alter material properties. Understanding of the atomic-scale mechanisms that produce the material properties, including response to environment, can be a powerful tool for the scientist bent on modifying a material's properties in desired ways, even including properties associated with health and disease.

The first step in developing this atomic-scale understanding is to determine the crystal structure (if the material is crystalline) of the model compound. The term crystal structure refers to the precise spatial arrangement of the constituent atoms of a material, and sometimes their root-mean-square thermal vibrations as well. The essence of crystallinity is spatial periodicity, i.e., repetition by translation, the repeating unit being the unit cell with edge lengths (axes) $\underline{a}$, $\underline{b}$, and $\underline{c}$ and with inter-axial angles α, β, and γ. Thus, the atom positions, x, y, z, need be specified only within a single unit cell; by definition, all other unit cells are identical. (In reality, we see that the specified unit cell and its contents represent an average, over the entire crystal, of the actual cells and their contents.)

The second step is to determine the structural locations and roles of the "impurities" and distortions that commonly occur in the natural materials of interest but not in the idealized model compound.

The third step is to relate differences in properties to the differences in the crystal structural details which occur naturally or can be imposed experimentally.

The fourth step, obviously, is to deduce the atomic scale mechanisms of the properties and processes of interest and to apply that knowledge to produce desired results, whether they be caries prevention, stimulation of bone regrowth, inhibition of formation of ectopic calcifications, improved biocompatibility of implanted or adherent restorative materials, or any other biological or nonbiological desiderata.

Thus, studies of the crystal structures of biological minerals and related compounds and of their structure-related properties are important areas of research in many disciplines. In the biomineralization context, some of the most important properties are growth and dissolution mechanisms, incorporation of impurities, presence of structural defects, chemical reactivity, nucleation mechanisms, and the thermodynamic properties which establish relative stabilities in the various biological fluids. More than 30 different "minerals" (inorganic crystalline deposits) occur in biological settings (Table 1).

In each, the atoms and molecules are fitted together in a specific periodic spatial arrangement, the crystal structure. The usefulness of knowing such detail about biological minerals will be illustrated with several discussions of atomic-scale mechanisms relating structural detail to differences in physicochemical properties of the crystals, including growth, change, and dissolution. In some cases, biological implications of these properties are addressed. An exhaustive, detailed description of all the known structures would be dull reading and not particularly enlightening, per se. Therefore, for the most part, detailed description is given for only a sampling of those structures which are most important in human biological mineralization and are best known in depth and variety of detail. The structures to be described are calcium phosphates, a urate, calcium carbonates, and calcium oxalates.

CALCIUM PHOSPHATES

Occurrence

The calcium phosphates comprise the largest group of biominerals in vertebrate animals, as judged with respect to numbers of compounds that exist (at least 11 different types), to the number that are present in hard tissue (at least 5 types), and to the amounts that are present in the body. The crystal

TABLE 1 - Diversity and phylum distribution of biomineralization products in extant organisms, adapted from "Minerals Formed by Organisms," by H.A. Lowenstam (30).

	KINGDOM				
	MONERA	PROTISTA	FUNGI	ANIMALIA	PLANTAE
CARBONATES:					
Calcite	+	+		+	+
Aragonite	+	+		+	+
Vaterite		+		+	+
Monohydrocalcite	+			+	
PHOSPHATES:					
Dahllite } Carbonate apatite	+	+	+	+	?
Francolite } with F, Cl, OH				+	
Mg-whitlockite				+	
Brushite				+	
Calcium Pyrophosphate Dihydrate				+	
Struvite				+	
Monetite				+	
HALIDES:					
Fluorite				+	
OXALATES:					
Whewellite		+	+	+	+
Weddelite			+	+	
SULFATES:					
Gypsum				+	+
Celestite		+			
Barite		+			
SILICA:					
Opal		+		+	+
Fe-OXIDES:					
Magnetite	+			+	
Maghemite	?				
Goethite				+	
Lepidocrocite				+	
Ferrihydrite			+	+	+
Mn-OXIDES:					
"Todorokite"	+				
Fe-SULFIDES:					
Pyrite	+				
Hydrotroilite	+				

structures of the calcium-phosphate biominerals can generally be modeled as derivatives (principally via atomic substitutions and disordering) of one or another of these five model compounds: $Ca_5(PO_4)_3OH$ (hydroxyapatite); $Ca_8H_2(PO_4)_6 \cdot 5H_2O$ (octacalcium phosphate); β-$Ca_3(PO_4)_2$ and its modification, whitlockite; $CaHPO_4 \cdot 2H_2O$ (brushite); and $CaHPO_4$ (monetite). Abbreviations used herein for these compounds are, respectively, OHAp, OCP, TCP, DCPD, and DCPA.

APATITES

Structure, Structural Variations and Properties

The biologically most important of these five compounds is OHAp because it is the starting model-compound for most of the mineral, "biological apatite," in hard tissues. OHAp is remarkable in how readily it accommodates a great variety, in both kind and degree, of substitutions and modifications, leading to an unusual degree of variability of properties. For both of these reasons, the OHAp derivatives are the most studied and have had the most information developed about them, yet are the most surrounded by controversies.

The apatites are notorious for their occurrence, in vitro and in vivo, in nonstoichiometric form (usually, but not always, with low Ca/P ratios). The other calcium phosphates are not nearly so inclined to be nonstoichiometric nor to incorporate lattice defects and impurities. A large number of proposals have been made to account for the nonstoichiometry of apatites, too many to discuss in detail here, and it is very unlikely that only one mechanism will account for all the cases of nonstoichiometry. Adsorption of calcium and phosphate ions on the surfaces undoubtedly accounts for much of the nonstoichiometry of very small particles. When such materials are washed to remove the adsorbed ions, the large surface area and the low solubility of the adsorbed ions require inordinate amounts of water to be used for their removal. It has been suggested (5) that an approximately half unit cell layer of OCP may act as the transition layer between the apatitic lattice and the aqueous phase. This would be a structural feature of the crystal and not an adsorbed layer. A simple calculation shows that if exactly a half unit cell of OCP were present on a hexagonal, needle-shaped crystal of hydroxyapatite 500 Å in diameter, the Ca/P ratio would be 1.64 instead of 1.67. It has been suggested that complexes with compositions Ca_2PO_4OH can form on the surface, and that these would alter the stoichiometry of the overall product. Reliable evidence for a discrete unit of this type, however, is lacking, and its existence appears improbable on structural grounds. However, the major cause of nonstoichiometry is simply incorporation of impurities (including vacancies), usually substitutionally, but also interstitially. Hydroxyapatite tends to incorporate much larger amounts of impurities than do the other calcium phosphates. Neither brushite nor monetite is known to incorporate significant amounts of impurities, except that brushite can form solid solutions with its isostructural analog, gypsum. It is noted later that whitlockite appears to prefer (a) the presence of certain monovalent or divalent cations in sites which have octahedral oxygen coordination, and (b) the incorporation of waters of constitution.

We will discuss the crystal structure of stoichiometric OHAp first, and then, the fact and consequences of various departures from the ideal until, finally, we arrive at a structural model that better approximates the most highly organized biological apatite, human tooth enamel (TE). To have an idea of how long that road is, the reader may wish to note that, compared to OHAp, TE has ~40% deficiency in OH, a Ca/P ratio closer to 1.58 than 1.67, 2-3 wt% CO_3 distributed between two sites, 3-5 wt% H_2O, of which at least one-half is structurally incorporated as molecular water, 2-5 wt% acid phosphate (HPO_4^{2-}), an expanded $\underline{a}$ axis in the unit cell, a few tenths wt% each of Cl, F, Na, and Mg, and a great variety of lesser substituents.

Crystals can have a great deal more symmetry than just the translation symmetry (spatial repetition) based on the unit cell dimensions. Other symmetry operations that can be present are face-centering and body-centering translations (F,A,B, C;I), mirror planes (m), rotation axes [2,3,4,6], inversion centers ($\bar{1}$), and translations of less than a unit cell dimension combined with a mirror plane (giving rise to a glide plane) or a rotation axis (giving rise to a screw axis). The space group of a crystal is the group of these symmetry elements that is present in it. The space group symbol for stoichiometric OHAp is $P2_1/b$, which means that there are no face- or body-centering translations, there is a screw axis which advances one-half a unit cell distance (in this case, along $\underline{c}$) with a 180° turn, and there is a glide plane perpendicular to the screw axis on which the glide is parallel to $\underset{\sim}{b}$ and equal to b/2. The monoclinic unit cell has parameters $\underline{a}$ = 9.419 Å,

$\underline{b} = 2\underline{a}$, $\underline{c} = 6.880$ Å, $\gamma = 120^{\circ}$, $\alpha = \beta = 90^{\circ}$, and it contains $4Ca_5(PO_4)_3OH$. This structure is shown in perspective in Fig. 1. As for all of the calcium phosphates, the structure consists of an array of PO_4 tetrahedra held together primarily by Ca ions interspersed among them. The Ca ions occur in two markedly different sites, in slightly staggered columns (Ca(I)) and in nearly equilateral triangles (Ca(II)) centered on the screw axis. The OH's occur in columns on the screw axes. These columns and triangles are more easily visualized in Fig. 2, which shows the projection of a portion of the structure onto the a, b plane.

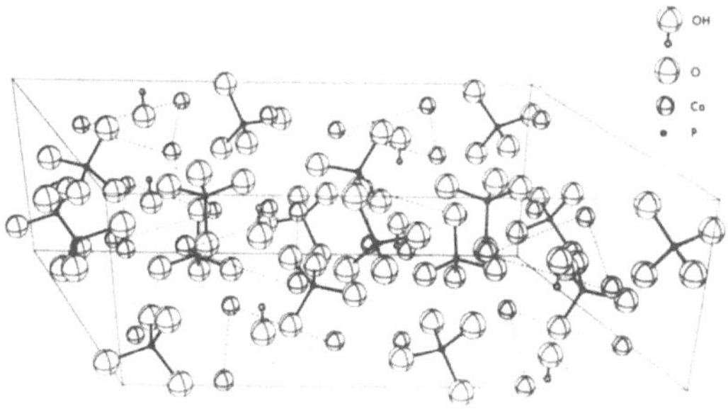

FIG. 1 - Monoclinic OHAp. Note how ordering of OH^- orientations has doubled the $\underline{b}$ axis (46).

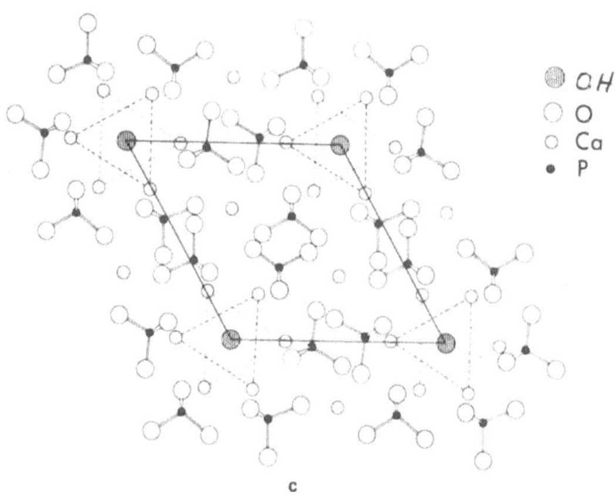

FIG. 2 - OHAp projected onto $\underline{a}$, $\underline{b}$ plane (46).

The parallelogram outlines one-half of the monoclinic unit cell. It can be seen in Fig. 1 that all of the OH's associated with the Ca triangles on the left point upward, while those on the right point downward. This relationship, imposed by the glide plane, is made clearer by Fig. 3.

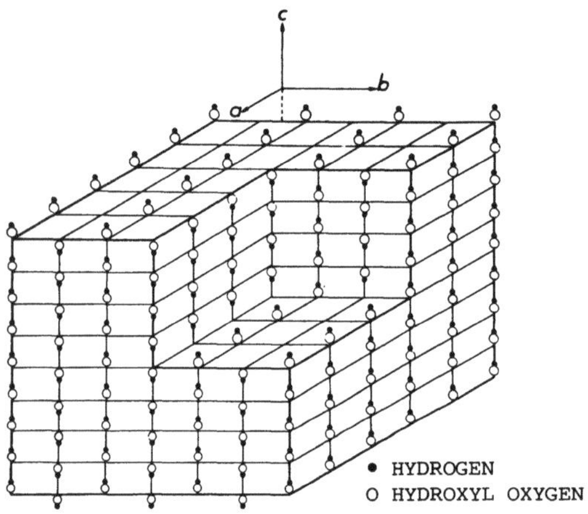

FIG. 3 - Schematic representation of the ordered arrangement of hydroxyl ions in monoclinic, $P2_1/b$, hydroxyapatite. The rest of the unit cell content, consisting of calcium and phosphate ions, has been omitted for clarity (22).

The O's do not lie exactly in the planes of the Ca triangles (represented by the horizontal lines in Fig. 3), but lie either above or below those planes by ~0.4 Å with the associated H being always ~1 Å farther away along the screw axis. This monoclinic form of OHAp is formed in bulk only at high temperatures. It is the most ordered and is the thermodynamically most stable form, even at ordinary temperatures.

The form of OHAp most frequently encountered is hexagonal. It can be regarded as being derived from the monoclinic form by allowing the OH's to become statistically disordered about their associated Ca triangles, as indicated in Fig. 4.

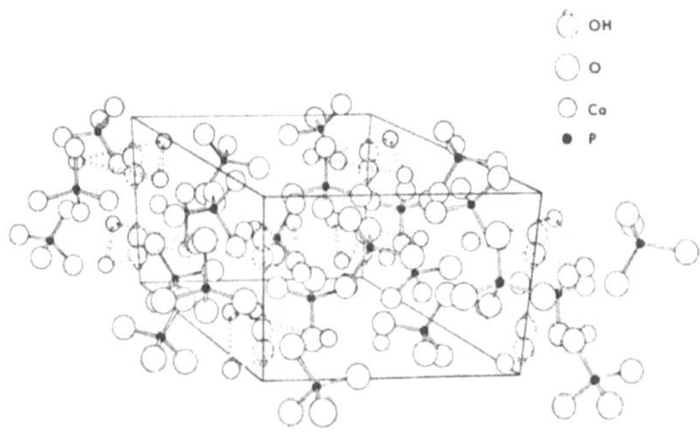

FIG. 4 - Hexagonal OHAp. a is directed to the right, c upward, and b into the page. The fact that each OH^- position is statistically only 50% occupied is indicated by the dotted outline of half of each OH group (46).

The glide plane becomes simply a mirror plane, and b = a = 9.419 Å. The twofold axis becomes sixfold, the space group is $P6_3/m$, the Ca (II) triangles are equilateral, and the Ca(I) ions are accurately aligned (no longer staggered) one above the other in columns at 1/3, 2/3, z and 2/3, 1/3, z. The actual changes in atom positions required to accomplish this are very small, <0.1 Å, except for the OH's; half of them must now be on the "other" side of the associated Ca triangle, i.e., the side opposite to where they would be if the structure were monoclinic.

As can be seen from Fig. 3, in monoclinic OHAp all of the OH's in a given column are pointed in the same direction. If one OH were reversed by being re-oriented and placed on the other side of its associated Ca triangle, there would be steric interference between it and the next adjacent OH. As far as is known, this steric hindrance is sufficient to prevent two OH's (pointing generally toward each other) from occurring between any two adjacent Ca triangles. Thus, a reversal of the OH direction within a column requires that some reversal point be provided by omission of an OH^-. This can be accomplished by replacement of an OH^- with a vacancy, or F^-, or Cl^-, etc.

Thus, one can conclude that hexagonal OHAp is probably never strictly stoichiometric. (The possible occurrence of an hexagonal form in which the columns of OH's are all ordered within themselves, but not with respect to one another, has not been entirely ruled out. However, neither has it ever received any experimental support, so it is now considered rather improbable.)

The differences between monoclinic and hexagonal OHAp (Fig. 4) are seen to be very small structurally, yet it is sufficient to affect some properties quite strongly, e.g., dissolution kinetics, deuteration kinetics (and presumably, kinetics of diffusion along the OH-column region). These differences seem to be associated primarily with the "X-ion channels."

If all OH's were removed, there would be left in the structure a tunnel, or channel, where each OH column had been. These channels appear to be the sites of a great deal of the interesting activity in apatites. They can accommodate a great variety of substitutions, and for this reason, they are sometimes called the X-ion channels, where X = OH, F, Cl, CO_3, □, etc., or almost any combination of these. Table 2 shows how most of these X-ion substitutions affect lattice parameters.

TABLE 2 - Effects of various substitutional or interstitial entities on lattice parameters.

	One TE unit* of H_2O	1 wt% CO_3 in B-site	1 wt% CO_3 in A-site	$(2\,OH^-)_x$ by $(O^{2-} + \square)_x$	$(OH)_x$ by Cl_x	$(OH)_x$ by F_x
Δa (Å)	+0.041	-0.0059	+0.026	-0.0268x	+0.223x	-0.052x
Δc (Å)	——	+0.0024	-0.0080	-0.0129x	-0.116x	0.00
Reference	Holcomb & Young (26)	LeGeros, Trautz, Klein & LeGeros (29)	Bonel (2) Elliott (19)	Verbeek, Heiligers, Driessens, & Schaeken (42)	Sudarsanan & Young (38)	Sudarsanan & Young (38)

* This is the amount of structurally incorporated H_2O rapidly lost in heating the TE between ~250 and 300°C (26).

Diffusion along the X-ion channel may be a particularly important property, perhaps even being associated with chemical dissolution. If the X-ion sites are filled with F or Cl instead of OH, one has fluorapatite ($Ca_5(PO_4)_3F$) or chlorapatite ($Ca_5(PO_4)_3Cl$). Fractional millimeter-sized single crystals of chlorapatite can be converted to single crystals of OHAp, all the while remaining the same single crystals, by being exposed to steam at 1200°C (21,23). This can only happen by diffusion of Cl^- ions out and OH^- ions in along these channels. Acid dissolution of synthetic OHAp and of TE tends to be very anisotropic, and in that, too, the X-ion channels are implicated.

Diffusibility can be assessed in part by deuteration kinetics for OH↔OD exchange in "OHAp," the model being that D_2O must be diffusing in and H_2O out along the channels. All hexagonal OHAp's tested deuterated much more readily (in most cases by orders of magnitude) than did any of the monoclinic OHAp's (Holcomb and Young, in preparation). Presumably, the fact that in the monoclinic OHAp each OH site was occupied with an OH oriented in a minimum free energy configuration with respect to its neighbors, left little opportunity for the diffusing species to pass each other in the channels. These channels are bordered by Ca(II) ions and by O(III) ions from the PO_4 groups. Any departure from stoichiometry resulting in either Ca(II) vacancies or PO_4 vacancies might be expected to assist diffusion by providing passing sites. Biological apatites tend to have such features and they do, in fact, deuterate rather readily compared to most synthetic apatites.

If an occasional F is substituted for an OH in OHAp (e.g., 1 in 12 to 15 OH groups), not only does this provide reversal points, forcing the structure to be hexagonal, but in this case it also inhibits diffusion. Figure 5 shows how the reversal can take place and that hydrogen bonding between F and O(H) occurs.

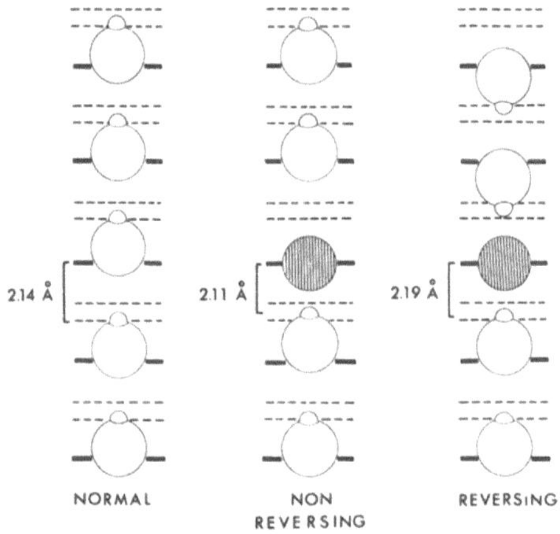

FIG. 5 - Models for F-OH interaction in (F, OH) apatite. The shaded ion is fluorine (46).

Thus the ability of an OH to participate in the deuteration exchange is inhibited if it is adjacent to an F^- (to $\lesssim$70%, according to one study). Our knowledge of the crystal structural details makes it possible for us to understand this in a mechanistic sense at the atomic level, and thereby, perhaps, to understand similarly the cariostatic effect of F in TE (if, indeed, it is the F in TE that is responsible).

Many other substitutions in the X-ion channel have been observed and studied. Each substituting species takes up its own particular location on the screw axis and each affects the lattice parameters (Table 2), among other things. In the pure forms of OHAp, FAp, and ClAp, OH is at 0,0,0.19, F is at 0,0,1/4, and Cl is at 0,0,0.06 (plus, in each case, others of the symmetry-related positions, such as 0,0,z; 0,1/2,1/2-z; 0,0,z+1/2; and 0,1/2,1-z in the monoclinic structure). But when two or more of these ions are present at once, interactions among them occur. The previously mentioned hydrogen bonding between F and OH tends to move both the F^- and the OH^- slightly. The large Cl^- ion not only affects the position of neighboring F^- or OH^- but, interestingly, when it is present in only minor amounts its own position is shifted nearly 0.5 Å along the screw axis (to z ~0.13). The effects of these interactions on physicochemical properties other than crystalline phase have not been studied in detail as yet. One might expect to find significant effects on many properties, for example, on diffusion along the X-ion channels.

CO_3^{2-} and O^{2-} are two X-ion substitutions which have been studied mostly in synthetic apatites prepared at high temperatures but which are, nonetheless, of biological interest.* Their effects on lattice parameters are included in Table 2. Each replaces 2(OH^-); only one occurs per unit cell. The precise location of the O^{2-} has not been determined; one would expect it to be at 0,0,1/4. The plane of the CO_3^{2-} group is inclined ~18° to the screw axis. One report, based on powder diffraction studies, places the C, of this CO_3^{2-}, near 0,0, 0.12 (47). The CO_3^{2-} ions in the X-ion channel tend to occur in an ordered fashion, eliminating the screw-axis translation and thus changing the space group (symmetry). When ~90% or more of the possible CO_3^{2-} A sites (i.e., in the X-ion channel) have been filled, the space group is changed again, to Pb (20).

CO_3^{2-} is also one of the many substituents for PO_4^{3-}. Charge balance is maintained by additional processes, of which several have been suggested (e.g., Na^+ for Ca^{2+}, [$CaOH^+$] loss, inclusion of an F^- with the CO_3^{2-} at the PO_4^{3-} site, substitution of a divalent ion such as O^{2-} for OH^-, etc.). The precise location(s) of the CO_3^{2-} within the confines of the PO_4^{3-} site

---

*CO_3^{2-} in this position (A-site) has been shown to be present in human tooth enamel. Apatites have been prepared at high temperatures (~1000°C) with essentially only O^{2-} in the X-ion sites, but on cooling, even in vacuum, there was some O^{2-} + $H_2O \rightarrow 2(OH^-)$, and the pure oxapatite could not be stabilized at room temperature. However, the conversion was not complete; some O^{2-} did remain in room temperature specimens stored in air (41). Also, a mineral apatite has been reported with X_2 = $F_{1.29}Cl_{0.26}(OH)_{0.14}O_{0.32}$. Therefore, it would seem that in biological apatite a few of the X sites could be taken up by O^{2-}.

are not known and may, in fact, differ with the charge compensation mechanisms.

Carbonate contents of both synthetic and biological apatites have been positively correlated with acid dissolution rate (25,27,35). The possible involvement of the X-ion channels in dissolution of carbonate-containing apatites is indicated by two recent observations. In the first (26), it was noted that, on thermal decomposition of human tooth enamel, much or all of the B-site CO_3^{2-} (i.e., in PO_4^{3-} sites) left by way of the X-ion channels, the actual diffusing species being CO_2 (some of it becoming, briefly, A-site CO_3^{2-} along the way). The second observation ((36), and Montel and Bonel, private communication) was that, when AB carbonate apatite (i.e., with CO_3^{2-} in both A and B sites) was subjected to high pressures, all vacancies were lost, the A sites were all filled, and the resulting product was stable at much higher temperatures than was the initial material. The implication is that, with no vacancies along the X-ion channels, diffusion even of CO_2 was strongly inhibited, locking in the B-site CO_3^{2-}. Again, we see the great influence exerted on properties by seemingly minor structural details.

Other than for X-ions, there are many substitutions and, perhaps, additions which are of biological interest. Substitutions for Ca, the occurrence of HPO_4^{2-}, and the structural incorporation of molecular water, H_2O, are high among these. Sr^{2+} for Ca^{2+} is of special interest because of ^{90}Sr and because in the normal turnover of bone the Sr is eliminated somewhat preferentially to Ca. One would like to find the atomic-scale mechanism. A detailed structure study (38) of a presumably adequate model compound $(Ca,Sr)_5(PO_4)_3Cl$ showed (a) that the Sr ion's position differed from the Ca position by ~0.1 Å, in accord with its larger radius, and (b) when the degree of substitution was small (7.2 and 2.3 atom % in the study), Sr replaced only Ca(II), and not Ca(I). The Ca(II) ions are those in the triangles surrounding the X-ion columns. Thus, when only small amounts of Sr ions are present, they are in positions bordering the X-ion channels. Possibly, this makes them more accessible for replacement than they would be in Ca(I) sites.

Both HPO_4^{2-} and structurally incorporated H_2O occur in some synthetic "hydroxyapatites" and in some biological apatites. Their precise structural locations are not known. Presumably, the PO_4 part of HPO_4 is in much the same location as is a non-protonated PO_4. Only little is known about the extent to which structural water affects lattice parameters. The most nearly definitive study correlated a 0.014 Å contraction of the a-axis in TE (human tooth enamel) with an abrupt (in temperature) loss at 270-300°C of 1/3 to 1/2 of the structural water initially present (26). This observation is the basis for the entry in Table 2. It may be that the structural water is present in more than one site, only one of which has a profound effect on the a lattice parameter. Possible locations are in the X-ion channel, in nearby locations corresponding to the positions of some of the H_2O molecules at an OCP-OHAp interface or an OCP interlayer which may have existed as the OHAp crystals grew from an OCP precursor.

Let us turn, now, to biological apatites. Insofar as is known, none of these is monoclinic. That is to be expected, because these natural materials contain many "impurity" ions and vacancies that can provide reversing points for the OH directions in the X-ion columns, thus enforcing hexagonal symmetry over the monoclinic symmetry of truly stoichiometric OHAp. The lack of observation of monoclinic biological apatite means only that the structures are hexagonal on the average over regions of the size sensed in X-ray diffraction (several tens to hundreds of unit cells). The possibility remains that smaller regions of monoclinic structure could exist. That would be interesting because the properties, including dissolution properties, of the monoclinic regions would differ from those of the regions with truly hexagonal structure. Thus, a considerable degree of heterogeneity in properties (e.g., diffusion and dissolution) could be produced in biological apatite by these minor structural differences.

Let us now turn to TE. TE deuterates very readily, almost as readily as the most easily deuterizable synthetic "OHAp" preparation tested. This fact together with the broadening of the 3570 cm^{-1} i.r. band (OH stretch) suggest that there is much disorder in the X-ion channels, with many places for the diffusing species to pass each other. These observations are consistent with the OH deficiency (up to 40%), the occurrence of a Cl instead of an OH in approximately 10% of the cells, the probable substitution of the few tenths wt% Na^+ for Ca^{2+} with additional (as yet) unspecified charge compensating events, the occurrence of a small percentage of F instead of OH, and the occurrence in these X-ion channels (A site for CO_3) of 10-20% of the total CO_3 present in TE. The remainder of the CO_3 replaces PO_4 (B sites). It is presumed that the presence of an H in HPO_4 does not disturb the location of the PO_4 very much, but neither that nor the location of the H has been determined experimentally. Neither has the precise location and orientation of the CO_3 in the B sites been determined, nor has the placement or the location of the structurally incorporated water. Nonetheless, it was possible for a preliminary crystal structure refinement of TE, per se, to be carried out with an OHAp-like model which simply omitted the B-CO_3, the H of HPO_4, and anything else not in OHAp and not on or near the screw axis in the X-ion channel (48). The result confirmed that TE (dense fraction) is, indeed, similar to OHAp in the placement of the main atoms. Differences in apparent site occupancies were observed which may later be used to help locate some of the omitted species.

OCTACALCIUM PHOSPHATE

OCP has been reported to be present in dental calculus, renal calculi, bone, and juvenile feline and canine tooth enamel (26). However, its most important role in biological mineralization is probably that of a precursor to other solid phases found in biological settings, especially defect OHAp. There is a considerable range of pH values and solution compositions over which OCP is a thermodynamically metastable phase and where it can be formed before OHAp (see Fig. 1 in Nancollas, this volume). Transformation of OCP to OHAp has been observed and accounts for many features of biological "OHAp," as will be discussed below. Therefore, the present discussion of the OCP structure is developed in the context of its relationships to OHAp.

Structure

OCP crystallizes in space group $P\bar{1}$ with $\underline{a}$ = 19.715 Å, $\underline{b}$ = 9.534 Å, $\underline{c}$ = 6.839 Å, α = 90.14°, β = 92.52°, and γ = 108.67°. It consists of alternating layers of PO_4^{3-} groups interspersed with Ca^{2+} ions, as in apatite (the "apatitic layer"), and layers of more widely spaced PO_4^{3-} and Ca^{2+} ions with H_2O interspersed (the "hydrated layer"), as indicated in Fig. 6 (4).

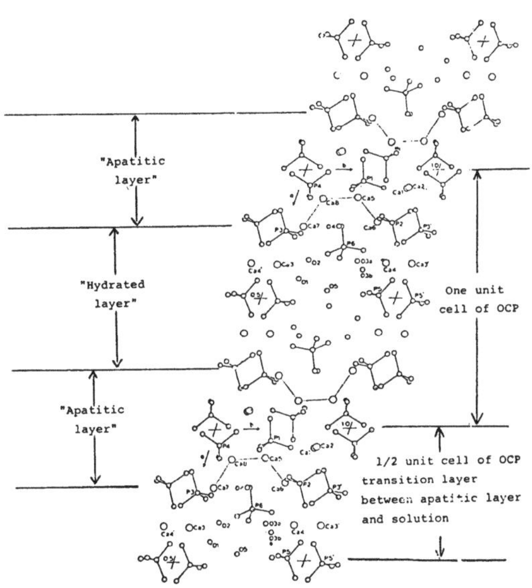

FIG. 6 - Crystal structure of octacalcium phosphate projected down c. Two unit cells are shown, and the "apatitic," "hydrated," and proposed transition layer between the apatitic and the aqueous phase are delineated.

The close relationship of the OCP and OHAp structures is shown in Fig. 7, where the atomic positions of OCP are seen superimposed on the structure of OHAp by means of the small filled circles. In the apatitic layer the two sets of positions are very close, whereas in the hydrated layer, although there is some similarity, the composition and atomic positions of OCP are quite different from those of OHAp. The closeness of fit in the apatitic layer accounts for the epitaxial, interlayered mixtures formed by these two compounds. Furthermore, since the hydrated layer of OCP is almost certainly hydrophilic in nature, and therefore more likely than the apatitic layer to form a low energy interface with aqueous phase, it has been suggested that the transition layer to the aqueous phase is equivalent to approximately a half unit cell of OCP as indicated in Fig. 6. The presence on each of an OCP hydrated layer would suggest that OCP can grow on OHAp and vice versa.

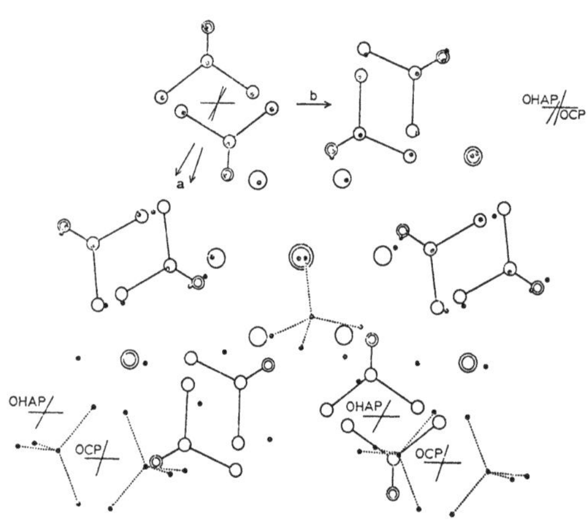

FIG. 7 - The crystal structure of hydroxyapatite projected down c with the atomic positions of octacalcium phosphate represented by small filled circles. The latter coincide very closely with those of hydroxyapatite in the "apatitic" layer but not in the "hydrated" layer.

OCP Role in Growth and Properties of Biological Apatites

Although interlayering is an important phenomenon in terms of the chemical properties and compositions of mixtures of these two salts, it is only one of the ways in which OCP can affect the growth and properties of OHAp. Dehydration, hydrolysis, pyrolysis, and fluoridation of OCP produce a variety of products, some poorly understood, which can have important effects on the properties of apatitic materials. Of these effects, we discuss here only those associated with the hydrolysis of OCP and crystal growth.

Hydroxyapatite and OCP (Figs. 1, 6, and 7) are the only calcium phosphates for which relatively detailed growth mechanisms have been proposed. These are of three different types: a spiral growth mechanism and two mechanisms involving OCP as a precursor which hydrolyzes in situ.

Mechanism I: A spiral growth mechanism has been observed (1) in minerals and in OHAp prepared under hydrothermal conditions. In this process the OHAp crystals can form directly from the supply materials without going through a precursor phase. There have been suggestions that such a mechanism is involved in enamel crystal growth. No detailed model has yet been developed which would reconcile a spiral growth mechanism with the ribbon-like morphology of the initial enamel crystallite.

Mechanism II: One process (5) by which OHAp can form from OCP is by hydrolysis of previously formed OCP crystals: This process takes place mostly in situ and produces crystals that have the same morphology as the original OCP. A single unit cell thickness of OCP (b_{100} = 18.68 Å) can hydrolyze to form a two unit cell thickness of OHAp ($2d_{100}$ = 16.32). The hydrolysis step is depicted in Figs. 8c and 8d.

In addition to a 2.36 Å contraction in the direction normal to the [100] plane, one apatitic layer must move relative to its adjacent apatitic layer by 1.6 Å in the <u>b</u> direction. Whether

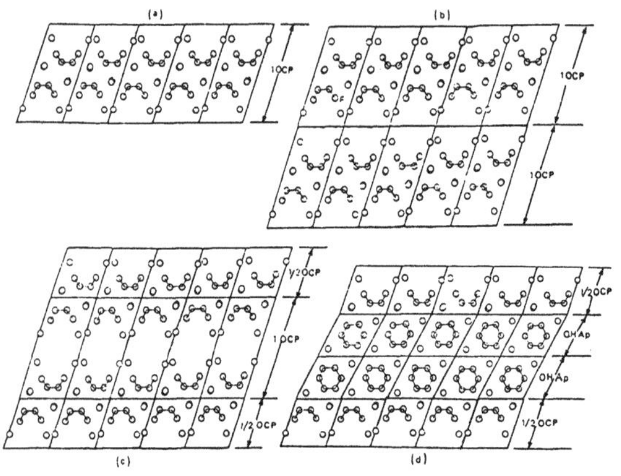

FIG. 8 - Mechanism of formation of sesquiapatite (6): a) single unit-cell thickness of OCP; b) double unit-cell thickness of OCP; c) same as b) visualized as a single unit-cell thickness of OCP bounded by two one-half unit-cell thicknesses of OCP; d) sesquiapatite layer formed by hydrolysis of the full unit-cell layer shown in c). Only Ca ions are shown.

this process involves translations of individual ions or clusters of ions, it is irreversible. Thus, this could be the crucial step in which imperfections and impurities may be incorporated into apatite.

Mechanism III: A second process (4) for the formation of OHAp involving OCP has been suggested. It is a layer-by-layer process (Fig. 8) in which the formation of a unit-cell thick layer of OCP on a [100] face alternates with its hydrolysis to a two unit-cell thick layer of OHAp. This process was suggested because of the close similarity of the structures of OCP and OHAp and is consistent with the suggestion that for both OHAp and OCP the transition layer from the lattice to the aqueous phase contains the hydrated layer of OCP. If these two processes (i.e., precipitation of an OCP layer followed by its hydrolysis to OHAp) occur, then their relative rates become critically important parameters. For example, if the rate of formation of the OCP layer is much more rapid than its hydrolysis, then the crystal would tend to be nearly pure OCP. On the other hand, if the rate of hydrolysis is quite rapid, but not so rapid that it prevents formation of the OCP layer, then the crystals would tend to be relatively pure apatite, and OCP would not be detected in the product. However, many effects due to the presence of OCP as a precursor (e.g., crystal morphology, incorporated impurities, and nonstoichiometry) would be fully displayed in the product. Variations in the rates of these two processes could account for differences observed in the properties of biominerals and in the maturation of amorphous calcium phosphates and other apatitic precipitates.

The mechanism which is presumed for the layer-by-layer process would involve nucleation of a "pill box" on a [100] face of the OCP or OHAp crystal, two-dimensional growth of that nucleus over the [100] face, followed by the possible hydrolysis step once it is initiated by some event. If the hydrolysis step is so rapid that the two-dimensional growth cannot take place fully on that face, the surface may form a series of island-like plateaus. Then each nucleation event would not produce a full layer of growth; this could cause the overall rate of crystal growth to decrease. It has been suggested (5) that fluorotic enamel is caused by a high fluoride concentration which accelerates excessively the hydrolysis step; fluorotic enamel is known to have a low density of mineral as compared to normal enamel, in accord with this view. It seems plausible that apatitic products produced by mechanism III would contain fewer structural impurities and incorporated impurities than those produced by mechanism II. Enamel may form by mechanism III.

There are a number of indications, besides the great structural similarity, that OCP participates in growth of OHAp. Apatitic products with low Ca/P ratios are commonly observed in these systems; these may be OCP-OHAp interlayered mixtures, or more probably, they may contain poorly crystallized products from the irreversible hydrolysis of OCP. The latter might be one or more of the following imprecisely defined materials: "calcium deficient apatite," "cryptocrystalline OCP," "collapsed OCP," "acidic apatite," "sesquiapatite," or a mixture of OHAp and $CaHPO_4$. Although such products may show only apatite-like X-ray diffraction patterns, they are probably much more complex. Even though they frequently do not give an OCP X-ray diffraction pattern, their chemical and optical properties resemble those of OCP. Unraveling the structural characteristics of the products that derive from the hydrolysis of OCP warrants high priority because it may provide answers to important long-standing controversies regarding the structural nature of nonstoichiometric "calcium deficient apatites." It appears highly significant that the apatitic crystals grown under conditions where OCP cannot act as a precursor to OHAp formation tend to be very nearly stoichiometric. Examples where OCP is not likely to be a precursor are hydrothermal conditions, high pH, presence of significant concentrations of fluoride, or very low supersaturations.

Epitaxially directed interlayering between OCP and OHAp is a well established phenomenon. In a Weissenberg X-ray diffraction study (9) of a "single crystal" which had optical properties intermediate between those of OCP and OHAp, it was shown that the OHAp and OCP parts of the crystal diffracted independently as though they were individual crystals with their b and c axes colinear. In a powder diffraction study of much smaller crystallites in which apparently the layers had random thicknesses (7, 18), it was shown that h00 reflections of OCP interacted and merged with those of OHAp and that the positions of the merged peaks would shift with Ca/P ratio. There is need for further study of this phenomenon; in particular, the reflections that are normally used to determine the unit cell dimensions of OHAp should be studied to see how they are affected by interlayering. This might account for some of the apparent variations in the unit cell dimensions of OHAp that are observed in biominerals.

Note that structures in Figs. 8a and 8d would be stabilized against further hydrolysis by the presence of the half unit cells of OCP on the [100] faces (6). Only OCP layers in the interior, as in Fig. 8c, would be susceptible to hydrolysis.

Materials with the structure of the type depicted in Fig. 8d have been given the name "sesquiapatite" because they would have a Ca/P ratio of about 1.5. The various combinations of units such as those shown in Figs. 8a, 8c, and 8d and higher growth products have been suggested (6) as prototypes for "amorphous calcium phosphate," for "tricalcium phosphate hydrate," for poorly crystallized apatites such as those found in bone, for the ribbon-like crystallites found in early enamel, and for partially hydrolyzed OCP. This unifying view is difficult to prove because of the small size of the crystals, but it is in accord with considerable structural, thermodynamic, kinetic, chemical, and physical information about these materials.

THE OTHER CALCIUM PHOSPHATES

Brushite

The sparingly soluble salt, $CaHPO_4.2H_2O$, is a member of a group of compounds (8) all of which contain a "corrugated sheet." These sheets comprise $CaPO_4$ chains in which calciums share PO_4 edges, $...Ca{=}O_2PO_2{=}Ca{=}O_2PO_2{=}Ca....$ These chains are arranged in a parallel fashion, alternating above and below the midplane (Fig. 9). A surprising variety of materials is generated by different choices of filling between the corrugated sheets. In $CaHPO_4.2H_2O$ it is H_2O molecules; in $Ca(H_2PO_4)_2.H_2O$ it is H_3PO_4 and H_2O molecules; in $Ca_2H_7NH_4(PO_4)_4.2H_2O$ it is H_3PO_4 water molecules and NH_4^+ ions, etc. With the exception of $CaHPO_4.2H_2O$, all the known members of this structural series are too soluble to be of physiological importance as crystals. However, its presence in so many compounds is strong evidence for the stability of the corrugated sheet moiety.

Despite the fact that $CaHPO_4.2H_2O$ is more soluble than OHAp, $Ca_8H_2(PO_4)_6.5H_2O$, and β-$Ca_3(PO_4)_2$, it forms under many physiological, geochemical, and laboratory conditions. Obviously this is a kinetic phenomenon, but the basic mechanisms which allow preferential formation of $CaHPO_4.2H_2O$ are not known. In general, precipitation of a hydrated compound from an aqueous system seems to be preferred, possibly because (a) the hydrated ions transpose more readily into a hydrated crystal,

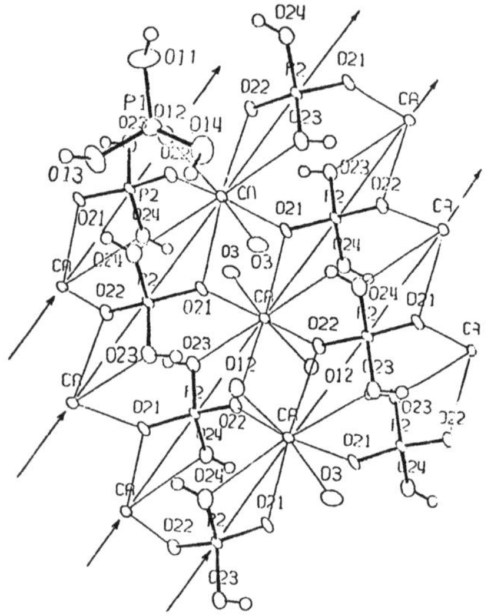

FIG. 9 - Corrugated sheet in $CaHPO_4.2H_2O$. The arrows show the direction of the $Ca-PO_4$ chains.

(b) a hydrated structure has a lower surface energy at the nucleation stage, and therefore, at its inception it is favored thermodynamically over the anhydrous crystal nuclei, or (c) hydrated crystals are less affected by growth poisons. In view of the importance of $CaHPO_4.2H_2O$ in biomineralization processes (it is known to participate in tooth, bone, calculus, and renal stone formation), more emphasis should be given to understanding the growth mechanism of brushite and why its formation is so favored. Both OCP and brushite hydrolyze to more basic calcium phosphates, but with one very important difference. Brushite seems to always dissolve first and then reprecipitate; OCP usually transforms topotactically. The latter incorporates more impurities and defects into the resulting structure.

Monetite

$CaHPO_4$ (15) is less soluble (i.e., more stable) than $CaHPO_4.2H_2O$ under all conditions of normal temperature and pressure. Yet, it forms only rarely under physiological conditions. However, the paucity of data on its occurrence as a biomineral may be due in part to the fact that it has a weak X-ray diffraction pattern and can be easily overlooked. On the other hand, because of its strong birefringence, it can be detected with the polarizing microscope in very small amounts. $CaHPO_4$ contains the corrugated sheet structure in highly distorted form. Apparently the absence of a "filling" between the corrugated sheets allows them to interact more strongly, thereby distorting the sheets.

Tricalcium Phosphates

There are four calcium phosphates with the formula $Ca_3(PO_4)_2$: α-$Ca_3(PO_4)_2$ (34) (the stable phase between 1120 and 1470°C); $\bar{\alpha}$-$Ca_3(PO_4)_2$ (34) (stable above 1470°C); β-$Ca_3(PO_4)_2$ (17) (stable below 1120°C); and β'-$Ca_3(PO_4)_2$ (37) (stable at high pressures). Two additional phase transformations have been indicated for β-$Ca_3(PO_4)_2$ at or below room temperature. The mineral whitlockite (11), which is also found in renal stones, in dental calculus, and as a dissolution-reprecipitation product in dental enamel, has a structure which is very similar to that of β-$Ca_3(PO_4)_2$. Generally it forms only in an aqueous environment. Finally, there are frequent references in the literature to "hydrated tricalcium phosphates" with apatitic properties. None of these has been shown to be a discrete crystalline compound.

$\bar{\alpha}$-$Ca_3(PO_4)_2$, a high temperature modification of α-$Ca_3(PO_4)_2$, presumably forms because of greater motion of the ions. It appears to have a structure related to that of the α-$Ca_2(SiO_4)$. The latter structure contains one Ca-Ca column for every two Ca-SiO_4 columns, producing a crystal with a cation-to-anion ratio of 2/1. In α-$Ca_3(PO_4)_2$, and presumably in $\bar{\alpha}$-$Ca_3(PO_4)_2$, every other calcium ion in the cation-anion column is missing, thus giving solids with cation-to-anion ratios of 3/2. Compounds of this type are not known to form in biological systems, but the existence of structures with single ion vacancies has relevance to the chemistry of other calcium phosphates.

β-$Ca_3(PO_4)_2$ has a structure which bears a formal resemblance to that of $Ba_3(PO_4)_2$, but apparently because the calcium ion is too small to provide ideal $Ba_3(PO_4)_2$-type packing, one out of every eight formula units is missing in β-$Ca_3(PO_4)_2$. Under high pressure these vacancies become filled, and the structure then mimics $Ba_3(PO_4)_2$ very closely. The $Ba_3(PO_4)_2$ structure contains columns of the type ...O_3PO-Ba-Ba-Ba-OPO_3... .

One calcium out of every 21 in the β-$Ca_3(PO_4)_2$ unit cell has unusual properties which affect the structure and composition in many ways:

1) Only one-half of this calcium's sites are occupied (in a statistical manner); thus the structure surrounding this site acquires either of two slightly different configurations, depending upon whether the site is occupied or vacant.
2) This calcium is octahedrally coordinated by its surrounding oxygens. This type of coordination provides a site which is somewhat too small for calcium but is ideal for magnesium, thereby accounting for the presence of part of the magnesium that tends to be incorporated into β-$Ca_3(PO_4)_2$.
3) Three of the six oxygens in this calcium's coordination sphere come from the face of a PO_4 tetrahedron; this is an unstable configuration because of repulsion between the Ca^{2+} ion and the P^{5+} ion in the PO_4. It is this PO_4 which is flipped 180° in the opposite direction in whitlockite.
4) A pair of small monovalent cations, one replacing the Ca^{2+}, the other filling the vacancy position, can fully substitute for this calcium. This may provide greater stability to the structure than does substitution by small bivalent cations such as Mg^{2+} or Fe^{2+} which can occupy only half of the sites. An example of this type of substitution may be the mineral merrilite found in meteorites and which then has a formula near $Ca_{18}(Mg,Fe)_2Na_2(PO_4)_{14}$.

There is another site in $Ca_3(PO_4)_2$ that has octahedral coordination and which seems to be preferred by small cations such as Mg^{2+} and Fe^{2+}. As a result, maximum substitution of Mg^{2+} into $Ca_3(PO_4)_2$ will produce a material with the formula $Ca_{18}Mg_3(PO_4)_{14}$.

Although it was once thought that β-$Ca_3(PO_4)_2$ prepared at high temperature might have the ideal formula $Ca_{20}H_2(PO_4)_{14}$, it was shown (43) in a chemical study that the formula was unquestionably $Ca_3(PO_4)_2$. It was then shown in a structure determination by Calvo and Gopal (11) that the mineral whitlockite (i.e., a material formed in an aqueous medium) had a formula which approximated $Ca_{18}(Mg,Fe)_2H_2(PO_4)_{14}$. They also found that the PO_4 group that previously shared a face with the unique calcium, had become inverted on the threefold axis and that the fourth oxygen on this PO_4 had a long P-O bond distance, indicating that it was bonded to a hydrogen. This calcium is replaced by two such protons to give the above formula.

Clearly the compositional and chemical properties of natural whitlockites are dependent upon structural details. β-$Ca_3(PO_4)_2$ is second only to OHAp in its stability under most physiological conditions. This was demonstrated in a thorough solubility study on a sample of β-$Ca_3(PO_4)_2$ prepared at a high temperature (24). It is probable that a material with formula $Ca_{18}Mg_2H_2(PO_4)_{14}$ would have an even lower solubility than β-$Ca_3(PO_4)_2$. The formation of $CaHPO_4.2H_2O$ and $Ca_8H_2(PO_4)_6.5H_2O$ under most physiological conditions instead of the more stable salt, whitlockite, is a clear indication of the importance of rates of precipitation in such processes. Although whitlockite does not seem to form under physiological conditions unless a small bivalent cation is present, it is conceivable that under conditions where such ions are present in high concentration, whitlockite might be more stable than OHAp, especially the fine grained hydroxyapatite found in most physiological precipitates. More information is needed on how chemical properties, such as solubility of the β-$Ca_3(PO_4)_2$-whitlockite series of compounds, are affected by the permissible variations in their structures and compositions.

Calcium Pyrophosphate

Triclinic calcium pyrophosphate dihydrate (CPPD) is of particular biological interest because of its involvement in chondrocalcinosis, pseudogout. Crystals of CPPD are found in vivo in knee joints. They produce inflammation by causing "the membrane of the digestive organelle, the lysosome, to rupture, allowing the lysosome's digestive enzymes to escape and begin the digestion of their parent white blood cell" (31). The most probable mechanism of this membranolysis (Mandel (32), who also cites Weissman and Rita (44)) involves the atomic composition of the most prominent crystal faces. In CPPD they are rich in both donor and acceptor hydrogen bonds and in alternating ribbons of ions of opposite charge. Thus, both hydrogen bonding and electrostatic attraction are available to bind to the membrane. This immobilizes it along the face of the crystal, producing breaking stresses at the edges of the crystal face as the rest of the membrane undulates. Let us examine this CPPD structure and see how these importantly active structural features arise, as viewed by Mandel (31,32).

CPPD has the formula $Ca_2P_2O_7.2H_2O$. It crystallizes in a triclinic form, space group $P\bar{1}$, with $\underline{a}$ = 7.365(4) Å, $\underline{b}$ = 8.287(4) Å, $\underline{c}$ = 6.691(4) Å, α = 102.96(1)°, β = 72.73(1)°, γ = 95.01(1)°, and with two formula units per unit cell. The crystals grow as rhomboids, slightly elongated along $\underline{c}$, with [100], [010], and [01$\bar{1}$] as the prominent faces.

The pyrophosphate ion $(P_2O_7)^{4-}$ is formed from two PO_4 tetrahedra sharing one vertex oxygen (Fig. 10). The torsion and bending angles at the bridge oxygen are not strongly fixed, as they vary over wide ranges with differing environments of the $(P_2O_7)^{4-}$ ion (i.e., in different compounds).

Each Ca ion is coordinated by seven oxygens at distances ranging from 2.29 Å to 2.67 Å. The coordination polyhedra have

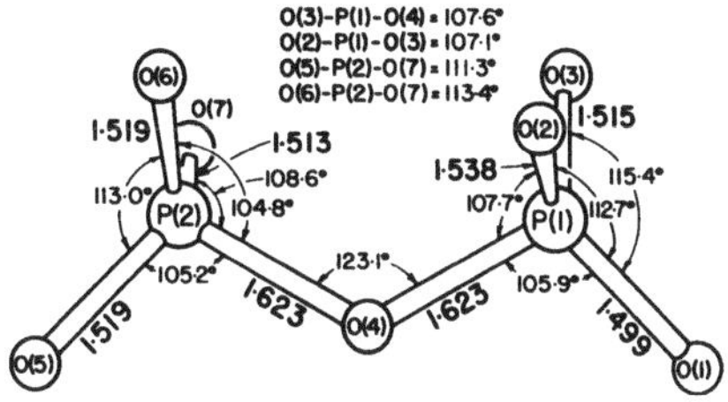

FIG. 10 - Bond distances and angles for the pyrophosphate anion (31).

somewhat different shapes about the two crystallographically nonequivalent Ca ions. About Ca(1), two O's are supplied by one PO_4 group, one O is supplied by each of 4 other PO_4 groups, and the 7th O is supplied from an H_2O molecule. About Ca(2), one O is supplied from the same H_2O molecule (i.e., Ca(1) and Ca(2) share one H_2O molecule), one O is supplied from an H_2O associated only with Ca(2), two O's are supplied by one PO_4 group, and the remaining 3 are supplied, one each, by 3 different PO_4 groups (Fig. 11).

On the prominent rhomboid surfaces, there are rows of Ca^{2+} ions, rows of O^- ions (the P-O bond is nearly 1/2 covalent), and rows of H_2O molecules. The dashed lines in Fig. 12, which

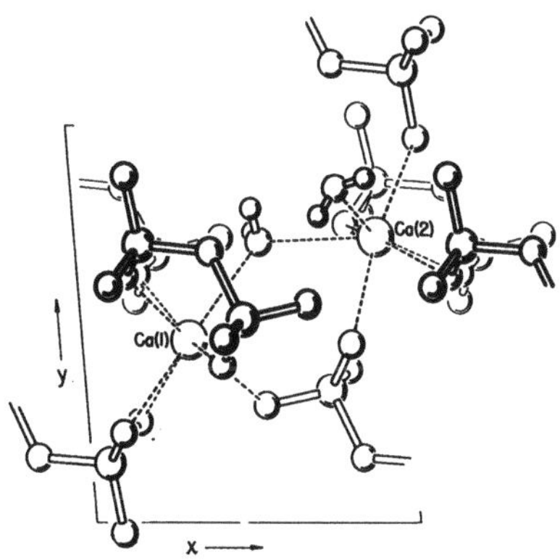

FIG. 11 - A view down the [001] direction in the crystal, indicating the two different types of calcium coordination exhibited in $Ca_2P_2O_7.2H_2O$ (31).

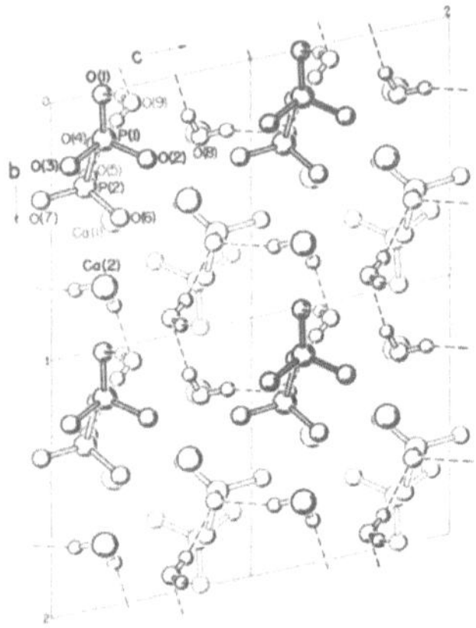

FIG. 12 - The calcium pyrophosphate dihydrate structure viewed onto the [100] face - the ribbons of Ca^{2+} ions, pyrophosphate-oxygen atoms, and water molecules (32).

is a view onto the [100] face (one of the dominant ones) show the richness of H bonds arising from the H_2O molecules. In each formula unit, there are 4 hydrogen bonds, three from water to phosphate oxygens and one between the two waters.

The zigzag rows (ribbons) of calcium and oxygen ions can best be seen in a view along the $[00\bar{1}]$ direction ((32), Fig. 2b). These alternate in charge and thus form an electrostatic surface which can interact with other slightly charged surfaces. On either surface, [100] or $[01\bar{1}]$, the hydrogen atoms of one water molecule (per formula unit) are available as H-bond donors. Thus, the possible mechanism for the membrane lysing property of CPPD can be understood at the atomic and molecular-structure level.

That CPPD has an inflammatory property is not in question. But whether that property arises from lysosome membrane rupture and whether the mechanism for the lysing set forth here is the correct one, are still open to question. Although plausible, satisfyingly based on atomic scale structural phenomena, and not in conflict with existing experimental data, these mechanisms are not yet proven.

Monosodium Urate Monohydrate

Crystals of monosodium urate monohydrate (MSUM), $NaC_5H_3N_4O_3 \cdot H_2O$, occur in vivo where, it is generally accepted, they are the pyrogens in acute gouty arthritis.

MSUM is thought to produce its pyrogenic effect in much the same way as CPPD, that is, through membranolysis caused by rigid attachment of sections of the membrane to the prominent crystal faces through H bond formation between the crystal surface and the O atoms of the phosphate head groups of the phospholipid membrane.

MSUM crystals, which tend to form as needles elongated along $\underline{c}$, have space group $P\bar{1}$ with $\underline{a}$ = 10.888(5) Å, $\underline{b}$ = 9.534(3) Å, $\underline{c}$ = 3.567(1) Å, α = 95.06(3)°, β = 99.47(5)°, γ = 97.17(3)°, and with two formula units per unit cell. Figure 13 shows the urate anion with dimensions found by Mandel and Mandel (33) in their crystal structure determination. They found that the MSUM structure could be viewed as stacks of slightly rippled layers, running perpendicular to $\underline{c}$, of nearly planar urate ions held together by Na^+ ions and hydrogen bonds from the H_2O. Thus, the crystal faces which are available to interact with the membrane are edge-on views of stacks of urate anions separated by stacks of Na ions and H_2O. Thus, again, as in the

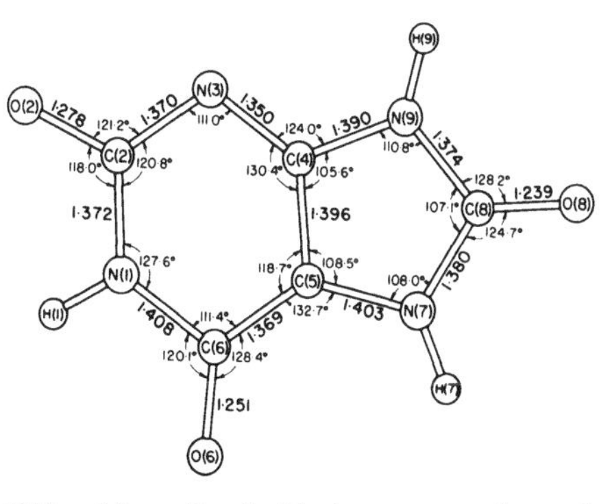

FIG. 13 - Bond distances and angles of the urate anion. The estimated standard deviations are 0.015 Å and 1.0° (33).

case of CPPD, one is provided with a very plausible structural basis for the membranolysis.

THE CALCIUM CARBONATES

The three anhydrous forms of calcium carbonate, calcite, aragonite, and vaterite, all occur as biominerals; the monohydrate also occurs as a biomineral (Table 1); and in accord with the principle suggested re OCP that hydrated crystals may act as precursors, the hexahydrate may participate in the incipient stages of crystal formation. Dolomite, $CaMg(CO_3)_2$, is a widely distributed mineral. It has the same structure as calcite except that every other calcium in a c-axis column is replaced by magnesium. On the basis of as yet very limited evidence, huntite, $Mg_3Ca(CO_3)_4$, may be present in the mineralizing front during formation of enamel. In view of the ubiquitous presence of sodium ion in biological systems, it is conceivable that the sodium calcium carbonates, shortite, $Na_2Ca_2(CO_3)_3$, pirssonite, $CaNa_2(CO_3)_2.2H_2O$, or gaylussite, $CaNa_2(CO_3)_2.5H_2O$, may participate in biological mineralization, at least in a transitory form. Calcite and vaterite are hexagonal crystals, and aragonite and the hexahydrate are pseudohexagonal crystals (45). The pseudohexagonal symmetry of aragonite is the basis for its extensive twinning. The plane of the carbonate ion lies perpendicular to the hexagonal or pseudohexagonal axis in calcite and aragonite, and parallel to this axis in vaterite and the hexahydrate.

The structure of calcite mimics that of the cubic close packed structure of NaCl, with Ca and CO_3 ions replacing, respectively, Na and Cl ions. Aragonite, on the other hand, resembles a hexagonal close packed lattice. The arrangement of calcium and carbonate ions in calcite and aragonite are compared in Fig. 14, where the numbers within the circles designate the heights of the atoms above the plane of the paper (in hundredths of the c axis length). It can be seen that in calcite each CO_3 ion is coordinated to two calcium ions (one above and one below), while in aragonite each CO_3 oxygen is coordinated to three calcium ions (two above and one below or vice versa).

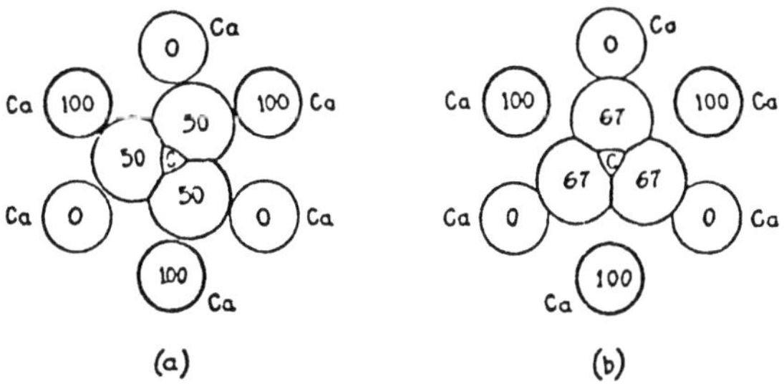

FIG. 14 - A comparison of the situation of the CO_3 group in (a) calcite and (b) aragonite. The upper layer of Ca atoms has the index 100, the lower layer the index 0. Each oxygen atom is linked to two calcium atoms in calcite and to three calcium atoms in aragonite. (Courtesy of Bragg (3).)

The structure of $CaCO_3.6H_2O$ (16) is unique in that it contains $CaCO_3$ ion pairs completely surrounded by water molecules so that there is no direct coordination between ion pairs. Each calcium shares an edge of the carbonate ion; the ion pair lies on a twofold axis, and these axes are arranged in a pseudo-hexagonal pattern. The structure of the hexahydrate provides a good starting model for the calcium-carbonate ion pair and its hydrated surroundings in aqueous solutions.

CALCIUM OXALATES

Calcium oxalates occur in plant tissues (especially profusely in rhubarb leaves), as minerals of organic origin, and in urinary stones. They normally occur either as whewellite, $CaC_2O_4.H_2O$, or as weddellite, $CaC_2O_4.(2 + x)H_2O$, where $0<x\leq0.5$. The trihydrate, $CaC_2O_4.3H_2O$, is also known and may be biologically important as a precursor to the lower hydrates (40). Typically, urinary calculi may be ~1/2 whewellite, ~1/6 weddellite, ~1/5 apatite, and the remainder organic material (28). Whewellite has two crystalline phases ($P2_1/c$ and I2/m, equivalent to standard group C2/m) related by an order-disorder transformation at or slightly below 45°C, i.e., near human body temperature (12). Unit cell data for these oxalates are given in Table 3.

TABLE 3 - Unit Cell data for calcium oxalate hydrates (10,12,13,39)

	a (Å)	b (Å)	c (Å)	α (deg)	β (deg)	γ (deg)	Space group	Z	Ref.
Whewellite									
$CaC_2O_4.H_2O$									
basic (55°C)	9.978(1)	7.295(1)	6.292(1)	90.0	107.04(2)	90.0	I2/m*	4	12
derivative (20°C)	9.9763(3)	14.5884(4)	6.2913(3)	90.0	107.03(2)	90.0	$P2_1/n$	8	12
Weddellite									
$CaC_2O_4.(2+x)H_2O$									
(x = 0.37)	12.371(3)	12.371(3)	7.357(2)	90.0	90.0	90.0	I4/m	8	39
Ca oxalate trihydrate									
$CaC_2O_4.3H_2O$	7.145(6)	8.600(7)	6.099(5)	112.30(5)	108.87(5)	89.92(5)	$P\bar{1}$	2	13

Deganello (12).

Tazzoli and Domeneghetti (39).

Deganello, Kampf, and Moore (13).

* This non-standard space group used by Deganello (12) is described by Buerger (10) and is transformable to C2/m.

In all of these oxalates, the Ca ions are surrounded (e.g., coordinated) by 8 oxygen atoms belonging either to oxalate (C_2O_4) ions or water (H_2O). The oxygen atoms are at distances of ~2.35 to 2.55 Å from the Ca ion, depending on the particular oxygen atom and oxalate form. The positions of these oxygen atoms can be visualized as being at the corners of a distorted square antiprism. (To visualize this, think first of a cube with an O atom at each of the 8 corners. Then rotate one of the faces 45° about a central line joining it to its opposite, parallel face, thus forming a square antiprism.) Drawing lines between nearest neighbor oxygens produces a polyhedron with 8 triangular and 2 "square" (actually distorted from a square) faces. In low whewellite ($P2_1/c$), there are two crystallographically nonequivalent Ca ions and, consequently, two nonequivalent, though very similar coordination polyhedra. By contributing O atoms to more than one Ca-coordination polyhedron, the oxalate ions (Fig. 15) hold the structure together with some help from hydrogen bonding associated with the H_2O molecules. The oxalate ions are accurately planar in high whewellite (I2/m) and probably so in weddellite. In low whewellite, one of the two crystallographically nonequivalent oxalate ions is planar within experimental errors, while the other is significantly, though slightly, nonplanar. In the trihydrate structure (13) the oxalate ion is distinctly nonplanar.

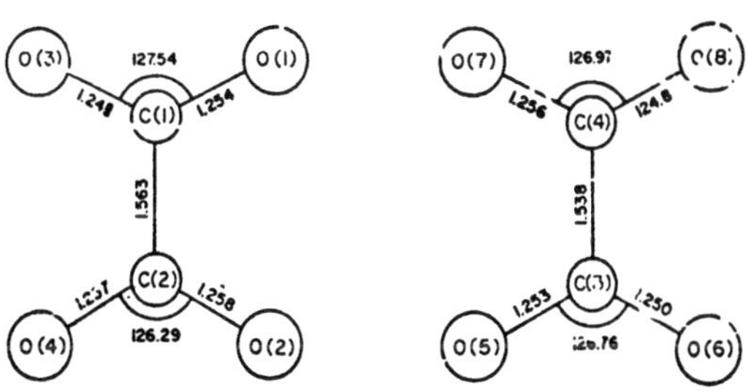

FIG. 15 - (a): The C(I) carboxylic group. Distances (Å) and (°) have average e.s.d.'s of 0.004 Å and 0.4°. (b): The C(II) carboxylic group. Same considerations as for (a) (14).

In weddellite, the fractional waters of hydration (i.e., x in $(2 + x)H_2O$) occur zeolitically and do not contribute to the Ca-coordination polyhedra. One sees, then, that progression from monohydrate to dihydrate to trihydrate is a one-to-one expression of the number of oxygen atoms contributed by H_2O to the 8-oxygen polyhedron around Ca. This logical progression might make it seem that the growth and transformation of each hydrate into the next lower one is simply a matter of successive replacement of a water molecule with an oxalate ion (each contributing one oxygen atom), even starting with the Ca ion and its hydration sphere in aqueous solution. However, the situation is not quite so simple. The three water molecules in the trihydrate occur at the three corners of one of the triangular faces of the coordination polyhedron, but the two coordinating water molecules in weddellite do not both occur on the same triangular face. Thus, transformation from the trihydrate to the dihydrate (really, 2 + x) requires not only replacement of a water oxygen by an oxalate oxygen, but also some rearrangement of at least one of the two remaining water oxygens in the coordination polyhedron.

Since H_2O is electrically neutral and $(C_2O_4)^{2-}$ is not, it is clear that replacing a water oxygen by an oxalate oxygen in the coordination polyhedron must be accompanied by some additional differences which maintain charge balance. These are the differences in edge sharing among polyhedra. In the trihydrate, each two pseudo-equivalent polyhedra share one edge, in the "di-hydrate" (weddellite) each coordination polyhedron shares two edges, one with each of two other polyhedra, and in the monohydrate (whewellite) each polyhedron shares three edges with other polyhedra. Thus, by this edge sharing, Ca-coordination polyhedra are linked to form sheets in whewellite, chains in weddellite, and separate dimeric units in the trihydrate.

These sheets in whewellite are parallel to [100], the good cleavage plane. (This refers to the cell setting for which the space group is $P2_1/c$.) For the cell setting yielding $P2_1/n$ used by Deganello and Piro (14), this plane is [$\bar{1}01$].

In weddellite, the chains run along c, the fourfold axis direction. They are linked by hydrogen-bonded "ribbons" composed of alternating sequences of oxalate and water. Obeying the fourfold axis symmetry, these ribbons and the Ca-coordination polyhedra chains generate channels, parallel to c, in which the zeolitic water occurs in two alternative, closely spaced (0.4 Å) positions. This channel and the chains of Ca-coordination polyhedra are shown in Fig. 16, taken from the work of Tazzoli and Domeneghetti (39). (Their work is also the principal source of up-to-date structural information on weddellite.) It appears physically impossible for these channels to accommodate more than 1/2 H_2O per cell, hence the otherwise mysterious upper limit on x in the formula, $CaC_2O_4.(2 + x)H_2O$, is explained.

Now that the zeolitic character of the x H_2O has thus been firmly established, one may next look purposefully for biological significance and possible function of the zeolitic character.

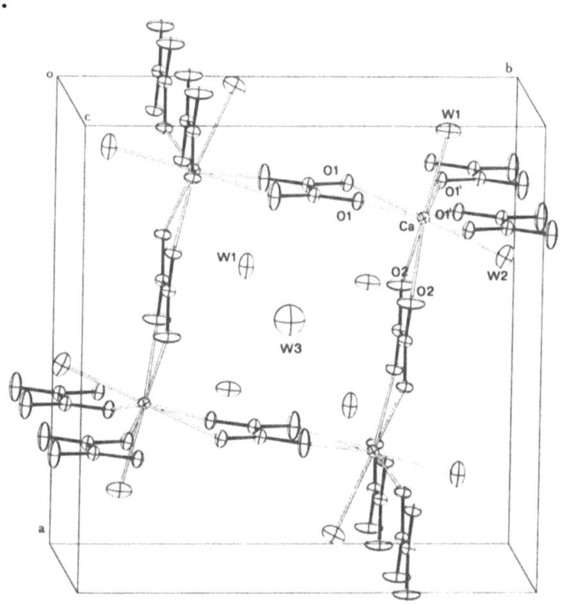

FIG. 16 - Weddellite, section almost parallel to [001] (39).

Acknowledgements. This work received financial support from the USPHS through NIH-NIDR Grants DE 01912 and DE 05030.

REFERENCES

(1) Arends, J., and Jongebloed, W.L. 1979. Ultrastructural studies of synthetic apatite crystals. J. Dent. Res. 58(B): 837-842.

(2) Bonel, G. 1972. Contribution de l'etude de la carbonatation des apatites. I. Syntheses et etude des proprietes physico chimiques des apatites carbonatees de type A. Ann. Chim. 7: 65-88.

(3) Bragg, W.L. 1965. The Crystalline State, vol. 4. Ithaca, NY: Cornell University Press.

(4) Brown, W.E. 1962. Crystal structure of octacalcium phosphate. Nature 196: 1048-1050.

(5) Brown, W.E. 1966. Crystal growth of bone mineral. Clin. Orthop. 44: 205-220.

(6) Brown, W.E.; Mathew, M.; and Tung, M.S. 1981. Crystal chemistry of octacalcium phosphate. In Prog. Crystal Growth Charact., vol. 4, pp. 59-87. Oxford, England: Pergamon Press Ltd.

(7) Brown, W.E.; Schroeder, L.W.; and Ferris, J.S. 1979. Interlayering of crystalline octacalcium phosphate and hydroxylapatite. J. Phys. Chem. 83: 1835.

(8) Brown, W.E.; Smith, J.P.; Lehr, J.R.; and Frazier, A.W. 1958. Crystallography of hydrated monocalcium phosphates containing potassium or ammonium. J. Phys. Chem. 62: 625-627.

(9) Brown, W.E.; Smith, J.P.; Lehr, J.R.; and Frazier, A.W. 1962. Crystallographic and chemical relations between octacalcium phosphate and hydroxyapatite. Nature 196: 1050-1054.

(10) Buerger, M.J. 1963. Elementary Crystallography, p. 306. New York: John Wiley.

(11) Calvo, C., and Gopal, R. 1975. The crystal structure of whitlockite from the Palermo Quarry. Am. Min. 60: 120-133.

(12) Deganello, S. 1980. The basic and derivative structures of calcium oxalate monohydrate. Z. Krist. 152: 247-252.

(13) Deganello, S.; Kampf, A.R.; and Moore, P.B. 1981. The crystal structure of calcium oxalate trihydrate: $Ca(H_2O)_3(C_2O_4)$. Am. Min. 66: 859-865.

(14) Deganello, S., and Piro, O.E. 1981. The crystal structure of calcium oxalate monohydrate (whewellite). N. Jb. Miner Mh. 2: 81-88.

(15) Dickens, B.; Bowen, J.S.; and Brown, W.E. 1972. A refinement of the crystal structure of $CaHPO_4$ (synthetic monetite). Acta Cryst. B28: 797-806.

(16) Dickens, B., and Brown, W.E. 1970. The crystal structure of calcium carbonate hexahydrate at ~-120°. Inorg. Chem. 9: 480-486.

(17) Dickens, G.; Schroeder, L.W.; and Brown, W.E. 1974. Crystallographic studies of the role of Mg as a stabilizing impurity in β-$Ca_3(PO_4)_2$ - I. The crystal structure of pure β-$Ca_3(PO_4)_2$. J. Solid State Chem. 10: 232-248.

(18) Eanes, E.D., and Meyer, J.L. 1977. The maturation of crystalline calcium phosphates in aqueous suspensions at physiologic pH. Calcif. Tiss. Res. 23: 259-269.

(19) Elliott, J.C. 1964. The crystallographic structure of dental enamel and related apatites. Ph.D. Thesis, University of London, London, England.

(20) Elliott, J.C.; Bonel, G.; and Trombe, J.C. 1980. Space group and lattice constants of $Ca_{10}(PO_4)_6CO_3$. J. Appl. Cryst. 13: 618-621.

(21) Elliott, J.C., and Mackie, P.E. 1975. Monoclinic hydroxyapatite. Physico-Chimie et Cristollographie des Apatites D'Interet Biologique. Colloques Internationaux C.N.R.S. (Paris) 230: 69-70.

(22) Elliott, J.C.; Mackie, P.E.; and Young, R.A. 1973. Monoclinic hydroxyapatite. Science 180: 1055-1057.

(23) Elliott, J.C., and Young, R.A. 1967. Conversion of single crystals of chlorapatite into single crystals of hydroxyapatite. Nature 214: 904-906.

(24) Gregory, T.M.; Moreno, E.C.; Patel, J.M.; and Brown, W.E. 1974. Solubility of β-$Ca_3(PO_4)_2$ in system $Ca(OH_2$-H_3PO_4-H_2O at 5, 15, 25 and 37°C. J. Res. Nat. Bur. Stand. (USA) 78A(6): 667-674.

(25) Hallsworth, A.S.; Weatherell, J.S.; and Robinson, C. 1973. Loss of carbonate during the first stages of caries attack. Caries Res. 7: 345-348.

(26) Holcomb, D.W., and Young, R.A. 1980. Thermal decomposition of human tooth enamel. Calcif. Tiss. Intl. 31: 189-201.

(27) Ingraham, G.S. 1973. The role of carbonate in dental minerals. Caries Res. 7: 217-230.

(28) Konjiki, T.; Sudo, T.; and Kohyama, N. 1980. Mineralogical notes of apatites in urinary calculi. Calcif. Tiss. Intl. 30: 101-107.

(29) LeGeros, R.Z.; Trautz, O.R.; Klein, E.; and LeGeros, J.P. 1969. Two types of carbonate substitutions in the apatite structure. Separatum Experientia 24: 5.

(30) Lowenstam, H.A. 1981. Minerals formed by organisms. Science 211: 1126.

(31) Mandel, N.S. 1975. The crystal structure of calcium pyrophosphate dihydrate. Acta Cryst. B31: 1731-1734.

(32) Mandel, N.S. 1976. The structural basis of crystal-induced membranolysis. Arthrit. Rheum. 19: 440-445.

(33) Mandel, N.S., and Mandel, G.S. 1976. Monosodium urate monohydrate, the gout culprit. J. Am. Chem. Soc. 98: 2319-2323.

(34) Mathew, M.; Schroeder, L.W.; Dickens, B.; and Brown, W.E. 1977. The crystal structure of α-$Ca_3(PO_4)_2$. Acta Cryst. B33: 1325-1333.

(35) Nelson, D. 1981. The influence of carbonate on the atomic structure and reactivity of hydroxyapatite. J. Dent. Res. 60(C): 1621-1629.

(36) Roux, P., and Bonel, G. 1980. Etude par diffraction des rayons X et par spectrometric d'assorption infrarouge des apatites carbonatees de type A phospho-calcique et arsenio-calcique "haute pression." Bull. Min. 101: 448-452.

(37) Roux, P.; Louer, D.; and Bonel, G. 1978. Chimie Minerale - sur une nouvelle forme cristalline de phosphate tricalcique. C. E. Acad. Sc. Paris 286: 549-551.

(38) Sudarsanan, K., and Young, R.A. 1978. Structural interactions of F, Cl and OH in apatites. Acta Cryst. B34: 1401-1407.

(39) Tazzoli, V., and Domeneghetti, C. 1980. The crystal structures of whewellite and weddellite: re-examination and comparison. Am. Min. 65: 327-334.

(40) Tomazic, B., and Nancollas, G.H. 1979. The kinetics of dissolution of calcium oxalate hydrates. J. Cryst. Growth 46: 355-361.

(41) Trombe, H.C., and Montel, G. 1978. Some features of the incorporation of oxygen in different oxidation states in the apatitic lattice - I. On the existence of calcium and strontium oxyapatites. J. inorg. nucl. Chem. 40: 15-21.

(42) Verbeek, R.M.H.; Heiligers, H.J.M.; Driessens, F.C.M.; and Schaeken, H.C. 1980. Effect of dehydration of calcium hydroxyl apatite on its cell perimeters. Z. anorg. allg. Chem. 466: 76-80.

(43) Wallace, B.M., and Brown, W.E. 1971. Stoichiometric composition of whitlockite. J. Dent. Res. 50: 343-346.

(44) Weissmann, G., and Rita, C.A. 1972. Molecular basis of gouty inflammation: interaction of monosodium urate crystals with lysosomes and lipsomes. Nature 240: 167-172.

(45) Wells, A.F. 1962. Structural Inorganic Chemistry, 3rd ed. Oxford, England: Oxford University Press.

(46) Young, R.A. 1975. Some aspects of crystal structural modeling of biological apatites. Physico-Chemie et Crystallographie des Apatites D'Interet Biologique. Colloques Internationaux C.N.R.S. (Paris) 230: 21-39.

(47) Young, R.A.; Bartlett, M.L.; Spooner, S.; Mackie, P.E.; and Bonel, G. 1981. Reversible high temperature exchange of carbonate and hydroxyl ions in tooth enamel and synthetic hydroxyapatite. J. Biol. Phys. 9: 1-26.

(48) Young, R.A., and Mackie, P.E. 1980. Crystallography of human tooth enamel: initial structure refinement. Mat. Res. Bull. 15: 17-29.

Biological Mineralization and Demineralization, ed. G.H. Nancollas, pp. 143-160.
Dahlem Konferenzen 1982. Berlin, Heidelberg, New York: Springer-Verlag.

Structural Properties of Cells Related to Tissue Mineralization

R. K. Schenk, E. Hunziker, and W. Herrmann
Anatomisches Institut der Universität Bern
3012 Bern, Switzerland

Abstract. Structural aspects of biological mineralization are discussed in view of a participation of cells and of the possible role of organized matrix components in cartilage and bone. In both tissues mineralization is confined to restricted compartments within the intercellular matrix. Structural properties of these compartments are analyzed in relation to a possible cellular control of calcification. Special attention is paid to the role of matrix vesicles, cellular derivatives which seem to be involved in the initial phase of mineral deposition in most calcifying tissues.

INTRODUCTION

Although mineralization of cartilage and bone are highly specialized processes, both in view of cellular activities and composition of their respective intercellular matrices, they share some structural properties which are important for a discussion of the mechanisms involved. The most striking feature is the fact that initial deposition of mineral never occurs in direct contact to the cell surface but always at some distance from it within the newly formed organic matrix. This distance ranges from 2 - 3 μm in cartilage and woven bone to about 10 μm in mature lamellar bone. It represents not only a spatial separation of the site of initial mineralization from the cell that is thought to mediate this process, but also a time interval of at least several hours up to roughly 10 days. This suggests that organic material deposited by the cells first undergoes some structural and/or biochemical alterations before it is ready to calcify. On the other hand, it is also possible that this capacity to undergo calcification is limited to a certain time span which is not exactly known. This assumption is based on observations about recovery of mineralization in pathological conditions where calcification is temporarily arrested. Explanations for this time lag and separation in space range from the presence and removal of mineralization inhibitors to the deposition of cellular derivatives which modify certain matrix components or ionic concentrations in the tissue fluid and so create conditions which finally trigger crystal formation.

MINERALIZATION IN CARTILAGE

Cartilage mineralization is discussed first, since this tissue is phylogenetically older, and it can be expected that the basic mechanisms involved are at least in part operating also in more elaborate tissues. Mineralization of cartilage is an essential step during embryonic development and throughout growth of the bony skeleton which is for its most part preformed as a cartilagenous model. In the adult cartilage, calcification may continue in articular cartilage at a very slow rate which can be accelerated in degenerative lesions. In fracture repair, a more primitive fibrocartilage is formed under unstable conditions between the fragment ends and also undergoes mineralization before its substitution by bone. Thus, in the adult, almost all calcified cartilage is replaced by bone, with the exception of the calcified cartilage layer permanently interposed between hyaline articular cartilage and subchondral bone underneath the joint surfaces.

Cartilage mineralization is best understood in the growth plate of growing long bones. There it plays a decisive role in an extremely well regulated concerted action of cell proliferation, matrix production, resorption of both unmineralized and mineralized cartilage, and finally bone formation. Retardation of inhibition of cartilage mineralization is a prominent feature in rickets and leads to a thickening of the growth cartilage, since regular vascular invasion and bony substitution of nonmineralizing growth cartilage are no longer possible.

Cell Proliferation and Matrix Production in the Growth Cartilage

Cartilage is formed by chondroblasts. These cells are derived from mesenchymal precursors and are still capable of mitotic divisions. After mitosis, the daughter cells immediately surround themselves with newly formed matrix and thus become completely separated. The cytoplasm of chondroblasts is well equipped with organelles. An abundant rough endoplasmic reticulum furnished with ribosomes is engaged in the production of collagen and other proteins, whereas glycosaminoglycans are mainly assembled before extrusion in a well developed Golgi-field (a system of smooth intracellular membranes near the cell nucleus). In fast growing animals such as rats, one single chondroblast produces about its own cell size in matrix volume in 24 hours. This matrix consists of collagen and proteoglycans in a ratio of about 60:40% (referring to the dry weight of the organic matrix). In growth cartilage collagen appears in specimens conventionally prepared for electron microscopy as fibrils with diameters of about 20 nm and no apparent cross banding. The latter, however, becomes visible after extraction of proteoglycans or in tissue after high pressure freezing and freeze substitution. Cartilage collagen has a specific chemical structure (type II) which is different from collagen in skin, ligaments, tendon and bone (type I), reticular fibers (type III), and basement membranes (type IV). The proteoglycans not only contribute to one third of the cartilaginous matrix, they are also decisive for its mechanical properties as well as for the diffusion of ions and nutrients into this avascular tissue. In both respects, the high negative charge density of the sulfated carbohydrate chains is of paramount importance: it creates electrostatic forces which try to expand the domain occupied by the giant molecules entrapped in the network of tensile resistant collagen fibrils (8,16).

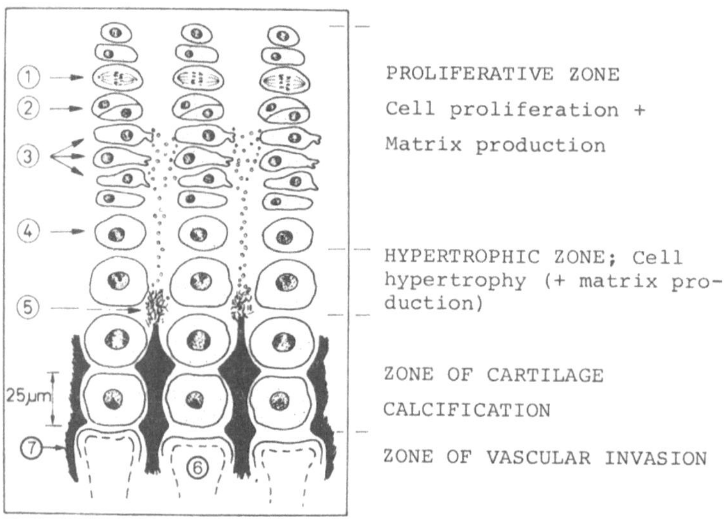

FIG. 1 - Structural and functional organization of the growth plate. 1) Mitotic cell division. 2) Postmitotic phase. 3) Matrix production and release of matrix vesicles. 4) Onset of cell hypertrophy. 5) Site of initial mineralization. 6) Invading blood vessels. 7) Calcified cartilage.

On the other hand, the negative charges favor diffusion of positively charged ions such as Ca^{2+} and of nutrients of comparably small molecular size which have to pass through the matrix to reach the cells (14). Unfortunately, routine fixation and embedding procedures extract up to 70% of the proteoglycans from the matrix. Preservation of these components for ultramicroscopy is possible by adding positively charged cationic dyes to the primary fixative (such as toluidine blue, alcianblue, ruthenium red, rutheniumhexamminetrichloride), or as recently shown, by high-pressure freezing and freeze substitution

((10), and Hunziker et al., submitted). But the problem of identification of various components within proteoglycan aggregates in such preparations is still unsolved. Among other matrix components chemically isolated from cartilage matrix, only lysozyme has been structurally located (11).

Cartilage Mineralization

In growth plates, cartilage mineralization is restricted to the lower part of the hypertrophic zone, and there again to a rather well-defined compartment called interterritorial matrix. This compartment lies between the typical cell columns, which are surrounded and bound together by the territorial matrix. This compartmentalization ultimately leads to a pattern of mineralization which was described in earlier papers as being confined to the longitudinal septa (running parallel to the direction of growth (19)). The transverse septa remain unmineralized. Describing mineralization as being limited to the interterritorial matrix is more adequate in one important aspect: there is always a distinct envelope of unmineralized matrix left, several microns thick, which separates the cells from the site of mineralization. This layer comprises the territorial and pericellular compartment. It remains unmineralized down to the level where the chondrocytes disintegrate and cartilage is finally degraded and invaded by capillaries.

For a long time electron microscopists have been attracted by the process of cartilage mineralization for several reasons: a) it is an extremely dynamic process, the tips of the mineralizing septa moving forward at a rate of up to 300 μm/day (as measured in the proximal tibia of rats weighing about 100 g); b) it is confined to a distinct cartilage compartment which can be selectively studied prior to the onset of initial calcification; and c) in the final stage, there is - in contrast to bone - no association of the apatite crystals with collagen fibrils, but rather a propagation of the mineral clusters into the interfibrillar space formerly occupied by proteoglycans only.

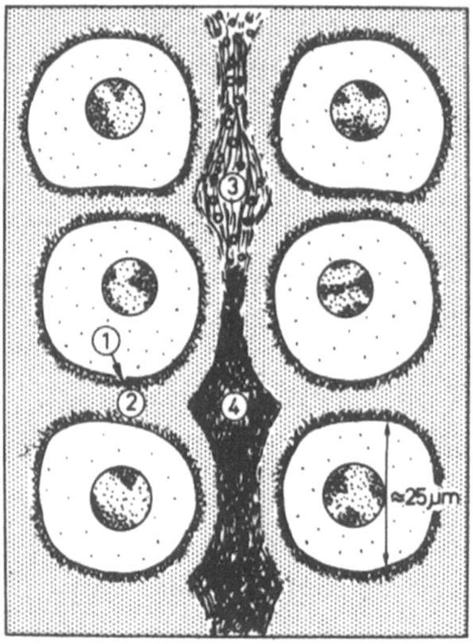

FIG. 2 - Matrix compartments in the zone of cartilage calcification. 1) Pericellular matrix (proteoglycans). 2) Territorial matrix. 3) Unmineralized interterritorial matrix. 4) Mineralized interterritorial matrix.

During the phase preceding cartilage mineralization, structural changes are seen both in chondrocytes and in interterritorial matrix. The most prominent feature is cell hypertrophy which has a stretching effect and increases the daily growth rate about 2 - 3 times. Hypertrophy is an active cell performance based on electrolyte and fluid transport and results in a 4 - 5 times volume increase in the rat. During hypertrophy, the volume ratio of cells to matrix changes from about 1:2 in the upper proliferative zone to 2:1 in the hypertrophic and calcifying zone. This gives the general impression that the cells expand at the expense of the surrounding matrix, which eventually could cause an increase in the concentration of ions in the extracellular fluid. But recent, more refined morphometric determinations, which also consider the stretching of the columns, show that the total matrix volume is also augmented by a factor of 1.4 (Hunziker and Schenk, in preparation). We therefore assume that there is none or only little change in the density of structural matrix components and especially in extracellular ion concentration.

Quantitation of cytoplasmatic organelles in hypertrophic cells also reveals a surprisingly high total volume of mitochondria per cell, which in part may reflect the energy requirement for osmotic work during swelling. On the other hand, mitochondria are also involved in intracellular calcium regulation, and it has been suggested that mitochondria in the hypertrophic cells might provide calcium or calcium phosphate packages which ultimately are released into the calcifying compartments of the matrix. In this respect, the dark mitochondrial granules, which readily appear after several fixation methods and have been shown to contain large amounts of calcium (by the pyroantimonate and other techniques), are especially noteworthy (4,5).

Role of Matrix Vesicles in Cartilage Calcification

So far these observations are in favor of a general participation of chondrocytes in providing some basic requirements for mineralization. But they cannot explain the initiation of this process in a clearly defined matrix compartment, as well as its confinement to the same area. The most striking but also most disputed contribution to the ultrastructural aspect of tissue mineralization is the discovery of matrix vesicles in the sites of initial calcification, first reported in cartilage and subsequently also in other tissues. In transmission electron micrographs of conventionally prepared specimens, the first mineral deposits are always visualized in close contact to spherical, rather electron-dense matrix inclusions with diameters of 100 - 200 nm (1). These vesicular structures are invested by a trilaminar membrane and contain a sometimes heterogeneously distributed, rather electron-dense material. The first deposits resembling apatite crystals are always found either inside the vesicles or in direct contact with their surface. As crystal formation proceeds, the membranous envelope ruptures and the mineral spreads out from these secondary nucleation centers in the form of clusters which virtually obscure the remnants of the former vesicular structure. Numerous studies deal with the origin, composition, and function of matrix vesicles, but some of the results are still confusing. At present, a group of "vesiculists" emphatically tries to establish a unified concept of the role of matrix vesicles in biological tissue mineralization, whereas others raise serious doubts about the postulated underlying mechanism (12) or even consider them as pure morphological artefacts.

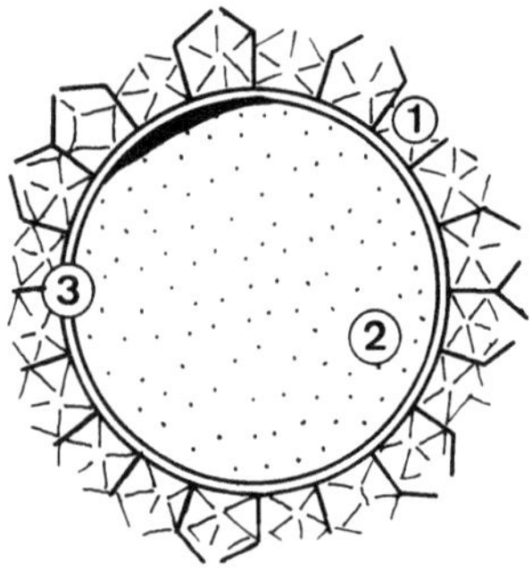

FIG. 3 - Structural integration of matrix vesicle. 1) Extravesicular domain (bound proteoglycans). 2) Intravesicular domain (phosphate-rich compounds?). 3) Trilaminar membrane (nucleation at PS, enzymes, transport).

In general, most authors agree with the following statements: 1) Matrix vesicles are derived from the cell surface, mostly from cell processes extending into the surrounding matrix. Their membrane shares certain properties with the plasmalemma of chondrocytes, i.e., it carries enzymes such as alkaline phosphatase, PPiase at ATPase (3). 2) Matrix vesicles have a high lipids content, most probably phospholipids, proteolipids, or calciumphospholipid-PO_4 complexes, which may also be involved in initial calcification. 3) Vesicles are released by chondrocytes in the proliferation zone, are then accumulated in the interterritorial matrix (passive translocation by additional matrix formation), and go through a maturation period before the onset of mineralization in the lower part of the hypertrophic zone. During this phase which, depending on the growth rate, lasts from 2 days to more than one week, they seem to accumulate calcium, and eventually phosphate, before the first crystals are formed. These observations are based on staining reactions for calcium (pyroantimonate) and on electron microprobe analysis. The microprobe data, however, gave contradictory results at the hands of different investigators, and some authors completely deny the presence of increased amounts of calcium and/or phosphate in matrix components which they consider to be matrix vesicles.

Recent investigations in growth cartilage processed by high pressure freezing and freeze substitution confirm some of the previous descriptions of the vesicles and also reveal a close structural relationship between them and the surrounding proteoglycans, which are locally attached to the outer surfaces of their membranes. This structural integration suggests that the membrane could provide an interface between compartments with different ion composition, since negatively charged proteoglycans facilitate diffusion of positively charged ions such as Ca^{2+} (3), and vesicles could be charged with compounds rich in phosphate (13). The same technique also reveals the first crystalline deposits to appear at the inner surface of the vesicular membrane (10).

It is evident that all isolation procedures will destroy this intimate structural interrelation, a fact which is worth consideration in view of the interpretation of in vitro nucleation tests with purified vesicle fractions.

The concept that inhibitors of mineralization could prevent precipitation until they are destroyed can only be examined morphologically in view of a possible role of proteoglycans. In this respect, most investigators state that there is no loss of proteoglycans during cartilage mineralization. But a change from highly aggregated proteoglycans into partly or completely dissociated units, which cannot be demonstrated by any morphological method currently available, is still possible. Such a regulation by a controlled dissociation of highly aggregated proteoglycans could also be important for the restriction of the mineralization to the interterritorial matrix compartment outlined above. The role of lysozyme as a possible regulator for proteoglycan aggregation merits further consideration, especially in view of its preferential accumulation in the pericellular compartment.

BONE MINERALIZATION

Specifications of Bone Cells

The bone surface is covered by a cell population consisting of bone-forming cells (osteoblasts), bone-resorbing cells (osteoclasts), and indifferent bone-lining cells. Within the bony intercellular substance, osteocytes are incorporated. They are interconnected by a system of numerous fine canaliculi containing cytoplasmic processes which make contact by specialized junctions, mostly gap junctions. This system is also connected to osteoblasts and lining cells and guarantees the transport of fluid and nutrients from the surface over a distance of about 100 µm into the deeper tissue layers. The fine, interdigitated net of lacunae and canaliculi which houses the osteocytes and their processes is directly surrounded by calcified bone and therefore provides an enormous surface for exchange of calcium between bone mineral, bone fluid, and the osteocyte-lining cell system which is thought to be essential for calcium homeostasis. Within certain limits, removal and deposition of calcium salts can structurally alter the lacunar and canalicular walls. This is taken as a proof of the osteolytic or osteoblastic capacities of osteocytes. But in this context, it is not considered as a true bone-forming or bone-resorbing activity, which is a privilege of specialized bone surface cells (15).

Origin and Structure of Osteoblasts

As with chondroblasts, osteoblasts stem from mesenchymal precursors called osteoprogenitor cells (17). Originally this cell was also thought to give rise to osteoclasts which to date are considered to be blood-borne elements (7,9,20). Contrary to cartilage, bone grows only by apposition along its surface. Cell division is restricted to preosteoblasts which then differentiate into osteoblasts. The osteoblasts form a continuous epithelial cell layer and exhibit a striking polarity: bone matrix is only secreted along the surface facing matrix. Adjacent osteoblasts are separated by narrow matrix-free intercellular clefts, and the surface opposed to bone is in contact with blood capillaries to a remarkable extent. This reflects a high metabolic activity which is further sustained by the abundance of cytoplasmic organelles, such as mitochondria, rough endoplasmic reticulum, and components of the Golgi complex. At regular intervals, osteoblasts are determined to become osteocytes, stop matrix production, and subsequently are buried by lateral and superimposed neighbors (6).

Formation and Mineralization of Lamellar Bone

The adult skeleton of higher mammals consists almost exclusively of lamellar bone. Formation of this bone type occurs in 2 phases: osteoblasts first deposit an organic matrix called osteoid which already shows the characteristic lamellar pattern. Each lamella corresponds to an about 2-3 μm thick layer of parallel, rather densely packed collagen fibrils with diameters of 60-100 nm. The orientation of the fibrils changes from lamella to lamella, but there is also a continuous interlamellar exchange of fibrils. These show a clear-cut periodicity of 64 nm and are chemically defined as type I collagen. There are only traces of interfibrillar substance visible which explains the low content in proteoglycans, amounting to only 5% of the dry organic matrix, 95% consisting of collagen. The only further structural components are cell processes and a small number of matrix vesicles.

In the light microscope this osteoid is clearly separated from the calcified bone by the calcification front. At this interface the osteoid undergoes mineralization. This second step in

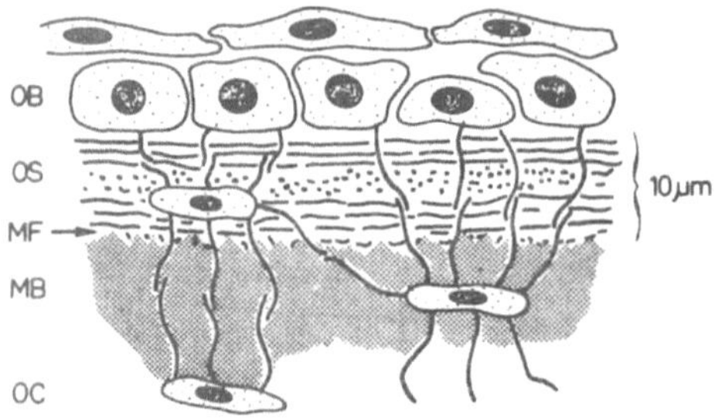

FIG. 4 - Lamellar bone formation. OB - osteoblasts (covered by an endosteal epithelium); OS - osteoid; MF - mineralization front; MB - mineralized bone; and OC - osteocyte.

lamellar bone formation ends with a close association of apatite crystals to collagen, which reflects or even accentuates the cross striation pattern of the fibrils. This preferential orientation has been explained by the macromolecular structure of microfibrils and the presence of holes between the tropocollagen molecules. It has also led to the assumption that lamellar bone mineralization is based on collagen as a nucleating substrate (15).

The dynamics of lamellar bone formation have been studied by various labelling techniques, conveniently by labelling with tetracycline or other fluorochromes. These labels are incorporated and permanently fixed in the mineralization front and allow precise determinations of its daily progress (18). In most adult mammals, a daily mineralization rate of 1 μm (human) to 2 μm (rabbit, horse) has been measured. On the other hand, osteoid seams (i.e., the osteoid interposed between osteoblasts and mineralization front) have a remarkably constant width (in humans about 10 μm) which indicates that osteoblasts also deposit new osteoid at the same constant rate of 1 μm/day (= apposition rate). The 10 μm thick osteoid seam thus represents a time interval of 10 days between osteoid deposition and its subsequent mineralization, which may be considered as a maturation or preparation time for mineralization.

Some staining properties of the osteoid, as well as the distribution of the rare interfibrillar ground substance under the electron microscope, give some indication of such changes, but this definitely needs further investigation. According to some authors, the collagen-associated mineralization along the calcification front is secondary to a vesicle-induced initial phase within the more superficial layers in the osteoid seam. Others deny any participation of vesicles in lamellar bone mineralization and consider this to be a secondary event, starting from the mineral already present in initially formed woven bone or even from underlying mineralized cartilage, upon which lamellar bone is deposited. Recently there has been increasing evidence from the healing of osteomalacia and other skeletal disorders with osteoid accumulation that active osteoblasts are a prerequisite for recovery of mineralization, and therefore that a messenger released by the cells into the matrix is also operative in lamellar bone mineralization.

Formation and Mineralization of Woven Bone

Woven bone is formed in embryonic and fetal life and also during growth, or in the adult organism in conditions of rapid and extensive new bone formation, as in certain bone diseases and in fracture repair. It is considered to be more primitive in its structural and chemical organization and is formed far more rapidly than lamellar bone. Besides the higher cellularity, its main characteristics are the collagen fibrils, which form a felt-like web without any preferential orientation. The staining properties suggest a higher content in proteoglycans, and the osteoid seams are much smaller and sometimes only detectable under the electron microscope.

In view of its mineralization, the most striking difference to lamellar bone is the presence of large numbers of matrix vesicles with all intermediate steps of vesicle-induced initial mineralization, which help to convert the freshly deposited

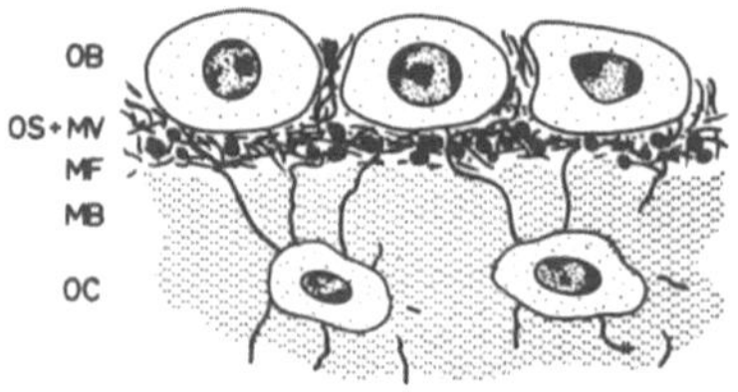

FIG. 5 - Structure and formation of woven bone. MV - matrix vesicles. For other abbreviations, see Fig. 4.

organic matrix into calcified bone within 24 - 72 hours (2). The initial clusters formed around vesicles thus increase rapidly in size and finally coalesce at the level of the calcification front. At this location the arrangement of the crystals also changes; they now appear to be tightly associated with the collagen fibrils and are spaced according to their cross banding. Mineralization in woven bone, therefore, clearly shows a sequence of two steps, the first being induced by vesicles, the second governed by collagen.

CELLS INVOLVED IN CALCIFIED TISSUE RESORPTION

Calcified tissues are uniformly resorbed by multinucleated giant cells with a highly specialized organization. According to their substrates they are called osteoclasts, chondroclasts, ev. dentinoclasts, and cementoclasts, but all are identical in both structure and function. We thus confine this description to osteoclasts.

Contrary to osteoblasts and chondroblasts, these cells are not derived from mesenchymal precursors but from monocytes originating from the bone marrow. These differentiate into mononuclear phagocytes and finally fuse to multinucleated "clasts." Besides the fusion, they also elaborate rather specific cytoplasmic features which seem to be essential for calcified tissue breakdown. Resorption always requires direct contact of the cells with the substrate. Traces of osteoclastic activity are therefore strictly limited to the osteoclast-bone interface and result in a depression in the bone surface called Howship's lacuna. Along this interface, the osteoclasts exhibit special surface characteristics: the marginal part, also referred to as clear zone or sealing part, consists of a rim of cytoplasm rich in filaments (actin?) and lacking other organelles. The plasmalemma of this area is closely attached to the bone and follows precisely its finest contours. Within the clear zone, filaments run perpendicular to the surface and may contribute to the adhesion. In the central part, the surface is complicated by a system of folds and indentations which cause a tremendous increase in surface. Furthermore, this "ruffled border" is in continuous motion. The cytoplasm of the remaining cell body contains numerous mitochondria, extensive Golgi fields, many lysosomes, vesicles, and vacuoles, but less endoplasmic reticulum and ribosomes. The mitochondria frequently contain dark granules rich in calcium, as already discussed in chondrocytes.

The area of the ruffled border is the proper site of bone destruction, which seems to start with dissolution of the mineral, followed by enzymatic degradation of the organic material, presumably collagen. The clear zone helps to maintain the concentration of cell products within the chamber where bone is decomposed. The dissolution of the mineral is most probably achieved by the release of H^+ ions. In this respect it is remarkable that parietal cells in the stomach which produce hydrochloric acid have a very similar organization with extensive plasmalemma folds along their surface associated with a mass of mitochondria. The matrix is most probably degraded by release of lysosomal and nonlysosomal proteolytic enzymes.

The activity of osteoclasts is impressive, they are able to penetrate 50 - 70 μm deep into compact bone within 24 hours. This activity, however, is influenced by a number of regulators, mostly hormones, and can also be reduced or completely arrested by other compounds, such as diphosphonates, which are used in the treatment of skeletal disorders where increased resorption dominates.

The question of calcified tissue resorption by mononucleated cells is often discussed. Besides misinterpretations of marginal sections through osteoclasts which reveal only a single nucleus and also miss the area of the ruffled border, there is evidence for participation of mononuclear phagocytes in locations where smaller calcified particles are engulfed, as in the growth plate, and also under experimental conditions where bone powder is offered to phagocytes in vitro (20). It is also possible that certain tumor cells can create conditions where bone is resorbed. But physiologically, when phagocytes are confronted with large particles and extensive surfaces, they see no other way to overcome the problem than by fusing into larger or even giant organizations and by elaborating the cytoplasmic equipment which is indispensable for a local attack.

REFERENCES

(1) Anderson, H.C. 1969. Vesicles associated with calcification in the matrix of epiphyseal cartilage. J. Cell Biol. 41: 59-72.

(2) Bernard, G.W. 1969. An electron microscope study of initial intramembraneous osteogenesis. Am. J. Anat. 125: 271-290.

(3) Boskey, A. 1979. Models of matrix vesicle calcification. Inorg. Persp. Biol. Med. 2: 51-92.

(4) Brighton, C.T., and Hunt, R.M. 1978. Role of mitochondria in growth plate calcification as demonstrated in a rachitic model. J. Bone Jt. Surg. 60: 630-640.

(5) Burger, E.H., and Matthews, J.L. 1978. Cellular calcium distribution in fetal bones studied with K-pyroantimonate. Calc. Tiss. Res. 26: 181-191.

(6) Fornasier, V.L. 1977. Osteoid: An ultrastructural study. Human Pathol. 8: 243-254.

(7) Göthlin, G., and Ericsson, J.L.E. 1976. The osteoclast. Clin. Orthop. 120: 201-231.

(8) Hascall, V.C. 1977. Interaction of cartilage proteoglycans with hyaluronic acid. J. Supramolecul. Struct. 7: 101-120.

(9) Holtrop, M.E., and King, G.J. 1977. The ultrastructure of the osteoclast and its functional implications. Clin. Orthop. 123: 177-196.

(10) Hunziker, E.B.; Herrmann, W.; Schenk, R.K.; Marti, T.; Müller, M.; and Moor, H. 1981. Structural integration of matrix vesicles in calcifying cartilage after cryofixation and freeze substitution. In Matrix Vesicles - Proceedings of the Third International Conference on Matrix Vesicles, eds. A. Ascenci, E. Bonucci, and B. de Bernard, pp. 25-31. Milano: Wichtig Editore srl.

(11) Kuettner, K.E.; Soble, L.; Günther, H.; Croxen, R.; and Eisenstein, R. 1970. Lysozyme in epiphyseal cartilage. I. The nature of the morphologic response of cartilage in culture to exogenous lysozyme. Calc. Tiss. Res. 5: 56-63.

(12) Landis, W.J. 1979. Application of electron probe X-ray microanalysis to calcification studies of bone and cartilage. Scanning Electr. Microsc. 11: 555.

(13) Landis, W.J.; Hammett, M.; and Glincher, M. 1979. Qualitative and quantitative X-ray microanalysis of ultrastructural components of rat cartilage prepared by ultra cryo microtomy. In Transactions of the 25th Annual Meeting, Orthopaedic Research Society, p. 158. Dependable Printing Co., Inc.

(14) Maroudas, A. 1980. Physical chemistry of articular cartilage and the intervertebral disc. In The Joints and Synovial Fluid II, ed. L. Sokoloff, p. 239. New York: Academic Press.

(15) Matthews, J.L. 1980. Bone structure and ultrastructure. In Fundamental and Clinical Bone Physiology, ed. M.R. Urist, pp. 4-44. Philadelphia, Toronto: J.B. Lippincott Co.

(16) Muir, H. 1977. Structure and function of proteoglycans of cartilage and cell-matrix interactions. In Cell and Tissue Interactions, eds. J.W. Lask and M.M. Burger, p. 87. New York: Raven Press.

(17) Owen, M. 1978. Histogenesis of bone cells. Calc. Tiss. Res. 25: 205-207.

(18) Schenk, R. 1980. Basic histomorphology and physiology of skeletal growth. In Treatment of Fractures in Children and Adolescents, eds. B.G. Weber, C. Brunner, and F. Freuler, pp. 3-19. Berlin, Heidelberg, New York: Springer-Verlag.

(19) Schenk, R.K.; Spiro, D.; and Wiener, J. 1967. Cartilage resorption in the tibial epiphyseal plate of growing rats. J. Cell Biol. 34: 275.

(20) Teitelbaum, S.L., and Kahn, A.J. 1980. Mononuclear phagocytes, osteoclasts and bone resorption. Min. Electrol. Metab. 3: 2-9.

Biological Mineralization and Demineralization, ed. G.H. Nancollas, pp. 161-177.
Dahlem Konferenzen 1982. Berlin, Heidelberg, New York: Springer-Verlag.

Structure of the Organic Matrix: Collagen Structure (Chemical)

D. J. Prockop and C. J. Williams
Dept. of Biochemistry, College of Medicine and Dentistry
of New Jersey - Rutgers Medical School
Piscataway, NJ 08854, USA

INTRODUCTION

Collagen is the major constituent of the organic matrix of bone in that it accounts for about 30% of its dry weight and 90 to 95% of its nonmineral content (see (14)). Of the proteins found in nature, collagen is one of the largest. In terms of structure and function, it is among the simplest. In contrast, its biosynthesis is complex in that the initial RNA transcripts of the genes must be extensively processed in order to generate the functional mRNAs which are used for translation. Also, the polypeptides first synthesized on ribosomes must undergo extensive post-translational processing before the protein can be assembled into an extracellular collagen fiber. (For reviews see (2,11,13); additional references cited here are to a few recent papers not included in these reviews.)

THE COLLAGEN MOLECULE

The function of collagen fibers is to hold together and give form to the tissues of the body. Collagen fibers have about the same tensile strength as steel wires and they consist of highly ordered polymers of a rod-like molecule. The molecule itself contains three polypeptide chains, called α chains (Fig. 1). Each of these α chains is coiled into a left-handed helix with about 3 amino acids per turn. The three helical

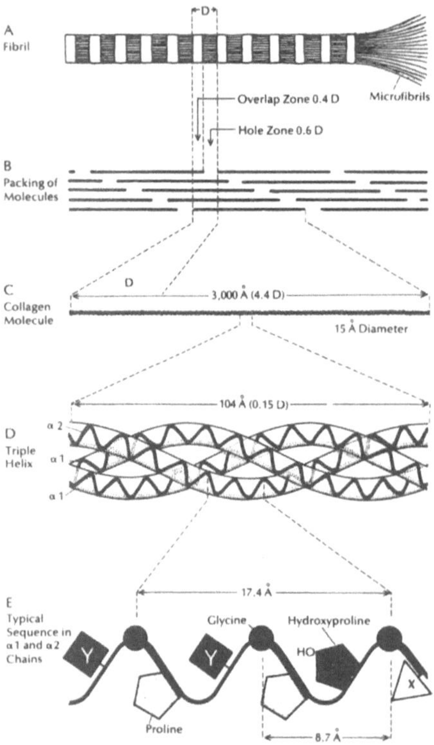

FIG. 1 - Schematic illustration of the structure of a collagen fibril and a collagen molecule (reproduced with permission (12)).

chains are then twisted around each other into a right-handed superhelix to form a rigid structure similar to a long, thin segment of a rope. Each α chain contains about 1000 amino acid residues and, with the exception of short sequences at the ends of the chains, every third amino acid in each chain is glycine. The molecular formula of an α chain can thus be approximated as $(X\text{-}Y\text{-}Gly)_{333}$ or $(Gly\text{-}X\text{-}Y)_{333}$, where X and Y represent amino acids other than glycine. The presence of glycine, the smallest amino acid, in every third position is crucial; the amino acid in this position occupies the restricted space in which the three helical α chains come together in the center of the triple helix. The triple-helical conformation also depends on the presence of proline and hydroxyproline in the α chains. In collagen from mammals and birds about 100 of the X-positions are proline, and about 100 of the Y-positions are hydroxyproline. Because both proline and hydroxyproline are rigid, cyclic amino acids, they limit rotation of the polypeptide backbone and thus contribute to the stability of the triple helix. Hydroxyproline in the Y-position has a special role in stabilizing the triple-helix, since collagen polypeptides lacking hydroxyproline do not fold into a triple-helical conformation at body temperature.

Studies with synthetic peptides demonstrate that α chains that consist only of glycine, proline, and hydroxyproline in the structure $(\text{Pro-Hyp-Gly})_n$ would form an extremely stable helix (see (13)). In mammalian collagen, however, about two-thirds of the X- and Y-positions are occupied by a variety of different amino acids. These other amino acids tend to be clustered in groups of hydrophobic and charged residues whose side chains point out from the center of the triple helix. Although these other amino acids decrease the stability of the triple helix, they are essential for organization of the triple-helical molecules into collagen fibrils. Collagen, in effect, is an "inside-out" protein in which the large number of amino acid side chains on the surface of the molecule direct precise self-assembly of the molecules. The specific amino acid sequences of α chains can, in fact, be used to predict that collagen molecules will aggregate so that each molecule is longitudinally displaced about one-quarter of its length relative to a nearest neighbor. This longitudinal displacement explains the characteristic "D spacing" seen by electron microscopy of collagen fibers (Fig. 1). The longitudinal displacement is, however, slightly less than one-quarter of the length of the monomer. This leaves a "hole" between the end of one triple helix and the beginning of the next. The hole probably provides a site for deposition of hydroxyapatite crystals in bone formation (see below). The fibrils appear to consist of smaller strands which may represent discrete "microfibrils" of four to eight collagen molecules in cross section. Recent evidence, however, suggests that the fibrils have no specific substructure and that the monomers are present instead in a uniform quasi-hexagonal lattice (7).

A variety of collagens are found in man and other animals (see (2)). All these collagens are composed of polypeptides with regions that have the same characteristic $(X\text{-}Y\text{-}Gly)_n$ structure. They also contain large amounts of proline and hydroxyproline. They differ, however, in the precise sequence of amino acids in the X- and Y-positions and, therefore, are the products of different genes (see Table I).

The different collagens and α chains have been classified on the basis of differences in amino acid sequences that can be directly attributed to differences in their messenger RNAs. The collagens also differ, however, in their contents of hydroxyproline, hydroxylysine, and glycosylated hydroxylysine, components that are synthesized by post-translational modifications of collagen polypeptides (see below). Most of the hydroxyproline is present in collagens as the trans-4-isomer.

TABLE I - Genetically distinct types of collagen.

Type	Chain Composition	Tissue	Characteristics
I	$[\alpha 1(I)]_2\alpha 2(I)$	Most connective tissues	Low glycosylated hydroxylysine
II	$[\alpha 1(II)]_3$	Hyaline cartilage	High glycosylated hydroxylysine
III	$[\alpha 1(III)]_3$	Skin, arteries and uterus	High hydroxyproline; interchain disulfide bonds
IV	α1, α2*	Basement membranes	High hydroxyproline and glycosylated hydroxylysine; large globular domains
V	α1, α2, α3*	Pericellular matrix	Similar to type IV

*Stoichiometry of chains uncertain.

All known collagens also contain a small amount of trans-3-hydroxyproline, and basement membrane collagens are relatively rich in 3-hydroxyproline. Most of the hydroxyl groups of the hydroxylysyl residues found in collagens are sites for the attachment, through O-glycoside linkage, of the monosaccharide galactose or the disaccharide glycosylgalactose. A number of physiologic and pathologic conditions can alter the efficiency of the post-translation reactions. Therefore, cells that use the same mRNA can produce collagens that differ markedly in their content of hydroxyproline, hydroxylysine, and glycosylated hydroxylysine.

THE COLLAGEN GENES AND mRNA PROCESSING

Collagen is first synthesized as a larger precursor known as procollagen and consisting of three proα chains (see below). The genes for type I procollagen have been cloned and, in part, characterized for chick embryo (6,19), sheep (1), and man (8,9). The proα2(I) gene in chick embryos, sheep, and man is about 40 kb in length and contains about 50 introns. The intervening sequences for the proα2(I) gene from chick embryos range in size from 50 to 2,000 bases. A few of the exons coding for the α chain domain of the chick proα2(I) are 45 and 99 bases but most are either 54 or 108 bases. It has been suggested that the collagen gene evolved by duplication of a 54 base fragment coding for 18 triplets of -Gly-X-Y- (18,19). The large number and the extent of the intervening sequences in the proα genes means that there must be 50 gene splicing events in converting the initial transcript of 40 kb to a functional mRNA (Fig. 2). Examination of polysomes from human fibroblasts showed that there are two functional mRNAs of about 7.2 and 5.9 kb for proα1(I) chains, and three functional mRNAs of about 6.2, 5.7, and 5.5 kb for proα2(I) chains (Chu et al., submitted for publication). By analogy with other systems, it is likely that these mRNAs differ only in the length of the nontranslated 3'-ends of the mRNAs. The reasons for such multiple size mRNAs is still unknown.

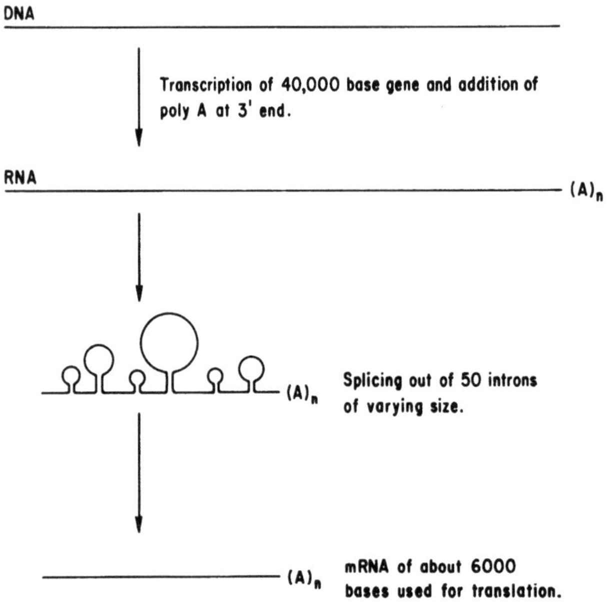

FIG. 2 - Scheme summarizing the transcription of a procollagen gene and the processing of the initial transcript.

Additional information recently obtained about the procollagen genes is that the human proα2(I) gene has been located on chromosome 7 (Junien et al., submitted for publication). Also, it has been shown that the structure of the 5'-end of the chick proα2(I) gene contains a TATA box and other sequences characteristic of the promoter regions of other eukaryotic genes (17).

THE INTRACELLULAR ASSEMBLY OF PROCOLLAGEN

The intracellular assembly of the procollagen molecule involves at least eight separate enzymes which modify the initial translation product at over 100 separate sites (Fig. 3).

Like most other proteins synthesized "for export," the amino-terminal ends of newly synthesized procollagen polypeptides contain "signal peptides" which are cleaved enzymatically as the chains pass into the cisternae of the rough endoplasmic

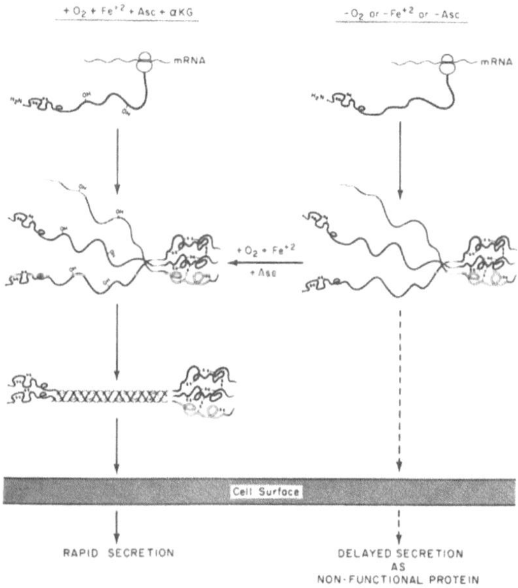

FIG. 3 - Scheme for the post-translational processing of procollagen which occurs before the protein is secreted. As discussed in the text, the rate constant for secretion depends on folding of the protein into a triple-helical conformation (reproduced with permission (13)).

reticulum. The signal peptides of proα1(I) and proα2(I) chains differ in their amino acid sequences. Both may be larger than the signal peptides found on other proteins. As the polypeptides enter the rough endoplasmic reticulum, they also encounter three hydroxylating enzymes located within this compartment. Two of these hydroxylases convert prolyl residues to 4-hydroxyproline or 3-hydroxyproline, and the third converts lysyl residues to hydroxylysine. The reactions catalyzed by these three hydroxylases all involve the same cofactors: ferrous ions, molecular oxygen, α-ketoglutarate, and ascorbate. The requirement for ascorbate in these reactions explains why wounds fail to heal, and why synthesis of bone matrix is deficient in scurvy.

All three hydroxylases act on substrates according to three unusual and specific requirements: (a) The residue hydroxylated must be located in a peptide and must occupy the correct position in the amino acid sequence of the peptide; (b) the peptide must have a nonhelical conformation; and (c) long peptides are more readily hydroxylated than are short peptides. Prolyl 4-hydroxylase and lysyl hydroxylase act only on proline or lysine residues in the Y-position of peptides containing -X-Y-Gly- sequences. In contrast, prolyl 3-hydroxylase acts on prolyl residues in the X-position but only if the Y-position is 4-hydroxyproline.

The requirement for a nonhelical conformation is just as rigid as the requirements for peptide linkage and for the correct X- or Y-position in the chain. This requirement for a nonhelical conformation has a critical role in the processing of the protein. The chains must have about 90 residues per chain of 4-hydroxyproline to form a triple helix that is stable at 37°C. On the other hand, after the chains have acquired the 4-hydroxyproline and the interchain disulfide bonds necessary for helix formation, they fold into a triple-helical conformation that prevents further hydroxylation of prolyl or lysyl residues (see (11,13)).

As lysyl residues in the newly synthesized proα chains are hydroxylated, the hydroxylysine residues are glycosylated by two specific enzymes, a galactosyltransferase and then a glucosyltransferase. Both these transferases use sugar in the form of a uridine diphosphate glycoside, and both reactions require the presence of a bivalent cation, preferably manganese. The enzymes, like the hydroxylases, require that the proα chains be in a nonhelical conformation (see (11,13)).

The mannose-rich oligosaccharides in the carboxy-terminal propeptides of procollagen (see below) are synthesized through dolichol intermediates, such as those involved in the synthesis of other glycoproteins. The sugars in the propeptides of procollagen can be added when the proα chains are still non-helical, and the addition of these sugars probably occurs in the cisternae of the rough endoplasmic reticulum.

Another important step in the intracellular processing of the propeptides is the synthesis of both intrachain and interchain disulfide bonds. As in other proteins containing disulfide bonds, it is not clear whether the synthesis of these linkages requires an enzymatic mediator or whether it occurs spontaneously during biosynthesis. Because the interchain disulfide links in type I procollagen are in the carboxy-propeptides, these links cannot be synthesized until most of the chains are translated. Indirect evidence suggests that the interchain bonds are not synthesized until after the chains are released from ribosomes (see (11,13)).

Most current evidence also suggests that the helix does not form until after proα chains are released from polysomes. In fact, the kinetics of helix formation in tendon fibroblasts are consistent with a model which assumes that the helix formation begins in a nucleus at the carboxy-terminal end of the α chain and then is propagated toward the amino-terminus at a rate which is controlled by the rate of isomerization of cis $\rightleftarrows$ trans peptide bonds (3). Also, most current evidence indicates that helix formation occurs in the rough endoplasmic reticulum, but the data do not completely exclude the possibility that the helix does not form until the chains reach the Golgi complex.

TRANSLOCATION AND SECRETION

Secretion of procollagen is similar to that of other proteins synthesized for export, in that the protein assembled in the rough endoplasmic reticulum passes through the Golgi complex before leaving the cell. One distinctive feature of procollagen, however, is that its secretion depends on intracellular processing of the protein and specifically, on folding of the collagen domains of proα chains into a triple-helical conformation (Fig. 3). If triple helix formation is prevented, the nonhelical proα chains accumulate within the cisternae of the rough endoplasmic reticulum and then are secreted with a slower rate constant than triple helical procollagen. This situation can be produced experimentally by incubating fibroblasts under an aerobic condition or with an iron chelator so that hydroxylation of prolyl residues is prevented. It is, in part, created in fibroblasts deficient in ascorbate. Also, it can be produced by incubating fibroblasts with proline analogs such as azetidine-2-carboxylic acid or cis-4-hydroxyl-L-proline which are incorporated into protein, and whose presence in proα chains prevents them from folding into a triple-helical conformation.

It is not clear why nonhelical forms of procollagen are slowly secreted and tend to accumulate in the cisternae of the rough endoplasmic reticulum. One intriguing hypothesis is that post-translational enzymes, such as prolyl hydroxylase and lysyl hydroxylase, may prevent secretion of incompletely processed procollagen. These enzymes have a high affinity for both nonhydroxylated and fully hydroxylated proα chains which are in a nonhelical conformation but do not interact with triple-helical chains (see (11,13).

THE CONVERSION OF PROCOLLAGEN TO COLLAGEN

The conversion of procollagen to collagen requires at least two enzymes (see (10,11,13)), a procollagen amino-proteinase that removes the amino-propeptides, and a procollagen carboxy-proteinase that removes the carboxy-propeptides (Fig. 4). Both the amino- and carboxy-proteinases are neutral proteinases which require calcium. They are not inhibited by inhibitors of serine or sulfhydryl proteinases. A procollagen amino-proteinase has been partially purified from calf and fibroblast culture and extensively purified from chick embryo tendons. These enzymes cleave the amino-propeptides from both the proα1 and proα2 chains of type I procollagen. The enzyme from chick embryo tendons also cleaves type II procollagen. It appears, therefore, that either the amino-proteinase has a broad specificity

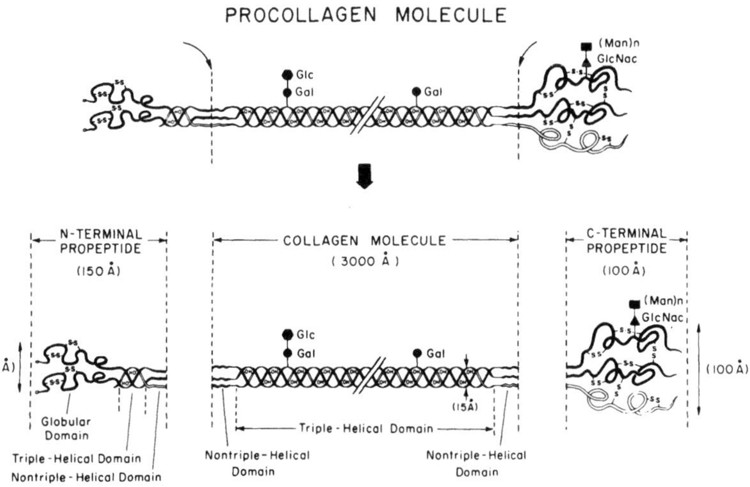

FIG. 4 - Structure of procollagen (reproduced with permission (13)).

or the cleavage sites in three different kinds of proα chains have similar amino acid sequences. A separate type III procollagen amino-proteinase has been identified. Procollagen carboxy-proteinases have not been as extensively studied as the amino-proteinases, but procollagen specific enzymes have been isolated from the medium of fibroblast cultures and from chick embryo calvaria. They appear to be smaller than the amino-proteinases but have a similar resistance to inhibitors of serine and sulfhydryl proteinase inhibitors.

ASSEMBLY OF COLLAGEN FIBRILS FROM EXTRACTED COLLAGEN AND FROM PROCOLLAGEN

Collagen monomers and polymers extracted from connective tissues with cold acidic or neutral buffers readily form characteristic fibrils when solutions containing the protein are adjusted to neutrality and warmed to 25°C to 37°C (see (11)). The fibrils which form have the typical cross-striated pattern of collagen fibrils synthesized in vivo. In part because of the ease with which such fibril reconstitution experiments can be carried out, the self-assembly of collagen solutions has been examined with a variety of physical techniques. However, the kinetics of fibril formation, and the morphology of the fibrils, are dependent on the history of the protein solution in that they are altered by the nature of the buffers used to extract the collagen, the loss of telopeptides from the ends of the molecule during extraction or purification, the presence of cross-links, the presence of aldehydes which can form cross-links in the protein (see below), the age of the solutions, and other variables such as the composition of the buffer used for fibril formation.

Recently, a new system has been developed for studying formation of collagen fibrils in vitro (Michinori, Njieha, and Prockop, in preparation). Partially purified procollagen amino-proteinase and carboxy-proteinase were used to initiate fibril formation by the enzymic removal of the amino- and carboxy-propeptides from purified type I procollagen in a physiological buffer at 35°C or 37°C. The system should make it possible to examine in detail the functional properties of both the propeptides of procollagen and of the collagen monomer.

CROSS-LINKING OF COLLAGEN FIBERS

To develop the necessary tensile strength, the collagen fibers first assembled in tissues must acquire intermolecular cross-links (see (5,11)). The cross-linking of collagen monomers in fibers occurs in two or three stages. The first step is oxidative deamination of ε-amino groups in certain lysyl and hydroxylysyl residues near the ends of the collagen molecule to produce reactive aldehydes. The enzyme catalyzing this reaction, lysyl oxidase, is a copper-containing protein and has been extensively purified from chick aorta and cartilage. The enzyme reacts poorly with isolated α chains, and it preferentially acts on collagen already assembled into collagen fibers.

After aldehyde groups are generated from collagen by the action of lysyl oxidase, cross-links are formed by one of two routes. The minor route involves reaction of two aldehydes in an aldol condensation reaction to join two α chains of the same molecule. The major route involves condensation between an aldehyde in one molecule and an ε-amino group of lysine, hydroxylysine, or glycosylated hydroxylysine in an adjacent molecule. The Schiff bases synthesized in this manner are the source of the major intermolecular cross-links in collagen. The Schiff bases, however, are not in themselves stable, but they serve as precursors for a variety of more stable cross-links. One of these more stable cross-links arises from a shift of the double bond in the Schiff base to yield a ketone. Another may be synthesized by hydration and oxidation to give a new peptide bond. A third may arise by a more complex rearrangement involving reaction with additional amino acid side chains of the collagen molecule, such as the imidazole group of histidine. A fourth, more recently discovered, intermolecular cross-link involves reaction of two Schiff base type of cross-links to form a trivalent aromatic cross-linking structure called 3-hydroxypyridinium (5). The latter cross-link may be the major cross-link of mature hard tissue collagens such as the type I collagen of bone.

THE ROLE OF COLLAGEN IN MINERALIZATION

One of the major questions about bone is, "What is the role of collagen in initiating or directing mineralization?"

One possible answer to this question is that collagen in fact has no direct role in mineralization and that it simply serves as a nonspecific scaffolding for the process. However, three suggestions for a specific role have been made.

One of these suggestions is that the "holes" in collagen fibers (Fig. 1) provide the primary sites for mineralization. This suggestion was first made about 15 years ago, but at least two objections have been raised. One is that the holes cannot accommodate more than two-thirds, and perhaps less, of the total mineral in bone. The other is that all collagen fibers have "holes" but all do not mineralize. Therefore, to accept this suggestion it seems necessary to make some further assumption such as that nonmineralizing collagen fibers are protected from mineralization by tissue components such as proteoglycans.

A second suggestion for a specific role for collagen is that bone collagen may contain phosphate and that the phosphate may serve as a site for initiating mineralization. Collagen α2(I) chains purified from the bones of chick embryos were found to contain about four atoms of organic phosphorus per mole (4). The phosphate is in the form of γ-glutamyl phosphate. It is not clear, however, whether the phosphate is linked directly to a glutamyl residue in the α2(I) chain or whether the γ-glutamyl phosphate is attached to collagen via a small covalently-linked peptide. Also, the biological role of the γ-glutamyl phosphate has not been clearly established.

A third suggestion as to the role of collagen is that it binds some protein which in turn provides a site for mineralization. Recently, a protein called "osteonectin" has been purified from periosteal bone (15). The protein, which is about 32,000 daltons, strongly binds to both hydroxyapatite and collagen fibers, and it accounts for about 25% of the noncollagenous protein in bone (15).

CONCLUSIONS

The information we have about collagen provides a broad base for understanding many critical features of bone. We are left, however, with a number of unresolved questions. In our opinion, some of the most critical of these unresolved questions are the following:

1. Is it correct to assume (see (14)) that bone formation always occurs in two discrete steps - deposition of a matrix of "osteoid" and then mineralization of this matrix?

2. What kinds of experimental data will resolve the question of whether collagen has a primary role in initiating or directing the mineralization of bone?

3. What are the principal signals which tell a bone-forming cell to synthesize more or less collagen, and more or less of the other components of matrix?

4. Is it correct to assume, as was done here (see above), that formation of collagen fibers is entirely an extracellular process of self-assembly controlled by the structural information found in monomer? Trelstad (16) and others have suggested fibers begin to be assembled intracellularly and are spun out in continuous threads, like cellulose from plant cells. How can this suggestion be ruled in or ruled out?

In considering how these questions might be resolved, we ourselves have developed the following opinions:

1. It will probably be highly fruitful to carry out further work on the isolation and chemical characterization of the proteins and other components of bone matrix.

2. Mutants, such as those represented by patients with osteogenesis imperfecta (see (11,13)), will probably provide critically important information about bone formation and structure.

3. Eventually, we will have to be able to make a bone in vitro from pure components isolated from the tissue.

REFERENCES

(1) Boyd, C.D.; Tolstoshev, P.; Schafer, M.P.; Trapnell, B.C.; Coon, H.C.; Kretschmer, P.J.; Nienhuis, A.W.; and Crystal, R.G. 1980. Isolation and characterization of a 15 kb genomic sequence coding for part of the proα2 chain of sheep type I collagen. J. Biol. Chem. 255: 3212-3220.

(2) Bornstein, P., and Sage, H. 1980. Structurally distinct collagen types. Ann. Rev. Biochem. 49: 957-1003.

(3) Bruckner, P.; Eikenberry, E.F.; and Prockop, D.J. 1981. Formation of the triple helix of type I procollagen in cellulo. A kinetic model based on cis-trans isomerization of peptide bonds. Eur. J. Biochem., in press.

(4) Cohen-Solal, L.; Cohen-Solal, M.; and Glimcher, M.J. 1979. Identification of a γ-glutamyl phosphate in the α2 chains of chicken bone collagen. Proc. Natl. Acad. Sci. 76: 4327-4330.

(5) Eyre, D.R. 1980. Collagen: Molecular diversity in the body's protein scaffold. Science 207: 1315-1322.

(6) Fuller, F., and Boedtker, H. 1981. Sequence determination and analysis of the 3' region of chicken proα1(I) and proα2(I) collagen messenger ribonucleic acids including the carboxyterminal propeptide sequences. Biochem. 20: 996-1006.

(7) Hulmes, D., and Miller, A. 1979. Quasi-hexagonal molecular packing in collagen fibrils. Nature 282: 878-880.

(8) Myers, J.C.; Chu, M-L.; Faro, S.H.; Clark, W.J.; Prockop, D.J.; and Ramirez, F. 1981. Cloning a cDNA for the proα2 chain of human type I collagen. Proc. Natl. Acad. Sci. 78: 3516-3520.

(9) Myers, J.C.; Sangiorgi, F.O.; Chu, M-L.; Ding, G.; Prockop, D.J.; and Ramirez, F. 1981. Isolation and characterization of part of the human proα2 collagen gene. Fed. Proc. 40: 1650.

(10) Njieha, F.; Morikawa, T.; Tuderman, L.; and Prockop, D.J. 1982. Partial purification of a procollagen C-proteinase. Inhibition by synthetic peptides and sequential cleavage of type I procollagen. Biochem. 21: 757-764.

(11) Prockop, D.J., and Champe, P.C., eds. 1980. Gene Families of Collagen and Other Proteins. New York: Elsevier/North-Holland.

(12) Prockop, D.J., and Guzman, N. 1977. Collagen diseases and the biosynthesis of collagen. Hospital Practice 12: 61-68.

(13) Prockop, D.J.; Kivirikko, K.I.; Tuderman, L.; and Guzman, N.A. 1979. The biosynthesis of collagen and its disorders. N. Eng. J. Med. 301: 13-23, 75-85.

(14) Sledge, C.V. 1981. Formation and resorption of bone. In Textbook of Rheumatology, eds. W.N. Kelley, E.D. Harris, S. Ruddy, and C.B. Sledge, pp. 277-293. Philadelphia: W.B. Saunders Co.

(15) Termine, J.D.; Kleinman, H.K.; Whitson, S.W.; Conn, H.M.; McGarvey, M.L.; and Martin, G.R. 1981. Osteonectin, a bone-specific protein linking mineral to collagen. Cell, in press.

(16) Trelstad, R.L., and Hayashi, K. 1979. Tendon fibrillogenesis: Intracellular collagen subassemblies and cell surface changes associated with fibril growth. Dev. Biol. 71: 228-242.

(17) Vogeli, G.; Ohkubo, H.; Sobel, M.E.; Yamada, Y.; Pastan, I.; de Crombrugghe, B. Structure of the promoter for the chick $\alpha2$ type I collagen gene. Proc. Natl. Acad. Sci., in press.

(18) Wozney, J.; Hanahan, D.; Morimoto, R.; Boedtker, H.; and Doty, P. 1981. Fine structural analysis of the chick pro$\alpha2$ collagen gene. Proc. Natl. Acad. Sci. 78: 712-716.

(19) Yamada, Y.; Avvedimento, V.E.; Mudryj, M.; Ohkubo, H.; Vogeli, G.; Irani, M.; Pastan, I.; and de Crombrugghe, B. 1980. The collagen gene contains evidence for its evolutionary assembly by duplication of a DNA segment containing an exon of 54 bp. Cell 22: 887-892.

Biological Mineralization and Demineralization, ed. G.H. Nancollas, pp. 179-198.
Dahlem Konferenzen 1982. Berlin, Heidelberg, New York: Springer-Verlag.

Components of the Organic Matrix: Proteoglycans

V. C. Hascall* and D. A. Lowther**
*National Institute of Dental Research, NIH, Bethesda, MD 20205, USA
**Dept. of Biochemistry, Monash University, Clayton, Victoria, Australia

Abstract. The structures and functions of proteoglycans are discussed, emphasizing those from cartilage. Various roles are considered for proteoglycans in the mechanisms of mineralization of cartilage matrix and osteoid during endochondral bone formation.

STRUCTURE OF PROTEOGLYCANS

Proteoglycans are macromolecules which contain one or more glycosaminoglycan chains covalently bound to a core protein. This definition encompasses many structures serving a variety of functions. Different classes vary in the nature and size of the core protein, in the type(s), number, average size, and degree of sulfation of the glycosaminoglycan chains, and in the type(s) of other oligosaccharides present on the core protein. The core proteins in different classes are almost certainly not related gene products the way that different α-chains of collagen are. Furthermore, the final structure of a proteoglycan depends upon the coordination of extensive post-translational steps using most of the known biosynthetic pathways involved in oligosaccharide and glycosaminoglycan synthesis. Thus, regulation or modulation of any of these pathways can greatly influence the final structures and functions of these macromolecules. A few examples will illustrate the range of structures and potential functions of proteoglycans. (Details and references will be found in recent reviews ((7,22,30), except when noted.)

The corneal stroma contains two classes of small proteoglycans, one with dermatan sulfate and the other with keratan sulfate. The dermatan sulfate proteoglycan (MW ≈ 150,000) contains 1-2 dermatan sulfate chains (MW ≈ 55,000). The keratan sulfate proteoglycan (MW ≈ 60,000) contains 1-3 keratan sulfate chains (MW ≈ 7,000). Each contains oligosaccharides, probably of the complex glycoprotein type. These proteoglycans have relatively small hydrodynamic sizes which just fit between the highly ordered lattice of collagen fibrils. This lattice would be severely distorted if larger proteoglycans such as those from cartilage were present. The corneal proteoglycans are important for extending the collagen network and for maintaining its ordered structure. For example, a deficiency in synthesis of the keratan sulfate proteoglycan in patients with corneal macular dystrophy leads to focal precipitates in the matrix which cause progressive cloudiness and loss of vision (7).

Glomerular basement membranes (13) and hepatocyte plasma membranes contain a class of heparan sulfate proteoglycans (MW ≈ 80-130 x 10^3) which have about 4 chains (MW ≈ 15-25 x 10^3) per macromolecule. In glomeruli, the proteoglycans form an ordered lattice on both surfaces of the basement membrane, providing a polyanionic barrier to filtration which is critical for retaining macromolecules in circulation (14). Heparan sulfate interacts with receptors in the hypatocyte plasma membrane (17) and with fibronectin (41), suggesting that heparan sulfate proteoglycans in plasma membranes may play roles in cell-cell and cell-matrix interactions.

The major cartilage proteoglycan has been widely studied because of its abundance in this tissue and its function in maintaining tissue resiliency and resistance to compressive load. It has a highly complex structure, and because of its postulated role in endochondral bone formation, its characteristics will be described in more detail (see Fig. 1). The large core protein (MW ≈ 300,000) contains 80-100 chondroitin sulfate chains (MW ≈ 20,000), 100-140 O-glycoside-linked oligosaccharides plus keratan sulfate chains, and

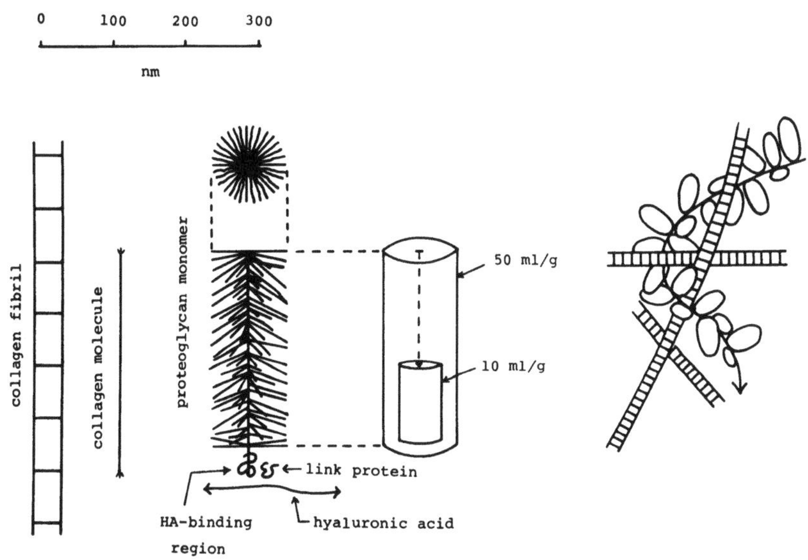

FIG. 1 - A schematic drawing of a typical cartilage proteoglycan with about 100 chondroitin sulfate chains attached to the core protein. In its extended structure, the proteoglycan would occupy a cylinder of sufficient size to contain about 50 ml/g. In the tissue, the concentration of proteoglycan is high, and the domain of an individual molecule would be no greater than 10 ml/g (the smaller cylinder). The interactions between link protein, hyaluronic acid, and the hyaluronic acid-binding region of the core protein necessary for aggregation are shown. For comparison, a single collagen molecule and a portion of a thin collagen fibril, such as those found in cartilage, are shown at the same scale. The drawing to the right shows a portion of an aggregate with many monomers bound to hyaluronic acid surrounding segments of collagen fibrils.

10-15 N-glycosylamine-linked oligosaccharides. Table 1 shows representative structures for each of these complex carbohydrates. The O-linked oligosaccharides are similar to those in mucins, while the N-linked oligosaccharides are similar to those in glycoproteins. The chondroitin sulfate chains are concentrated along 60% of the polypeptide at one end, while the keratan sulfate chains and O-linked oligosaccharides are concentrated in an adjacent middle region. A portion of the polypeptide (MW ≃ 60,000) at the other end end with few or no glycosaminoglycans contains a highly specialized binding site which interacts specifically with hyaluronic acid to form aggregates. The N-linked oligosaccharides are concentrated in this region. The overall MW of a cartilage proteoglycan, then, is about 2.5×10^6.

The hyaluronic acid-binding site interacts specifically with a length of hyaluronic acid extending over 10 sugars

TABLE 1 - Glycoconjugates in cartilage proteoglycans (7,21)

Chondroitin sulfate (n ≃ 50)

```
               OSO3
                | 3
(GlcUAβ1,3GalNacβ1,4)nGlcUAβ1,3Galβ1,3Galβ1,3Xylβserine
```

Keratan sulfate (n ≥ 1) and O-glycoside oligosaccharide (n=0)

```
                          OSO3
                           | 3
NANAα2,3(Galβ1,4GlcNacβ1,3)nGalβ1,4GlcNacβ1,6
                                        |
                     NANAα2,3Galβ1,3GalNac-serine (threonine)
```

N-glycosylamine oligosaccharide

```
        NANAα2,3Galβ1,4GlcNacβ1,4
                                |
NANAα2,3Galβ1,4GlcNacβ1,2Manα1,6(or 3)    Fucα1,6
                                |          |
                               Manβ1,4GlcNacβ1,4GlcNac-N-asparagine
                                |
NANAα2,3Galβ1,4GlcNacβ1,2Manα1,3(or 6)
```

GlcUA = glucuronic acid; GalNac = Nacetylgalactosamine; Gal = galactose; Xyl = xylose; NANA = Nacetylneuraminic acid; GlcNac = Nacetylglucosamine; Man = mannose; Fuc = fucose.

(glucuronosylβ1,3Nacetylglucosaminylβ1,4)$_5$, with a high affinity ($K_d \simeq 10^{-7}$ M). Many proteoglycans interact with a single hyaluronic acid molecule to form large aggregates. A minimum of about 100 sugars is required between adjacent proteoglycans since their large hydrodynamic sizes inhibit closer approach. Native aggregates contain a third component, the link protein (MW $\simeq$ 40-45 x 10^3). The link protein binds to an adjacent stretch of hyaluronic acid and interacts with a portion of the hyaluronic acid as well as with a portion of the hyaluronic acid-binding region of the proteoglycan, thereby preventing the proteoglycan from dissociating from the hyaluronic acid.

Proteoglycan biosynthesis begins with the synthesis of the core protein in the rough endoplasmic reticulum. The core protein precursor prior to adding chondroitin sulfate chains has been isolated from chondrocytes prepared from a rat chondrosarcoma (16). It has an apparent MW of ∿370,000 and contains a functional hyaluronic acid-binding site. High mannose type oligosaccharides would be added from a dolichol-diphospho-oligosaccharide donor at this stage; these would be modified to the complex type present in the final structure (Table 1), later in the biosynthetic sequence (see (18) for a review). After release from the ribosomes, the core protein migrates to the Golgi region where the chondroitin sulfate and O-linked oligosaccharides (and keratan sulfate, if present) are added. This appears to be the rate-limiting step since the chondrocytes contain a large quantity of core protein (∿500 ng/10^6 cells), and the addition of the chondroitin sulfate is very rapid. The half-life of a core protein is about 1 hour before adding chondroitin sulfate (16). Completed proteoglycans are rapidly exocytosed with an intracellular half-life of 5-7 minutes (7). Newly secreted proteoglycans form aggregates in the extracellular matrix within the interstices of the collagen fibrillar network. First, a proteoglycan-link protein complex forms, either in the cell or shortly after secretion, and then this complex binds to an available stretch of hyaluronic acid (7). How premature aggregation inside the cell is prevented is unknown; perhaps hyaluronic acid is segregated from the other two components.

Proteoglycans occupy large domains, up to 50 ml/g dry weight, when fully extended in solution, and they exhibit reversible compressibility in centrifugal fields (6). In hyaline cartilage, their concentration is up to 10% w/v, and they are constrained to far smaller volumes. At these concentrations, they form a visco-elastic gel in which the glycosaminoglycans exhibit a large degree of segmental mobility as assessed by ^{13}C-nmr. This exerts an extensive pressure on the collagen network which is able to resist equivalent compressive loads with less total deformation than possible at lower proteoglycan concentrations. The ability of the chondrocyte to maintain a high concentration of proteoglycan in its immediate matrix by regulating biosynthetic and catabolic pathways, while poorly understood, is critical for normal tissue function (see Fig. 1).

Overall, then, the cartilage extracellular matrix consists of a network of collagen fibrils, which provide tissue shape and tensile strength, surrounded by a concentrated proteoglycan phase, which provides swelling pressure and resistance to compressive forces. The proteoglycan phase contains a large concentration of fixed anionic groups, carboxyls and sulfate esters, on the glycosaminoglycan chains. At 10% (w/v), a typical concentration in the matrix, the proteoglycans contain a fixed sulfate ester content of about 0.2 molar. Thus, the proteoglycan phase will greatly influence the ionic environment of the matrix and the flow of molecules through it. For example, nutrients must diffuse through this phase to the chondrocytes as must the ions and precursors which eventually form the mineral deposits at the sites in the longitudinal septa where mineralization occurs (see discussion below). It is quite possible that the fixed charge in the proteoglycan phase can facilitate movement of cations such as Ca^{2+} through the matrix but would at the same time present a barrier to movement of anions such as phosphate. The mechanisms by which the chondrocytes regulate the flow of material through the proteoglycan phase, then, are likely to play key roles in the process of matrix mineralization.

PROTEOGLYCANS AND ENDOCHONDRAL BONE FORMATION

Growth cartilages must have sufficient structural strength to withstand compressive load during normal skeletal function and must provide a proper spatial scaffold for bone formation. Proteoglycans are critical for both of these functions. Their role in resisting compression is outlined above, and their involvement in defining volume and shape is shown in several skeletal growth anomalies. Cartilages in nanomelic chickens (24) and cartilage-matrix-deficient mice (15), both homozygous lethals, are almost devoid of proteoglycans. Chondrocyte morphology and the amount and organization of the collagen fibrils appear normal. However, the extracellular matrix is greatly reduced in volume. Although they are short, thickened, and deformed, long bones do form. Thus bone formation must occur and some aspects of growth cartilage function may be retained. A study of the morphology of growth plates from such animals might reveal clues for the role, if any, of proteoglycans in cartilage mineralization. In the absence of proteoglycans, structural support would be weak and columnar organization in the growth plate would undoubtedly be abnormal, probably the major factor contributing to the deformation of the bones. Chondrocytes in brachymorphic mice, a recessive nonlethal, are not able to synthesize normal amounts of phosphoadenosylphosphosulfate (32), the donor required for sulfation of glycosaminoglycans. In growth cartilages, where proteoglycan synthesis is high, this donor becomes limiting and the proteoglycans are undersulfated by 15%. With less charge, the glycosaminoglycan chains would be more flexible and less extended, giving reduced volumes for proteoglycans under equivalent load when compared to normal. This would account for shorter growth plates and the subsequent foreshortened limbs characteristic of these animals (23).

BIOLOGICAL PROCESSES IN ENDOCHONDRAL BONE FORMATION

Several distinct biological processes occur in defined locations between the region where cartilage mineralizes in the hypertrophic zone and the metaphysis where mineralized cartilage is replaced with bone. Each involves removal and/or deposition of inorganic matrix, including proteoglycans, and each must be considered separately in evaluating possible direct role(s) of proteoglycans in mineralization mechanisms.

The territorial matrix, including the transverse septa of the hypertrophic chondrocytes, does not mineralize, whereas the interterritorial matrix in the longitudinal septa, frequently at the level of the penultimate hypertrophic chondrocyte, does mineralize. The transverse septa of the last layer of chondrocytes in the hypertrophic zone are penetrated by the metaphyseal capillaries as they enter the lacunae of the degenerating chondrocytes. During this process, advancing endothelial cells and/or perivascular cells erode the matrix, including proteoglycans, of the septa (31). This leads to a net loss of cartilage proteoglycans at the junction of the growth plate and the metaphysis. However, since the territorial matrix does not mineralize, this loss cannot be directly involved in the mineralization process occurring in the longitudinal septa.

Mineralization is initiated in or near extracellular matrix vesicles in the longitudinal septa surrounding the last two layers of hypertrophic chondrocytes (see (34) for review). It then spreads throughout the intercellular matrix until it is uniformly and densely mineralized at the level of the metaphysis just distal to the last transverse septa, thus forming spicules of calcified cartilage. Two processes can then be distinguished: initial focal mineralization and subsequent growth of the mineral phase. Many hypotheses have been advanced for the role of matrix macromolecules in these processes; those relating to proteoglycans are discussed below.

Just inside the metaphysis, portions of the newly mineralized cartilage spicules are removed by chondroclasts before being covered by bone. Both mineral and the residual organic phase, including any entrapped proteoglycans, are removed. This would also contribute to a net loss of proteoglycans which would be unrelated to the mechanism of mineralization.

Slightly further into the metaphysis, osteoblasts cover the residual cartilage spicules and deposit a thin layer of organic matrix (osteoid) which is subsequently mineralized to bone, forming the trabeculae of the primary spongiosa. Thus, two distinctly different mineralized tissues are juxtaposed in this region. The composition of the organic matrix in these two tissues is quite distinct. For example, bone tissue contains type I collagen while the cartilage spicule contains type II collagen (36), and the structure of the proteoglycans in osteoid and bone differs from that for cartilage proteoglycans (see below). It is, therefore, likely that the mechanisms involved in mineralizing the organic matrices of these two tissues differ.

Deeper in the metaphysis, the bone trabeculae and their cartilage cores are remodelled. This involves erosion and removal of their surfaces by osteoclasts with subsequent deposition of new bone on these surfaces by osteoblasts to form the secondary spongiosa.

ROLE OF PROTEOGLYCANS IN CARTILAGE MINERALIZATION

Do proteoglycans have a direct role in the mechanism for mineralizing cartilage in addition to their structural and template functions described above? It is widely thought that the proteoglycans are degraded and at least partially removed from the cartilage matrix just prior to mineralization. This has led to the hypothesis that intact proteoglycans inhibit either initiation or growth of the mineral phase. Evidence for a loss of proteoglycans in the hypertrophic zone where mineralization occurs is derived from chemical, histological, and immunological studies. For examples: (a) Zonal analyses of growth plate by Wuthier (40) suggested a net loss of proteoglycans near the hypertrophic zone-metaphysis junction. (b) Lohmander and Hjerpe (20) used powdered growth plate and separated mineralized cartilage particles from non-mineralized cartilage particles. They found a lower ratio of hexuronic acid (proteoglycan) to total organic material in the mineralized tissue. (c) Jibril (12) reported that glycosaminoglycans diffused more rapidly from "ossifying" cartilage in acidic solvents (pH 4) than from non-mineralized cartilage of growth plates and suggested that the proteoglycans in the mineralized cartilage had been degraded. (d) Thyberg et al. (35) observed in electron micrographs of growth cartilage that the size and number of proteoglycan matrix granules decreased from the proliferating to the hypertrophic zone, suggesting removal of some proteoglycans. (e) Hirschman and Dziewiatkowski (9) prepared antibodies to a proteoglycan preparation which were then conjugated to fluorescein. Immunolocalization of the tagged antibodies suggested that there was less proteoglycan in the hypertrophic zone of mineralization than in the proliferating zone, and that they were absent in the metaphysis.

Other investigators have interpreted their studies within the context of this hypothesis. For examples: (a) Granda and Posner (4) found that the activity of cathepsin D was almost twice as high in the hypertrophic zone as in the proliferating zone and suggested that the increased activity was related to the removal of "ground substance," mainly proteoglycans, during mineralization. (b) Woodward and Davidson (39) suggested that proteoglycans could bind Ca^{2+} ions and that the degradation of proteoglycans, presumed to occur at the onset of mineralization, could release Ca^{2+} ions, thereby increasing their activity and initiating precipitation of calcium phosphate salts. (c) Thyberg and Friberg (34) provided evidence that one of the two classes of matrix vesicles in the longitudinal septa is derived from, or closely related to, lysosomes. They suggest that disruption and leakage of the vesicle contents into the adjacent matrix could create a local acidity, thereby initiating degradation of proteoglycans by cathepsins and subsequent mineralization adjacent to the vesicle.

Nevertheless, there are major problems with this hypothesis, and there is some evidence that proteoglycans may not be removed or degraded extensively during mineralization. Because several distinct biological processes occur within a very small distance of each other near the junction of the hypertrophic zone and the metaphysis, the results of gross chemical studies may not reflect the changes that occur in the specialized area where mineralization is actually occurring. A net loss of proteoglycan could be accounted for by removal of the matrix of the transverse septa and of some newly mineralized cartilage in the primary spongiosa as described above, and some biochemical studies have failed to detect significant decreases in glycosaminoglycan content until entering the metaphyseal region (19,37). Furthermore, few methods for processing tissue for histological studies are satisfactory for proteoglycans; losses from the tissue frequently occur in fixation, dehydration, or demineralization steps. Differential loss or diffusion of proteoglycans from the matrix can make quantitative interpretation of histochemical and immunolocalization studies difficult.

Poole et al. (27) have recently reinvestigated the immunolocalization of proteoglycan related antigens in bovine fetal growth plates. Fab fragments were prepared from primary antibodies directed against proteoglycan core protein and against link protein. Fluorescein-tagged Fab fragments directed against the primary antibodies were then used to localize the immune Fab in sections of growth plate and metaphysis after demineralization. Staining intensity for both antigens remained the same or increased from non-mineralized cartilage to mineralized cartilage of both the hypertrophic zone and the cartilage spicules in the metaphysis. Safranin, O, which interacts with glycosaminoglycans, stained similar sections uniformly in the cartilage regions. The antibodies did not stain osteoid or bone matrix, suggesting that proteoglycans in bone tissue did not share antigenic determinants with cartilage proteoglycans. The authors suggest that proteoglycans are entrapped without appreciable loss when cartilage mineralization occurs.

In a similar study, these investigators studied a model for endochondral bone formation induced by implanting powdered bone matrix in rats (28). This triggers a sequence in which the implant is populated by mesenchymal cells (days 0-3) which undergo chondrogenesis (days 5-8) with subsequent cartilage mineralization (days 9-11) and eventual remodeling by replacement with bone (days 11 and following). Immunofluorescent staining for proteoglycan and link protein occurred at the onset of chondrogenesis and persisted with no loss of intensity in the cartilage spicules; osteoid and bone did not stain. The authors again suggest that prior removal of proteoglycans is not required for mineralization. While such studies are not quantitative and do not provide information on the degree of aggregation or structure of proteoglycan in the mineralized cartilage, they clearly indicate that proteoglycan aggregate antigenic determinants remain in the matrix after mineralization occurs.

Accurate quantitative data for proteoglycan content of newly mineralized cartilage are hard to obtain because of difficulties in cleanly separating the different tissues, including bone tissue, and in identifying a basis of comparison. Lohmander and Hjerpe (20) described a procedure for separating acetone dried, pulverized growth plate cartilage into subfractions with no-, low-, and high-mineral contents using acetone-bromoform density gradients. The hexuronic acid content per dry weight of organic matter of non-mineralized cartilage was almost twice that of highly mineralized matrix, suggesting that mineralization was accompanied by a significant loss of proteoglycan. However, differential distribution of organic matter between the different matrix areas could influence these results, and it would be instructive to compare proteoglycan content to a component such as type II collagen which is not removed during mineralization (36). While this procedure enriches mineralized matrix relative to non-mineralized matrix and facilitates quantitative analyses, it does partially degrade proteoglycan structure, making it less useful for studies of structure and function of proteoglycans in mineralized cartilage.

Alternative procedures are being developed to extract macromolecules from mineralized tissues. Termine et al. (33) developed a sequential extraction procedure using first 4 M guanidine HCl and then 4 M guanidine HCl plus EDTA, which they applied to dentin and enamel. Several macromolecules were selectively extracted by the second extract, which dissolves the mineral, and they were presumably tightly associated with the mineral phase. Application of this methodology to bone tissue is discussed below.

Although proteoglycans may not be extensively degraded and removed prior to mineralization, evidence from the work of Pita, Howell, and their collaborators (11,25,26) suggests that they may have to be disaggregated at least partially before mineralization will occur. These investigators have used a micropuncture technique to aspirate nl quantities of fluid from the matrix of the hypertrophic region of rat growth plates and ultramicro techniques to analyze the content of the fluid. The fluid contains $\sim$10 mg/ml proteoglycan with 60% or more present as aggregates; this reflects their physiological state at the time of isolation since no prior disaggregation step was used. The concentrations of

Ca^{2+} and phosphate in the fluid are sufficient for mineral formation if incubated at 37°, but this does not occur unless the original volume is diluted to 1/6 with a synthetic "lymph" containing the same electrolyte concentrations as the fluid. After centrifugation to remove ~30% of the proteoglycans, predominantly the largest aggregates, mineral formation will occur in the remaining fluid (Pita, personal communication).

Additional studies on rachitic rats, induced by maintaining animals on a low phosphate, vitamin D deficient diet, provided evidence that the inhibitory factor preventing mineral formation residues in the largest aggregate fraction. The hypertrophic zones of rachitic growth plates are greatly expanded due to inability of the cartilage to mineralize and to be replaced by metaphyseal bone. When animals are restored to a normal diet, mineralization occurs throughout the expanded hypertrophic zone over 5-6 days well above the metaphysis. Proteoglycans from the expanded hypertrophic zone contain a higher proportion of large aggregates in fluid isolated from rachitic growth plates than from fluid isolated from the same region during healing and initial stages of mineralization. The investigators suggest that cartilage lysozyme facilitates the breakdown of the large aggregates, perhaps by affecting hyaluronic acid, thereby removing their inhibitory capacity (25). Incubation of fluid from rachitic cartilage with purified lysozyme decreased aggregate size without altering monomer size after which the fluid could undergo mineral formation. The lysozyme content almost doubled in tissue isolated after restoring rachitic animals to a normal diet for 72 hours. When rachitic growth plate is incubated in organ culture, mineralization occurs in the hypertrophic region; the presence of N-acetylchitotriose, which selectively binds to lysozyme, prevented mineralization in the same conditions (Howell, personal communication).

While these investigations provide circumstantial evidence for a regulatory role for proteoglycans in the mineralization process, it seems likely that the actual mechanism is more complex. For example: (a) Proteoglycans, whether aggregate or monomer, have been shown to inhibit hydroxyapatite formation in a model system in vitro (2). (b) It has been difficult to show that cartilage lysozyme disaggregates purified proteoglycan aggregates in vitro in an analogous way to that observed with cartilage fluid, nor is hyaluronic acid a substrate for this enzyme (5). (c) It is difficult to rationalize why only a portion of the aggregates, namely, the very largest, exert an inhibitory effect on mineralization. (d) The cartilages in animals with the genetic anomalies nanomelia and cartilage-matrix-deficiency described above are not mineralized throughout the matrix even though they lack proteoglycans. If cartilage mineralization in the growth plate of these animals occurs at the normal time and location, proteoglycans alone would be unable to act in a regulatory role. Perhaps there is another component, such as the matrix vesicles, to which proteoglycans bind. Lysozyme could degrade this component, releasing the bound proteoglycans as well as initiators for mineral formation, such as the phospholipid fraction recently identified as a potential nucleation factor in cartilage fluid (11). In this case proteoglycans would play only a passive or secondary role after initiation has occurred.

PROTEOGLYCANS IN BONE

It has been known for some time that cortical bone contains small amounts of glycosaminoglycans (8,10) and that osteoid stains with dyes that have affinity for glycosaminoglycans (1). Relatively little is known about the structure and function of the proteoglycans in this tissue, however. Reddi et al. (29) identified a proteoglycan with ony a few chondroitin (or dermatan) sulfate chains (MW ≃ 50,000), which was synthesized by osteoblasts during the active bone forming stage after implanting powdered bone in rats. Recent work (Termine, personal communication) used sequential extraction of 4 M guanidine HCl without, and then with, EDTA to isolate proteoglycans from calf bone tissue. The predominant proteoglycans in both extracts had similar properties to those described above. Electrophoresis on SDS-PAGE yielded a broad band (MW ≃ 2-3 x 10^5). Glycosaminoglycan analyses indicated the presence of dermatan sulfate chains (MW ≃ 40,000), and oligosaccharides with sialic acid were also present. It is not known if proteoglycan in the osteoid is partially removed when bone subsequently forms, although ion probe analyses of sections suggest that this might be the case (38). Assuming that the second extract with EDTA contains proteoglycans primarily associated with the mineral phase of bone matrix, the results suggest that the structure of the proteoglycans in the mineralized matrix may not be appreciably different from that of proteoglycans in the osteoid. The role, if any, for proteoglycans in bone mineralization remains unknown. It is quite possible, however, that they are important for defining the proper spatial matrix required for the subsequent deposition of mineral.

Less is known about proteoglycans in other mineralizing tissues, dentin, and enamel. Because of its similarity to bone, dentin proteoglycans are probably closely related to bone proteoglycans and will play similar roles. A recent study (3) has identified an unusual keratan sulfate proteoglycan which is selectively synthesized during synthesis of medullary bone induced by injecting male Japanese quail with estrogens. This proteoglycan may play a major role in the storage of Ca^{2+} in medullary bone for subsequent eggshell formation in the egg-laying cycle.

CONCLUDING REMARKS

Proteoglycans undoubtedly play prominent roles in the organization, structure, and function of extracellular matrices. While there have been many experimental data in the past suggesting that they play an active role in the mineralization process in cartilage, most of this evidence is circumstantial. It remains to be proven convincingly that their role in this process is an active one.

Acknowledgements. The authors are extremely indebted to L. Rosenberg, J. Pita, J. Termine, H. Reddi, M. Yanagishita, and R. Poole for generously sharing their ideas and many unpublished observations upon which parts of this position paper are based.

REFERENCES

(1) Baylink, D.; Wergedal, J.; and Thompson, E. 1972. Loss of protein polysaccharides at sites where bone mineralization is initiated. J. Histochem. Cytochem. 20: 279-292.

(2) Blumenthal, N.C.; Posner, A.S.; Silverman, L.D.; and Rosenberg, L.C. 1979. Effect of proteoglycans on in vitro hydroxyapatite formation. Calcif. Tiss. Intl. 27: 75-82.

(3) Fisher, L.W., and Schraer, H. 1980. The glycosaminoglycans of estrogen induced medullary bone in Japanese quail. Arch. Biochem. Biophys. 205: 396-403.

(4) Granda, J.L., and Posner, A.S. 1971. Distribution of four hydrolases in the epiphyseal plate. Clin. Orthop. 74: 269-272.

(5) Greenwald, R.A. 1976. Lack of effect of lysozymes on cartilage proteoglycans. Arch. Biochem. Biophys. 175: 435-437.

(6) Hascall, V.C. 1977. Interactions of cartilage proteoglycans with hyaluronic acid. J. Supramol. Structure 7: 101-120.

(7) Hascall, V.C. 1981. Proteoglycans: structure and function. In The Biology of Carbohydrates, eds. V. Ginsburg and P. Robbins, vol. 1, pp. 1-49. New York: John Wiley.

(8) Herring, G.M. 1968. Studies on the protein-bound chondroitin sulfate of bovine cortical bone. Biochem. J. 107: 41-49.

(9) Hirschman, A., and Dziewiatkowski, D.D. 1966. Protein polysaccharide loss during endochondral ossification: immunochemical evidence. Science 154: 393-395.

(10) Hjertquist, S.-O., and Vejlens, L. 1968. The glycosaminoglycans of dog compact bone and epiphyseal cartilage in the normal state and in experimental hyperparathyroidism. Calcif. Tiss. Res. 2: 314-333.

(11) Howell, D.S.; Blanco, L.; Pita, J.C.; and Munez, O. 1978. Further characterization of a nucleational agent in hypertrophic cell extracellular cartilage fluid. Metab. Bone Dis. Rel. Res. 1: 155-159.

(12) Jibril, A.O. 1967. Proteolytic degradation of ossifying cartilage matrix and removal of acid mucopolysaccharides prior to bone formation. Biochim. Biophys. Acta 136: 162-165.

(13) Kanwar, Y.S.; Hascall, V.C.; and Farquhar, M.G. 1981. Partial characterization of newly synthesized proteoglycans isolated from the glomerular basement membrane. J. Cell Biol. 90: 527-532.

(14) Kanwar, Y.S.; Linker, A.; and Farquhar, M.G. 1980. Increased permeability of basement membrane to ferritin after removal of glycosaminoglycans (heparan sulfate) by enzyme digestion. J. Cell Biol. 86: 688-693.

(15) Kimata, K.; Barrach, H.J.; Brown, K.S.; and Pennypacker, J.P. 1981. Absence of proteoglycan core protein in cartilage from the cmd (cartilage matrix deficiency) mouse. J. Biol. Chem. 256: 6961-6968.

(16) Kimura, J.H.; Thonar, E.J.-M.A.; Hascall, V.C.; Reiner, A.; and Poole, A.R. 1981. Biosynthesis of the core protein of proteoglycans by chondrocytes from the Swarm rat chondrosarcoma. J. Biol. Chem. 256: 7890-7897.

(17) Kjellen, L.; Oldberg, A.; and Höök, M. 1980. Cell surface heparan sulfate: mechanism of proteoglycan-cell association. J. Biol. Chem. 255: 10407-10413.

(18) Kornfeld, R., and Kornfeld, S. 1980. Structure of glycoproteins and their oligosaccharide units. In Biochemistry of Glycoproteins and Proteoglycans, ed. W. Lennarz, pp. 1-34. New York: Plenum Press.

(19) Lindenbaum, A., and Kuettner, K.E. 1967. Mucopolysaccharides and mucoproteins of calf scapula. Calcif. Tiss. Res. 1: 153-165.

(20) Lohmander, S., and Hjerpe, A. 1975. Proteoglycans of mineralizing rib and epiphyseal cartilage. Biochim. Biophys. Acta 404: 93-109.

(21) Lohmander, L.S.; Nilsson, B.; De Luca, S.; and Hascall, V.C. 1981. Structures of N- and O-linked oligosaccharides from chondrosarcoma proteoglycan. Sem. Arthr. Rheum. 11 supplement: 12-13.

(22) Muir, I.H.M. 1980. The chemistry of the ground substance of joint cartilage. In The Joints and Synovial Fluid II, ed. L. Sokoloff, pp. 27-94. New York: Academic Press.

(23) Orkin, R.W.; Williams, B.R.; Cranley, R.E.; Poppke, D.C.; and Brown, K.S. 1977. Defects in the cartilaginous growth plates of brachymorphic mice. J. Cell Biol. 73: 287-299.

(24) Pennypacker, J.P., and Goetinck, P.F. 1976. Biochemical and ultrastructural studies of collagen and proteochondroitin sulfate in normal and nanomelic cartilage. Dev. Biol. 50: 35-47.

(25) Pita, J.C.; Muller, F.J.; and Howell, D.S. 1979. Structural changes of sulfated proteoglycans of the growth cartilage of rats during endochondral calcification. In Glycoconjugate Research II: Proceedings of the 4th International Symposium on Glycoconjugates, eds. R. Jeanloz and J. Gregory, pp. 743-746. New York: Academic Press.

(26) Pita, J.C.; Muller, F.J.; Morales, S.M.; and Alarcon, E.J. 1979. Ultracentrifugal characterization of proteoglycans from rat growth cartilage. J. Biol. Chem. 254: 10313-10320.

(27) Poole, A.R.; Pidoux, I.; and Rosenberg, L. 1982. The role of proteoglycans in endochondral ossification: immunofluorescent localization of link proteins and proteoglycan monomer in bovine fetal epiphyseal growth plate. J. Cell Biol., in press.

(28) Poole, A.R.; Reddi, A.H.; and Rosenberg, L.C. 1982. Persistence of cartilage proteoglycan and link protein during matrix-induced endochondral bone development: an immunofluorescent study. Dev. Biol. 89: 532-539.

(29) Reddi, A.H.; Hascall, V.C.; and Hascall, G.R. 1978. Changes in proteoglycan types during matrix-induced cartilage and bone development. J. Biol. Chem. 253: 2429-2439.

(30) Rodén, L. 1980. Structure and metabolism of connective tissue proteoglycans. In Biochemistry of Glycoproteins and Proteoglycans, ed. W. Lennarz, pp. 267-371. New York: Plenum Press.

(31) Schenk, R.K.; Spiro, D.; and Wiener, J. 1967. Cartilage resorption in the tibial epiphyseal plate of growing rats. J. Cell Biol. 34: 275-291.

(32) Sugahara, K., and Schwartz, N.B. 1979. Defect in 3'-phosphoadenosine-5'-phosphosulfate formation in brachymorphic mice. Proc. Natl. Acad. Sci. (USA) 76: 6615-6618.

(33) Termine, J.D.; Belcourt, A.B.; Christner, P.J.; Conn, K.M.; and Nylen, M.U. 1980. Properties of dissociatively extracted fetal tooth matrix proteins. J. Biol. Chem. 255: 9760-9768.

(34) Thyberg, J., and Friberg, U. 1978. The lysosomal system in endochondral growth. Prog. Histochem. Cytochem. 10: 1-45.

(35) Thyberg, J.; Lohmander, S.; and Friberg, U. 1973. Electron microscopic demonstration of proteoglycans in guinea pig epiphyseal cartilage. J. Ultrastruct. Res. 45: 407-427.

(36) von der Mark, K., and von der Mark, H. 1977. The role of three genetically distinct collagen types in endochondral ossification and calcification of cartilage. J. Bone Joint Surg. 59-B: 458-464.

(37) Weatherell, J.A., and Weidman, S.M. 1963. The distribution of organically bound sulfate in bone and cartilage during calcification. Biochem. J. 89: 265-267.

(38) Wergedal, J.E., and Baylink, D.J. 1974. Electron microprobe measurement of bone mineralization rate in vivo. Am. J. Physiol. 226: 345-352.

(39) Woodward, C., and Davidson, E.A. 1968. Structure function relationships of protein polysaccharide complexes: specific ion binding properties. Proc. Natl. Acad. Sci. (USA) 60: 201-205.

(40) Wuthier, R.L. 1969/70. A zonal analysis of inorganic and organic constituents of the epiphysis during endochondral calcification. Calcif. Tiss. Res. 4: 20-38.

(41) Yamada, K.M.; Kennedy, E.W.; Kimata, K.; and Pratt, R.M. 1980. Characterization of fibronectin interactions with glycosaminoglycans and identification of active proteolytic fragments. J. Biol. Chem. 255: 6055-6063.

Biological Mineralization and Demineralization, ed. G.H. Nancollas, pp. 199-218.
Dahlem Konferenzen 1982. Berlin, Heidelberg, New York: Springer-Verlag.

Three Systems of Biomineralization in Plants with Comments on the Associated Organic Matrix

H. J. Arnott
Dept. of Biology, The University of Texas
Arlington, TX 76019, USA

Abstract. Three systems of biomineralization are common in higher plants. The formation of intracellular deposits of crystalline calcium oxalate takes place in a crystal vacuole. The individual crystal (one to many) develops within a crystal chamber bounded by a membrane and is often associated with tubules and membranes in the crystal vacuole. Other than the presence of polysaccharide and protein, the chemical nature of these elements is not well established. Exactly which elements constitute an organic matrix is not clear. In the formation of calcium carbonate in cystoliths, however, the structure that constitutes the organic matrix is clear. The carbonate is deposited in association with a siliconized cellulose and callose framework. Demineralized cystoliths clearly show this organic "skeleton." The third system is the formation of hydrated silicon dioxide as deposits of silica gel or opal. Silicon may impregnate primary walls where it is associated with the hemicellulose-pectic fraction of the wall; it may form as intercellular bodies, or it may form internal casts of various cell types. The organic matrix associated with mineralization in higher plants is not well defined.

INTRODUCTION

The hard tissues of animals often involve mineralization; however, in plants the most obvious hard tissue, wood, gains its strength and hardness from two organic polymers, cellulose and lignin. Plant scientists have been intensely concerned with the importance of minerals in nutrition, but although mineral deposits in plants have been known since the 17th Century (7,

14), the nature and meaning of the deposits has not received broad interest. Even application of the concepts of "calcified tissue" and "biomineralization" to higher plants is relatively recent (2,7). Most vascular plants have mineralized tissues; some may be strongly mineralized, e.g., certain cacti are reported to have 85% of their dry weight composed of calcium oxalate (9). Many nonvascular plants, especially marine algae and certain fungi, produce mineralized tissues or mineralized cells. Some lichens may have as much as 66% of their dry weight formed by calcium oxalate (10), and some algae as much as 85% dry weight formed by calcium carbonate (27). The general conclusion that mineralized tissues are common in plants is sound. However, a complete understanding of the distribution, development, and especially the physiological significance of these mineral deposits has not yet been achieved.

There are three systems of mineralization in higher plants which deserve consideration because of their broad distribution. The most widely distributed mineralized system is crystalline calcium oxalate (Figs. 1-11); McNair (24) found that over 74% of the families of seed plants (218/295) had been reported to have calcium oxalate deposits. In the taro the corms have been reported to contain 120,000 crystals per cm^3 (34). The second system is calcium carbonate in the form of cystoliths (Figs. 12-17). The cystolith system, which is found commonly in the genera of eight to ten families of seed plants, is more narrowly distributed than the previous system. Calcium carbonate is also found in noncystolith deposits, especially as calcite deposits in wood (33). The final mineralized system involves hydrated silicon dioxide, either as silica gel or opal (Figs. 16-19). As with calcium carbonate, this system is also restricted in family distribution. However, these deposits are common in one very important flowering plant family, the grasses, and in the lower vascular plant, Equisetum. Deposits of calcium carbonate in coccolithophorids and red algae and silicon in diatoms and other algae are, of course, extremely important but are outside the scope of this article (see (27)). For each of these three systems the nature of the mature deposits will be shown, and some understanding of their variation and development as well as some indication of their functional nature will be provided. When possible, comments on the organic matrix associated with the deposits will be made. Pentecost (27) has defined matrices as "substances which assist in the initiation, and control the direction of, crystal growth"; he also points to the difficulty in deciding the extent to which substances are active and the exact site of their activity.

CALCIUM OXALATE DEPOSITS

Despite its wide occurrence in vascular plants, calcium oxalate always occurs as an intracellular product and is often found in specialized cells called crystal idioblasts (Figs. 8,10,11) (2,3,7,14). Calcium oxalate occurs in two hydration states in plants, calcium oxalate monohydrate (whewellite) (Figs. 4,6) and, less commonly, calcium oxalate dihydrate (weddellite) (Figs. 5,7) (7,14,16). A number of crystal morphologies have been seen for both hydration states (compare Figs. 4 and 6 with 5 and 7). Botanists have traditionally classified these crystal morphologies as: raphide - one or more bundles of acicular crystals per cell (Figs. 1,8,9); styloid - an elongate single or twin crystal in a cell (Figs. 2,3); druse - an aggregation of crystals with a roughly spherical shape (Figs. 6,7); prismatic crystals - rhombohedral, single or twin crystals which usually occur one per cell (Figs. 4,5, 10); and crystal sand - numerous small single and/or twin crystals in a single cell. Raphide idioblasts, in which it was sometimes possible to count over 2,000 needle-like crystals, have been examined with transmission electron microscopy (TEM) (2,7,12,14,25,31,35,38). With isolated crystals (4,6), it is possible to correlate studies with light microscopy (LM), scanning electron microscopy (SEM), and X-ray diffraction. For example, the persimmon has prismatic crystals of calcium oxalate monohydrate in the leaf. When studied with SEM, five morphological types of contact twins (Fig. 4), as well as interpenetrant twins, were described (Fig. 5) (6). In plants with druses, SEM and LM show that many druses are often at least partly composed of contact (6) or interpenetrant (14) twins. Twinning has been demonstrated by polarization microscopy and SEM in each of the traditional crystal classifications except raphides (6,16). Twinning must be considered an important feature of calcium oxalate deposits in higher plants because of its common occurrence. At present, the role of the cell and/or the crystal-forming apparatus in this twinning process is not understood. However, it is clear that certain plant species have genetic information which somehow causes (or allows) not only the formation of twins, but twins of specific size and shape. It should also be pointed out that Kim and Johnson (21) believe that twinning, along with aggregation and epitactic crystal intergrowth, "plays a significant role" in the formation of calcium oxalate in crystals in urinary stone.

When a wide spectrum of plants is surveyed, a substantial variation will be found in the size, shape, number, and distribution of crystals. However, when individual species or tissues are examined, in normal individuals of a species the pattern of distribution, size, shape, number, and hydration state of the calcium oxalate deposits is very consistent. This consistency leads to the obvious conclusion that many crystal parameters

---

FIGS. 1-7 - Isolated calcium oxalate crystals as viewed by SEM.
FIG. 1 - Five raphide packets from Eichhornia crassipes. Each packet of needle-shaped crystals is enclosed in a covering that holds them together. Bar = 20 μm.
FIG. 2 - Styloid of Eichhornia crassipes. Bar = 20 μm.
FIG. 3 - Styloid of Agave americana. Bar = 20 μm.
FIG. 4 - A prismatic crystal of Diospyros virginiana which is a contact twin of the monohydrate. Bar = 10 μm.
FIG. 5 - Prismatic crystal of Begonia sp. which is an interpenetrant twin of the dihydrate. Bar = 5 μm.
FIG. 6 - Druse from Ricinus communis. This druse is a multiple interpenetrant twin of the monohydrate. The druse still retains the cellulose covering characteristic of Rosanoffian crystals. Bar = 10 μm.
FIG. 7 - Druse from Begonia sp. formed of the dihydrate. Note that many of the interpenetration twins are partly formed dodecahedra. Bar = 10 μm.

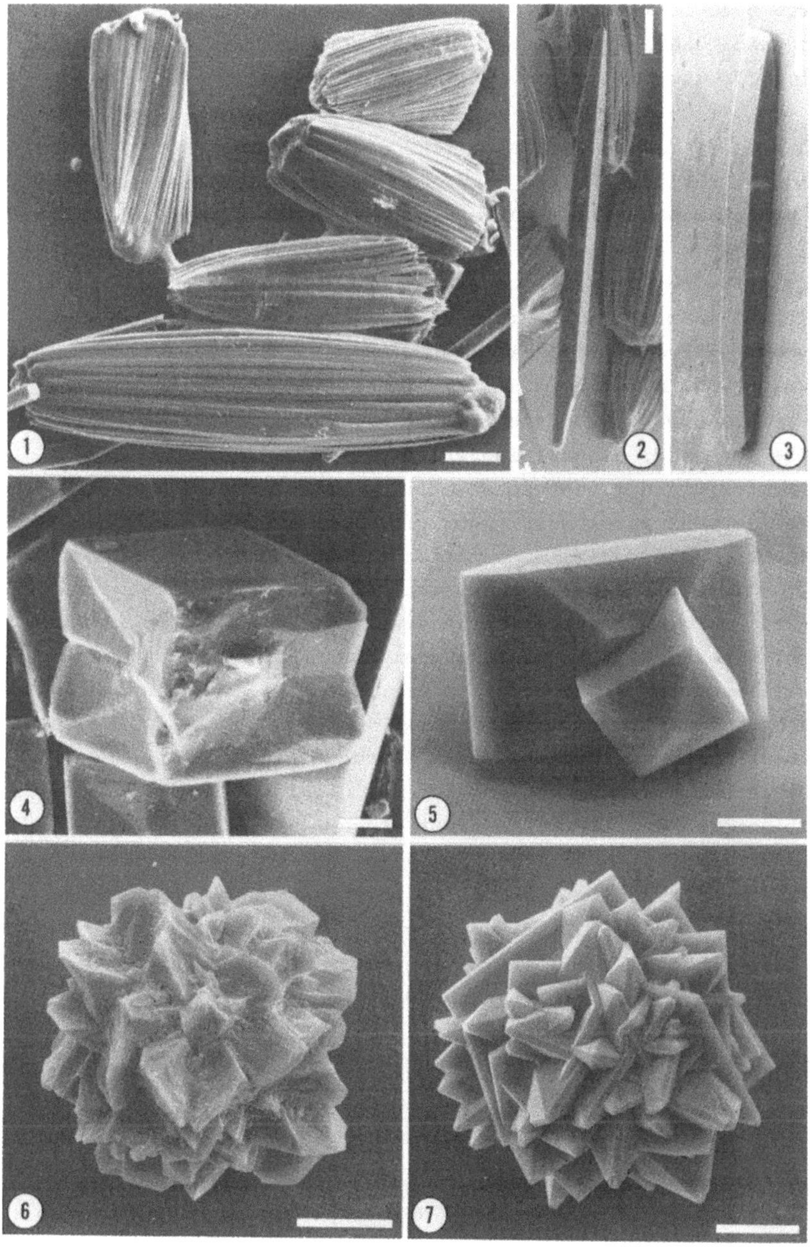
1
2
3
4
5
6
7

are genetically fixed (3,7,14); that is, the cell controls physical and chemical parameters in such a way that a specific type of crystal is formed (7).

How do Citrus or persimmon leaf cells develop into crystal idioblasts that always form calcium oxalate contact twins of about the same size, shape, and positional distribution?

An obvious suggestion to explain this is that a compartment is formed in which a specific type of crystal is allowed to develop (2,3,7,14). In the sense used here a compartment is a volume delimited by a "membrane-like" boundary (2); during crystal growth, the compartment may also increase in volume. The factor(s) determining which of the cell(s) of a tissue form crystals is not known, but in some examples, crystal cell development is clearly genetically controlled, e.g., cases in which every cell of a particular layer or every cell of a file of cells becomes a crystal cell clearly must have a genetic origin. Crystal cells develop from meristematic cells which have the following general construction: A primary cell wall encloses a nonvacuolate protoplast with a prominent nucleus, dictyosomes, mitochondria, endoplasmic reticulum, ribosomes, proplastids, and other standard plant organelles (26). At an early stage, the developing idioblast is similar to its neighbors (2,7,14). The first stage in idioblast development is the formation of a central vacuole; it is this vacuole in which the crystals develop. As idioblasts develop, changes occur in the cytoplasm, and these are undoubtedly important (12,15,20,35,37), but it is in the crystal vacuole where further development takes place (7,14).

The Crystal Vacuole

The crystal vacuole is delimited from the cytoplasm by the tonoplast (Figs. 9,11), a membrane with a dark-light-dark

---

FIG. 8 - A cryofractured raphide idioblast of Eichhornia crassipes as viewed by SEM. Note crystals in crystal vacuole with peripheral cytoplasm surrounded by the cell wall. Bar = 10 μm.

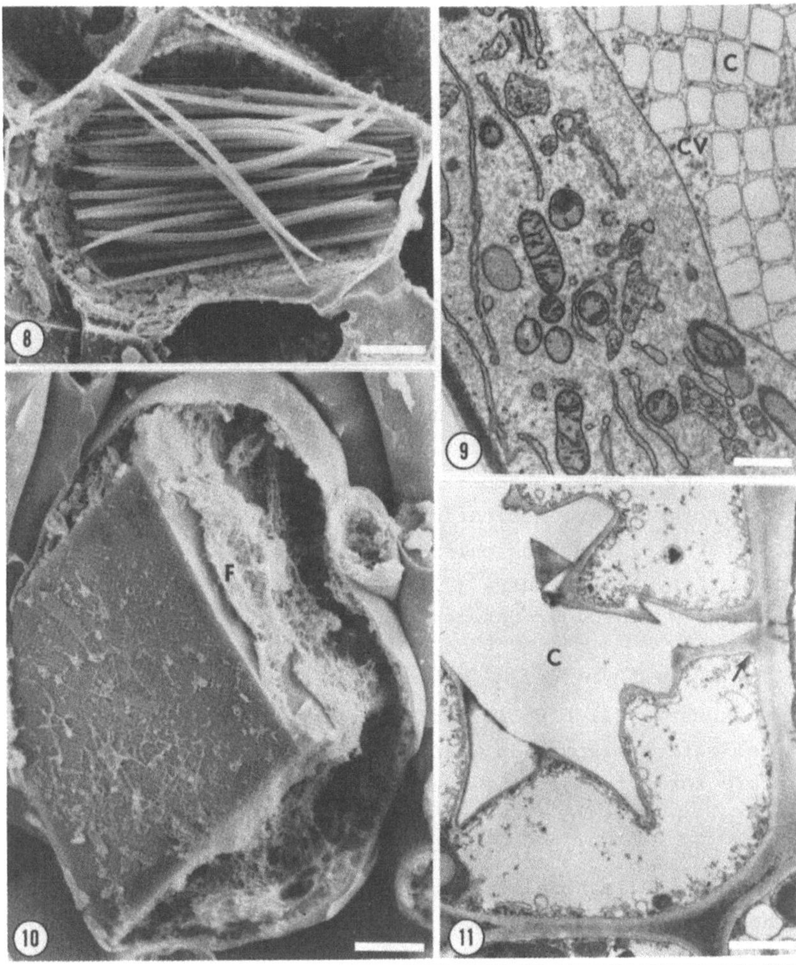

FIG. 9 - Section of raphide idioblast of Eichhornia crassipes showing portion of crystal vacuole (CV) and the peripheral cytoplasm by TEM. Crystal chambers (C) have oxalate removed by processing. Bar = 1 µm.
FIG. 10 - A cryofractured prismatic crystal of Diospyros. Note closely associated fibrous material (F) which is present in the crystal vacuole. Bar = 10 µm.
FIG. 11 - TEM view of a Rosanoffian calcium oxalate crystal (C) of Morus alba. Note extension of the cell wall around the crystal (arrow). These cells are living and the wall surrounding the crystal is produced after the crystal has developed. Bar = 1 µm.

structure (26); sometimes the membrane may be asymmetric with a thicker dark inner layer (30). A number of components have been observed in crystal vacuoles before or simultaneously with the formation of crystals: crystal chambers (2,3,7,12,14, 15,20,30,31,35); tubules (7,12,15,35); membranes (2,7,12,15,35); mucilage (25); amorphous or fibrillar Periodic Acid Schiff (PAS) positive material (31); fields of tubules which connect the crystal surface and the tonoplast (15); and a fibrillar matrix with spherical, circular, and tubular (5.4 μm in diameter) components (35). The matrix in the central vacuole stains for protein, carbohydrate, and RNA (35). In Capsicum, zones called nucleation sites have been identified (20). The sites occur with crystal membrane complexes in the crystal vacuole. The crystal chambers and the crystals develop from these paracrystalline nucleation sites. In Lemna a series of pleated parallel membrane pairs are responsible for the generation of crystal chambers (7); the crystal originates from dense areas which occur between the points of the pleated membranes.

Most authors (7,14) have identified a crystal membrane (sheath) that delimits the space in which the crystal will grow and therefore forms the crystal chamber (compartment). Whether a cell is a large idioblast with 2,000 crystals or a single crystal, the mechanism is the same. When crystals are isolated and then demineralized, the sheath can be seen by LM. The sheath stains positively for polysaccharides (PAS+) and mucopolysaccharides (35), but according to Wattendorff (38), it is formed by suberin (a waxy material) in Agave. In many raphide-forming cells the sheath may become multilayered (12,35,38). A complex system is associated with the production of calcium oxalate crystals in the crystal vacuole.

There is some evidence that the oxalate is present in the vacuole and that calcium is supplied from the cytoplasm (23). It is also clear that crystal formation does not occur as a random event in the vacuole (7,14). Rather the process is highly controlled so that from 1 to 2,000 or more entirely separate crystals can develop within a single vacuole. Assuming that oxalate ions were free in the vacuole when calcium is pumped or released into the vacuole, crystal development would occur, perhaps, at nucleation site(s); with multiple nucleation sites, the crystals would soon form an irregular mass.

As an alternative, one could consider a case in which oxalate is confined to the crystal chambers rather than the vacuole at large (39). As calcium enters the chamber, crystallization would occur, and thus the size, number, and orientation of the crystal chambers provide a mechanism to control crystal development (2,7,14). In such a system calcium would enter the vacuole and, subsequently, the crystal chambers, and the vacuole could serve as a part of a mechanism to control the concentration of calcium in the cytoplasm. Instead of ridding the plant of oxalate, this process would function in the detoxification of Ca^{+2} as has been suggested as a general mechanism of biomineralization (22). In such a hypothetical mechanism the oxalate could be continually supplied by the cell's metabolism and pass through the tubules and membranes that connect to the crystal chambers. A critical point in this hypothesis is the connection of the tubular or membrane system with the tonoplast or its passage through the tonoplast. Frank and Jensen (15) have already described such a system, and Eilert (12) reports the same small tubules in the cytoplasm and the vacuole but has not shown a clear connection. In that regard it has been experimentally shown that fewer crystals are formed in the dark; a situation that would reduce the oxalate but not the calcium available to the plant. With either hypothesis the size, number, and orientation of the crystal chambers provides a mechanism in which the size, number, and orientation of the crystals could be controlled.

In some crystal idioblasts, a further stage of development occurs such that an extension of the cell wall grows around the crystal. These crystals, first described by Rosanoff, have been termed Rosanoffian crystals. The details of this development have been followed by LM, and stages have been seen in TEM (30,37,39,40) (Fig. 11). Through the developmental stages the cells remain alive, but some may die at maturity. In the study of Rosanoffian crystals of Larix (37), Wattendorff found that the cell wall surrounding the crystal was composed of cellulose and suberin, and he was not able to demonstrate the crystal sheath (membrane) between the wall and the crystal. A similar situation was found in Cercidium (30). In Rosanoffian crystals of okra seeds, portions of the surrounding cell wall do not contact the crystal and an unidentified material intervenes (40). Because of their cell wall attachment, Rosanoffian crystals may be more difficult to isolate (Fig. 6). The role of this sheathing wall material in oxalate calcification deserves further study.

Clearly, we are at a primitive stage in the understanding of calcium oxalate crystal development and a long way from understanding the associated organic components. What constitutes the organic matrix of the calcium oxalate system? Is it the crystal membrane, the associated membranes and tubules within the vacuole, the nucleation sites (paracrystalline bodies), or the entire vacuole? It appears that the crystal vacuole is an integrated structure; perhaps the best answer may be that the whole vacuole (or at least its entire contents) is the organic matrix, the site in which crystallization occurs. In terms of the matrix definition given previously (27), the crystal chamber along with associated nucleation sites would constitute an organic matrix. At present we are unable to give clear chemical statements about the organic matrix of calcium oxalate mineralization other than the histochemical staining mentioned earlier (35).

CALCIUM CARBONATE DEPOSITS

A major deposition of calcium carbonate in higher plants occurs through the production of cystoliths. These intracellular deposits occur in a large number of genera but are confined to a limited number of plant families. Amorphous calcium carbonate (28) is the usual mineral found in cystoliths, but vaterite has been reported from Morus (5). As amorphous calcium carbonate is unstable in water, cystoliths must be isolated from dry leaves or through the use of absolute alcohol (28). In Ficus elastica leaf powder yielded less than 1% cystoliths on a dry weight basis (13). Cystoliths were composed of 91.5% calcium carbonate; 4.4% callose; 2.2% cellulose; and 1.8% inorganic components (13). Cystoliths vary in shape from almost spherical in Morus (Fig. 13) to elongated cylinders such as in Beloperone and may even be branched. They may have smooth or warty surfaces; their structure has been used by taxonomists to separate species of certain genera (5). Cystoliths may reach a length of 0.5 mm, and are produced in a cell called a lithocyst (Fig. 12). The cystolith is attached by a stalk to the lithocyst's external (apical) cell wall (Fig. 12). The mature lithocyst of Morus is a large (120 μm x 90 μm) living cell that contains a thin layer of cytoplasm appressed to the cell wall and to the surface of the cystolith; the cytoplasm surrounds the vacuole. The cystolith develops outside the plasma membrane, as was first shown by Hiltz and Pobeguin (19) through the use of plasmolysis and as has recently been confirmed by TEM (Arnott, unpublished). In many places in the mature lithocyst the cytoplasm is extremely thin; the plasma membrane and the tonoplast are separated by only a few nm. Organelles, such as chloroplasts, can be seen in this cytoplasm as it lines the peripheral wall. The vacuole contains fibrous material which also occurs between the cystolith and the plasma membrane.

In Ficus elastica cystolith development begins with the formation of a cell wall process which grows into the cell (1). As growth continues, the terminal part of the process expands and forms the young cystolith body. The growth of the cystolith in Ficus and Morus presses it into the cytoplasm of the lithocyst, which rapidly expands in size and becomes many times the size of neighbor cells (5) (Fig. 12). Since the growth begins outside the plasma membrane and the cystolith retains this position during development, it is extracytoplasmic and most closely related to the cell wall. Scott (33) studied Beloperone cystolith development and found that the first sign of a cystolith was "a small centrally located crystal" from which the remainder of the cystolith developed. She also reported that the cystolith was connected to the cell wall by many protoplasmic connections which passed through the cell wall into the surrounding cells. In Beloperone a stalk is not readily visible. As cystolith growth continues in Ficus, calcium carbonate is deposited in the matrix formed by the cystolith. At the same time, the stalk and the cell wall to which the stalk is attached (Figs. 12,16,17) become silicified. An energy dispersive X-ray analysis of Morus (Arnott, unpublished) showed a very strong peak for silicon in the cystolith stalk but none in the body of the cystolith. The degree of silicification varies even in a single plant; a surface layer or even the entire cystolith may become silicified (8,28). The exact details of cystolith growth at the ultrastructural level have not been elucidated. The external cell wall and the cystolith stalk undergo silicification by a process that has not been studied.

When a cystolith is demineralized an internal skeleton is revealed. In Ficus and Morus this skeleton consists of radial components extending from the terminal process and interconnected by concentric bands of similar material (Figs. 12,14, 15). The concentric bands follow the external contours of the

---

FIGS. 12-15 - Cystoliths of Morus alba.
FIG. 12 - Longisection showing stalk (S), cystolith body (B), and siliconized apical cell wall (W). Bar = 5 µm.
FIG. 13 - SEM of isolated cystolith showing stalk (S) and body (B). Bar = 20 µm.
FIG. 14 - Transverse section of a demineralized cystolith showing concentric banding of the organic matrix. Central region would be cellulose rich. Bar = 2 µm.
FIG. 15 - Transverse section showing fibrous reticulum that forms the organic matrix. Spaces represent areas where calcium carbonate was removed by demineralization. Bar = 1 µm.

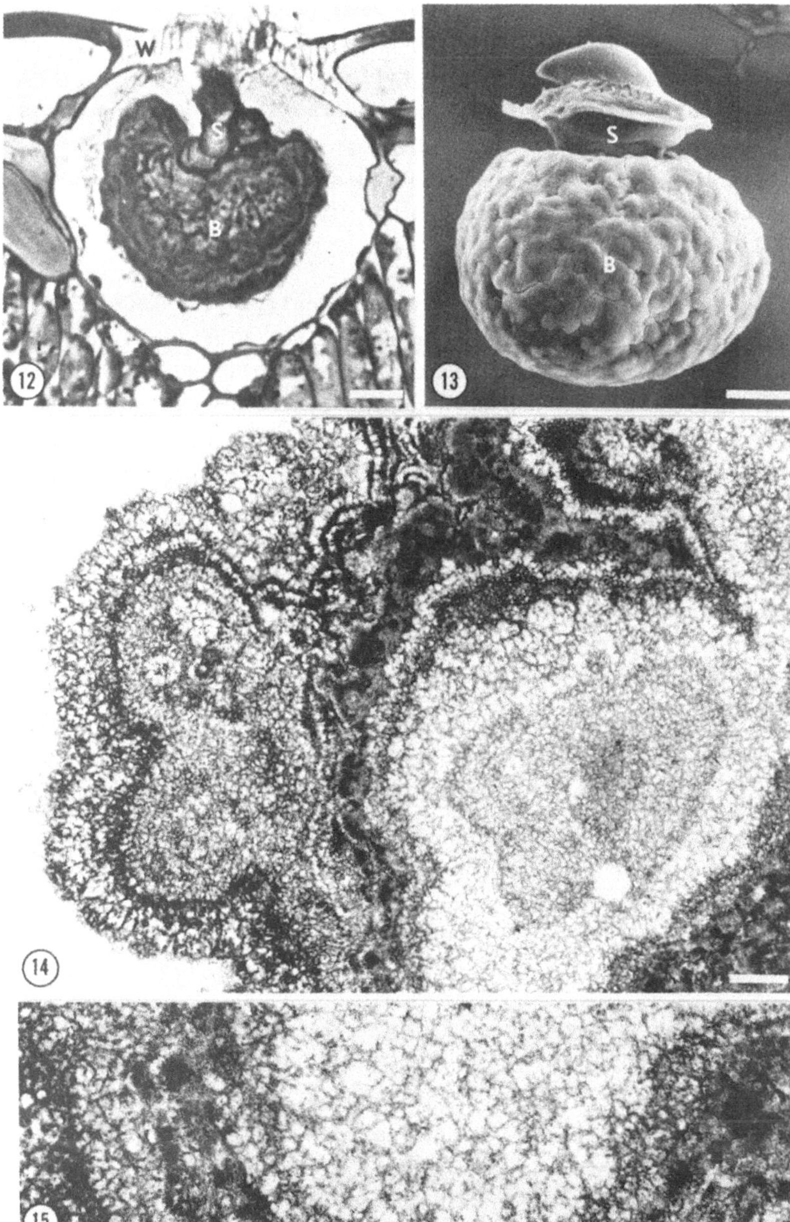
W
S
B
12
S
B
13
14
15

cystolith body, and these layers have been interpreted as alternate layers due to the changes in the concentration of calcium carbonate and the organic matrix (8). The deposition of new layers is by apposition, and several layers are deposited in one day - the layers not being part of a daily rhythm (8). Different authors feel that the radial strands may be either high or low in cellulose content.

The skeleton can be stained histochemically. In Ficus the entire cystolith stains blue in resorcinol blue (test for callose) (13). When the cystolith is demineralized, resorcinol blue staining can be produced on the cystolith skeleton. When similar skeletons are stained with zinc-chloride iodine (test for cellulose), only the central part of the cystolith stains strongly, but delicate radial threads may also be seen (13). The stalk does not stain in either case and is probably not calcified (13,19). The work of Eschrich (13) showed that the organic matrix of the cystolith is composed of cellulose and callose. By hydrolytic studies he determined that the callose in Ficus cystoliths was a polysaccharide synthesized only from d-glucose (13).

The organic matrix of calcium carbonate in cystoliths is clear and appears to be analogous to that found in some animal systems. The amorphous calcium carbonate spherules found in Pomacea paludosa show a similar radial and concentric banding of the organic matrix when demineralized (36).

SILICON DIOXIDE DEPOSITS

Deposits of silicon dioxide ($SiO_2 \cdot nH_2O$) represent the third system of biomineralization in higher plants. Unlike calcium, silicon is not necessary for normal growth in most plants except for Equisetum arvense. Silica deposits occur within the cellular structure in many plants during normal growth, and they are especially common in grasses. The particles produced are termed phytoliths (17,29), biogenetic opal (17), and silicon bodies. They are found as persistent microfossils in soils after the death and decay of the plant parts in which they occur (17).

Three types of silicon deposition are known (18,32). The cell lumen may be infilled resulting in an opaline phytolith, or the silicon may be deposited in cell walls or in intercellular spaces. Postek (29) has recently described several types of phytoliths in Magnolia grandiflora (Figs. 18,19), phytoliths derived from vein sheath cells, tracheary elements, epidermal cells, and guard cells. These phytoliths form by the deposition of monosilicic acid in the cell lumen. The phytolith is formed of spherical particles 35-66 μm in diameter which form a cast of the cell. The cast may occur in conjunction with a fibrillar material which may act as an organic matrix for silicification.

Silicon occurs in the walls of lithocysts (5,19) and in other "normal" cells (32) in which the cell wall constitutes the organic matrix. The primary cell wall of plants is not homogeneous, but rather contains varying proportions of cellulose, hemicellulose, and pectins in several layers (11). According to Sakai and Thom (32), the pectin-hemicellulose regions of the wall are the sites in the wall where silicon deposition occurs, and they consider hemicelluloses and pectins as representing the matrix for silicification. Pectic polysaccharides have galacturonosyl residues as their most characteristic component.

The hemicelluloses are composed of xyloglucans, xylans, and heteroxylans (11). There are many references to silicon in the walls of lithocysts (1,5,8,19,28); there are fewer references to silicon in the cell walls of other plants. Monosilicic acid enters the plant through the transpiration stream. When the water is lost from the monosilicic acid the silica solidifies (29). This solidification may occur in the

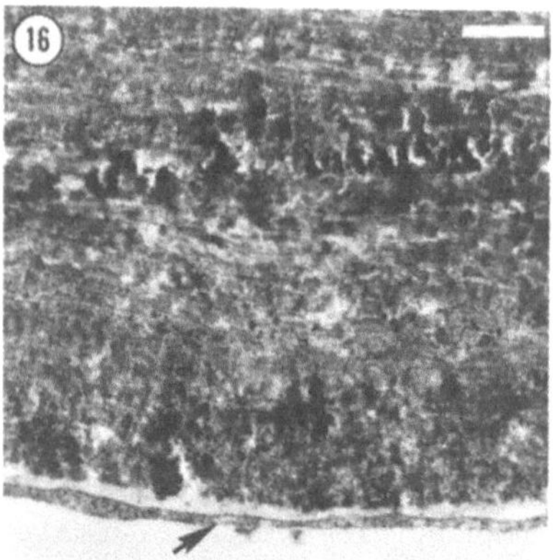
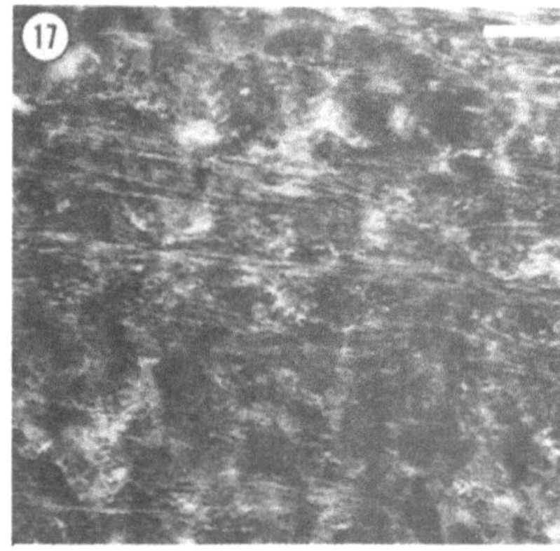
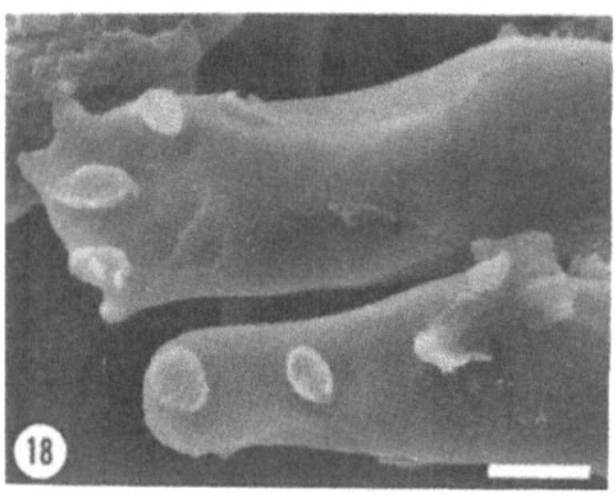
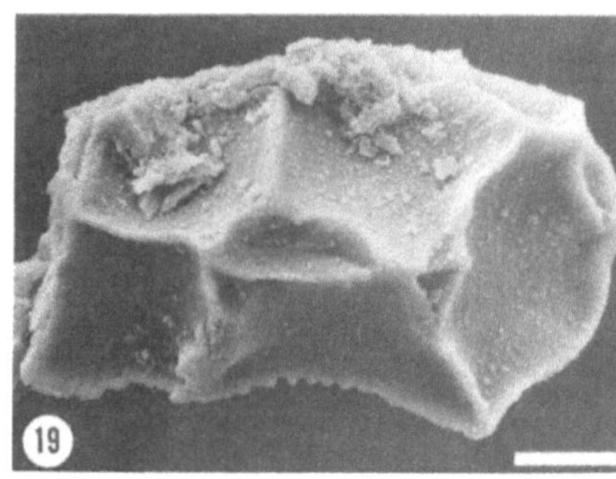

FIG. 16 - Silica impregnated apical wall of Morus alba lithocyst. Note thin layer of cytoplasm of the lithocyst (arrow). Bar = 1 µm.
FIG. 17 - Microfibrils in silicon impregnated lithocyst apical wall. Bar = 0.2 µm.
FIGS. 18-19 - Phytoliths of Magnolia grandiflora.
FIG. 18 - Tracheary element cast. Bar = 5 µm.
FIG. 19 - Terminal tracheid cast. Bar = 10 µm.

lumen or an intercellular space, in which case a phytolith is formed. Alternatively, the monosilicic acid may enter the cell wall where water loss may cause its deposition.

CONCLUSION

Calcium carbonate in cystoliths and silicon dioxide in silicification in higher plants appear to have clear associations with an organic matrix. In either case cellulose appears to be involved along with callose in cystoliths and hemicelluloses and pectins in silicification. As yet, little specific information is available on the presence of proteins in association with these mineral systems.

In the calcium oxalate system cellulose is not involved except in a secondary sense. The latter system appears to be more variable in terms of the materials and structures found in association with the crystal development. In calcium oxalate calcification it is difficult to define "the organic matrix" at our present stage of knowledge; however, polysaccharides and proteins have commonly been found associated with crystal development. The chemistry of the organic matrix in these higher plant calcification systems is clearly in need of careful study and represents an open area for research.

Acknowledgements. I thank M.J. Goad and M.A. Webb for their help in preparing this manuscript. Thanks to M. Kelly for the use of the micrograph in Figure 7. Research supported in part by the Graduate School, The University of Texas at Arlington.

REFERENCES

(1) Ajello, L. 1941. Cytology and cellular interrelations of cystolith formation in Ficus elastica. Am. J. Bot. 28: 589-594.

(2) Arnott, H.J. 1966. Studies of calcification in plants. In Third European Symposium on Calcified Tissues, eds. H. Fleisch, H.J.J. Blackwood, and M. Owens, pp. 152-157. New York: Springer-Verlag.

(3) Arnott, H.J. 1976. Calcification in higher plants. In The Mechanisms of Mineralization in the Invertebrates and Plants, eds. N. Watabe and K.M. Wilbur, pp. 55-78. Columbia, SC: South Carolina Press.

(4) Arnott, H.J. 1980. An SEM study of crystal idioblasts in pecan. Scanning Electron Microsc. 1980(III): 563-569.

(5) Arnott, H.J. 1980. Carbonates in higher plants. In Mechanisms of Biomineralization in Animals and Plants, eds. M. Omori and N. Watabe, pp. 211-218. Tokyo, Japan: Tokai University Press.

(6) Arnott, H.J. 1981. An SEM study of twinning in calcium oxalate crystals of plants. Scanning Electron Microsc. 1981(III): 225-234.

(7) Arnott, H.J., and Pautard, F.G.E. 1970. Calcification in plants. In Biological Calcification; Cellular and Molecular Aspects, ed. H. Schraer, pp. 375-446. New York: Appleton-Century-Croft.

(8) Beyrich, H. 1944. Neue Beiträge zur Entwicklungsgeschichte und Morphologie der Zystolithen. Protoplasma 33: 287-313.

(9) Cheavin, W.H.S. 1938. The crystals and cystoliths found in plant cells. Part I. Crystals. Microscope 2: 155-158.

(10) Crombie, W.M.L. 1960. Metabolism of extracyclic organic acids. In Encyclopedia of Plant Physiology, ed. W. Ruhland, vol. 12, part 2, pp. 890-933. Berlin: Springer-Verlag.

(11) Darvill, A.; McNeil, M.; Albersheim, P.; and Delmer, D. 1980. The primary cell walls of flowering plants. In The Biochemistry of Plants, ed. N.E. Tolbert, vol. 1, pp. 91-120. New York: Academic Press.

(12) Eilert, G.B. 1974. An ultrastructural study of the development of raphide crystal cells in the roots of Yucca torreyi. Ph.D. Dissertation, University of Texas at Austin (Libr. Congr. Card No. Mic. 74-24, 855). Ann Arbor, MI: University Microfilms.

(13) Eschrich, W. 1954. Ein Beitrag zur Kenntnis der Kallose. Planta 44: 532-542.

(14) Franceschi, V.R., and Horner, Jr., H.T. 1980. Calcium oxalate crystals in plants. Bot. Rev. 4: 361-427.

(15) Frank, E., and Jensen, W.A. 1970. On the formation of the pattern of crystal idioblasts on Canavalia ensiformis DC, IV. The fine structure of the crystal cells. Planta 95: 202-217.

(16) Frey-Wyssling, A. 1981. Crystallography of the two hydrates of crystalline calcium oxalate in plants. Am. J. Bot. 68: 130-141.

(17) Geis, J.W. 1978. Biogenic opal in three species of Gramineae. Ann. Bot. 42: 1119-1129.

(18) Hayward, D.M., and Parry, D.W. 1980. Scanning electron microscopy of silica deposits in the culms, floral bracts and awns of barley (Hordeum sativum Jess.). Ann. Bot. 46: 541-548.

(19) Hiltz, P., and Pobeguin, T. 1949. Sur la constitution des cystolithes de Ficus elastica. Comptes Rendus 228: 1049-1051.

(20) Horner, Jr., H.T., and Wagner, B.L. 1980. The association of druse crystals with the developing stomium of Capsicum annuum (Solanaceae) anthers. Am. J. Bot. 67: 1347-1360.

(21) Kim, K.M., and Johnson, R.B. 1981. Calcium oxalate crystal growth in urinary stones. Scanning Electron Microsc. 1981(III): 147-154.

(22) Krampitz, G., and Witt, W. 1979. Biochemical aspects of biomineralization. In Topics in Current Chemistry, ed. F.L. Boschke, pp. 58-144. Berlin: Springer-Verlag.

(23) Mazia, D. 1938. The binding of Ca, Sr and Ba by Elodea protoplasm. J. Cell Comp. Physiol. 11: 193-203.

(24) McNair, J.B. 1932. The interrelation between substances in plants: essential oils and resins, cyanogen and oxalate. Am. J. Bot. 19: 255-271.

(25) Mollenhauer, H.H., and Larson, D.A. 1966. Developmental changes in raphide-forming cells of Vanilla planifolia and Monstera deliciosa. J. Ultr. Res. 16: 55-70.

(26) Newcomb, E.H. 1980. The general cell. In The Biochemistry of Plants, ed. N.E. Tolbert, vol. 1, pp. 1-54. New York: Academic Press.

(27) Pentecost, A. 1980. Calcification in plants. Int. Rev. Cytol. 62: 1-27.

(28) Pobeguin, T. 1954. Contribution a l'étude des carbonates de calcium précipitation du calcaire par les végétaux comparaison avec le monde animal. Ann. des Sci. Nat. Bot. 15: 29-109.

(29) Postek, M.T. 1981. The occurrence of silica in the leaves of Magnolia grandiflora L. Bot. Gaz. 142: 124-134.

(30) Price, J.L. 1970. Ultrastructure of druse crystal idioblasts in leaves of Cercidium floridum. Am. J. Bot. 57: 1004-1009.

(31) Sakai, W.S., and Hanson, M. 1974. Mature raphid and raphid idioblast structure in plants of the edible aroid genera Colocasia, Alocasia, and Xanthosoma. Ann. Bot. 38: 739-748.

(32) Sakai, W.S., and Thom, M. 1979. Localization of silicon in specific cell wall layers of the stomatal apparatus of sugar cane by use of energy dispersive x-ray analysis. Ann. Bot. 44: 245-248.

(33) Scott, F.M. 1946. Cystoliths and plasmodesmata in Beloperone, Ficus, and Boehmeria. Bot. Gaz. 107: 372-378.

(34) Sunnell, L.A., and Healy, P.L. 1979. Distribution of calcium oxalate crystal idioblasts in corms of taro (Colocasia esculenta). Am. J. Bot. 66: 1029-1032.

(35) Tilton, V.R., and Horner, Jr., H.T. 1980. Calcium oxalate raphide crystals and crystalliferous idioblasts in the carpels of Ornithogalum caudatum. Ann. Bot. 46: 533-539.

(36) Watabe, N.; Meenakshi, V.R.; Blackwelder, P.L.; Kurtz, E.M.; and Dunkelberger, D.G. 1976. Calcareous spherules in the gastropod Pomacea paludosa. In The Mechanisms of Mineralization in the Invertebrates and Plants, eds. N. Watabe and K.M. Wilbur, pp. 283-308. Columbia, SC: South Carolina Press.

(37) Wattendorff, J. 1969. Feinbau und Entwicklung der verkorkten Calciumoxalate-Kristallzellen in der Rinde von Larix decidua Mill. Z. Pflanzenphysiol. 60: 307-347.

(38) Wattendorff, J. 1976. A third type of raphide crystal in the Plant Kingdom: six-sided raphides with laminated sheaths in Agave americana L. Planta 130: 303-312.

(39) Wattendorff, J. 1978. Feinbau und Entwicklung der Calciumoxalate-Kristallzellen mit suberinähnlichen Kristallscheiden in der Rinde und in sekundären Holz von Acacia senegal Willd. Protoplasma 95: 193-206.

(40) Webb, M.A., and Arnott, H.J. 1981. An ultrastructural study of druse crystals in okra cotyledons. Scanning Electron Microsc. 1981(III): 285-292.

Biological Mineralization and Demineralization, ed. G.H. Nancollas, pp. 219-232.
Dahlem Konferenzen 1982. Berlin, Heidelberg, New York: Springer-Verlag.

Structure of the Organic Matrix in Mollusc Shells and Avian Eggshells

G. P. Krampitz
Abteilung für Biochemie, Institut für Anatomie, Physiologie und Hygiene der Haustiere der Universität Bonn
5300 Bonn 1, German Federal Republic

Abstract. The organic matrices of mollusc shells as well as of avian eggshells include organic components soluble in aqueous solutions and constituents not soluble in water and organic solvents. The matrix components are regarded as being closely involved in biomineralization processes. The concept of an organic matrix depends on the hypothesis that macromolecular organic constituents largely control the deposition of solid inorganic matter. Concerning the chemical nature of matrices of mollusc shells and avian eggshells, macromolecules are thought to be important in binding various ions at anionic or cationic sites. These macromolecules are usually predominantly proteins. Occasionally the matrix is rich in highly basic polyanions, polyuronic acid, or sulfated polysaccharides. Ca-binding macromolecules have been identified among the soluble constituents of the matrix. The binding sites for calcium by soluble macromolecules are probably acidic building blocks of the polymers. The exact function of Ca-binding macromolecules in calcification processes is not clear. Among the soluble biopolymers of the matrix, also the enzymes have been detected as polyphenol oxidases and carbonic anhydrase. The chemical structure of the insoluble matrix and its biological significance on the molecular level need more elucidation, although studies on the secondary structure revealed in some cases antiparallel β-pleated sheet conformation and Ca-binding properties.

INTRODUCTION

It has generally been accepted that the organic matrix plays an important role in the processes leading to the formation of both mollusc shells and avian eggshells. The organic matrix of mollusc shells and avian eggshells can be obtained by dissolution of the mineralized layers of shell. This procedure yields two fractions, one that is soluble and the other insoluble in aqueous solutions (61). The content of organic substances of molluscan shell varies with species and is commonly between 0.01% and 10% (62). In analyses of shell of 14 species of estuarine bivalves a range from 1.4% to 21.4% has been found (42). Reported values for a particular species may vary considerably (42). The organic components of mollusc shell may be divided into two regions; a sklerotized, non-mineralized layer covering the outer surface of the shell, termed the periostracum, and the organic material associated with the mineralized layers of the shell, termed the organic matrix. The content of organic substances in the mineralized layer of the avian eggshell (biocrystalline layer) has been estimated to be in the range of 2-3% (23).

ORGANIC MATRIX OF MOLLUSC SHELLS

Soluble Matrix

The soluble matrix has been regarded as intracrystalline (5,38, 54), while the insoluble matrix has been thought to be intercrystalline (14). The concentration of soluble matrix components varies widely among species, ranging from 14% to 64% (61).

The composition of the soluble matrix of molluscan shells seems to be far from being simple with respect to the number of components observed within a single species (22,27,46,55,58), between species (21,27,38,58,62), and between different crystal layers of a single bivalve shell (43).

The first report on the extraction and at least partial purification of a soluble matrix component came from Crenshaw (5). This author isolated a water-soluble glycoprotein which was evenly distributed throughout the shell. The apparent molecular size was in the range of 160,000 daltons. It has been shown that in this substance negatively charged groups predominate due to ester sulfate which provides the negative groups associated with hexosamine residues (60,61). Its amino acid composition is very similar to that of insoluble mollusc shell matrices except that the isolated sulfated glycoprotein has a higher aspartic acid content. All the dicarboxylic amino acid residues are present in the protein as their amides. A soluble fraction extracted from Crassostrea virginica shells exhibited a molecular weight of 1 million daltons. Soluble proteins from Crassostrea irrescens, Mercenaria mercenaria, and Nautilus pompilius have been isolated and analyzed for amino acid composition. It has been presumed that a significant proportion of the soluble protein of the organic matrix of molluscan shells is composed of a repeating sequence of aspartic acid separated by either glycine or serine. This regularly negatively charged aspartic acid molecule may function as a template upon which mineralization occurs (56). However, the amino acid composition of the soluble protein of Mercenaria mercenaria (56) does not correspond with that purified by Crenshaw (5).

Fossil glycoproteins of the soluble matrix are present in an 80 million year old mollusc shell from the Late Cretaceous period. Discrete molecular weight components are preserved. A particular repeating amino acid sequence $(\text{- Asp-Y-})_n$ found in contemporary mollusc shell proteins was identified in fossil glycoproteins (57).

A more recent study of the amino acid composition of the major subfractions of the soluble matrix of Nautilus repertens, Strombus gigas, and Plicatula plicata confirms earlier findings that the proteins of all subfractions are enriched in acidic and polar amino acids. The subfraction which contains the major portion of total protein also contains the highest concentration of aspartic acid (55).

An important property of the soluble matrix is its ability to bind calcium. The sulfated glycoprotein isolated from Mercenaria mercenaria shells (5) has been demonstrated to bind calcium selectively even in the presence of excess sodium, potassium, and magnesium ions. Ca^{2+}-binding may be accomplished by the spacing of negatively charged ligands such as sulfate esters (5,7). There is a difference between the findings of Crenshaw (5,7) and Weiner (55,56). Crenshaw has clearly shown that all of the dicarboxylic amino acids are present as their amides in the Ca-binding glycoprotein, while Weiner's hypothesis of functions of soluble matrix proteins is based on the existence of carboxylic groups of aspartic acid. Another approach has been made to answer the question as to which building blocks of a Ca-binding protein or polypeptide are essential for that property (Ca-binding). By stepwise degradation, as commonly used in amino acid sequence analysis, fragments of Ca-binding polypeptides from shells of Crassostrea gigas exhibit the presence of glutamine and asparagine as well as sulfated serine residues which bind calcium. Similar results were obtained with Ca-binding peptides extracted from fossil oyster shells (Ostrea carrissima, Lopha diluviana, Lopha marshi) (49). The prevailing amino acid sequence from recent and fossil oyster shells were nearly the same except for the exchange of some amino acid residues probably due to evolutionary as well as diagenetic processes (22,46).

Polyphenol oxidases have been found in the soluble matrix of Crassostrea gigas, Strombus aureus (37,47), and Mercenaria mercenaria (13). Substrates for these enzymes, such as tyrosine and other phenolic substances, have been identified in shell matrices (13,43). It has been concluded that polyphenol oxidases and phenolic components may play a role in polymerization of the insoluble matrix.

Insoluble Matrix (Conchiolin)

The insoluble matrix is primarily a group of proteins with a sometimes significant polysaccharide content which has been detected in many species (7,38,59). The insoluble matrix has also an affinity for lipoidal stains such as Sudan black, even after thorough extraction with solvents used for lipids. The preponderance of aliphatic amino acids may explain this properly (6).

The insoluble fraction has cross-linked proteins which may be quinone-tanned as indicated by the high tyrosine content (7,38), suggesting that polyphenol oxidases may cross-link proteins after they have been secreted by the mantle.

Examination of the X-ray diffraction patterns given by the matrix derived from molluscan shells revealed that either an α- or β-keratin pattern was obtained, depending on the species studied (63). Also the molecular conformation of proteins present in the mollusc shell and those of the decalcified films of the nacreous layer was analyzed in the range of 4000-700 cm^{-1}. The decalcified films were found to consist primarily of proteins. The structure of proteins constituting the decalcified films of the nacreous layer is coiled and also has a β-conformation (20). More recently X-ray diffraction patterns of the proteinaceous components of insoluble matrices from nacreous shell of species representing the three major orders of molluscs show an antiparallel β-pleated sheet conformation similar to that of silk fibroin (59). The polysaccharide components, where present, exhibit a structure equivalent to that of chitin. Proteins of the insoluble matrix of prismatic layers of some species, however, show no well oriented pattern (61).

Functions of the Organic Matrix

Three functions of matrix presumed but not definitively established are crystal nucleation, limitation of crystal growth, and crystal orientation. A fourth function, the binding together of crystals and crystalline layers, is important in the physical behavior of shell (61).

ORGANIC MATRIX OF AVIAN EGGSHELLS

The biocrystalline layer of an avian eggshell (hen) contains organic material usually referred to as organic matrix. The organic substances of the biocrystalline layer include primarily protein-polysaccharide complexes, with low contents of water and lipids (50). Polysaccharides composed of glucosamine, galactosamine, mannose, and fucose have been detected. Chondroitin sulfate A and dermatan sulfate have also been found (1), and sialic acid is present only in small amounts, the highest concentration being in the mammillary layer, and then gradually dropping in the external zone (4).

The evidence for the presence of the proteinaceous material in the organic matrix was based on amino acid analysis (1). Systematic studies of the amino acid composition of the organic matrix of eggshells from Galliformes, Gruiformes, Anseriformes, and reptiles suggested evolutionary developments within orders of the classes Aves and Reptilia (24,29,31). From the amino acid composition data of ratite eggshells, affinities could be derived (32). A much better indicator for evolutionary tendencies could be the distribution of matrix proteins or their complexes with other substances.

Soluble Matrix

The organic matrix of eggshells can also be separated into a soluble and an insoluble fraction. Extraction and fractionation of proteinaceous macromolecules from the calcareous layer of chicken eggshells have recently been reported (25,26). Typical biopolymer profiles of eggshell matrix extracts have been obtained for a given species. Even race-dependent protein distributions have been revealed (Rhode Island Reds, White Leghorns, HNL-hybrids) (29).

A given pattern of soluble matrix macromolecules can be influenced by environmental poisons, e.g., Cd, Pb, Hg, Be (16,17, 29,35,37,51).

In the same context the phenomenon of eggshell reduction has been discussed since thin eggshells are correlated with changed protein profiles of the biocrystalline layer. Animals with a high egg production usually exhibit a protein pattern lacking one or more fractions compared with those having a low egg production.

Chemical analyses of isolated macromolecules from the soluble matrix revealed a proteinaceous character of most of these substances. Some can be described as protein-polysaccharide aggregates.

The biological functions of proteins and protein derivatives isolated from the soluble matrix have been studied. One of the matrix substances has been identified as active carbonic anhydrase after extraction of soluble matrix components. Crude eggshell extracts did not exhibit that enzyme activity. However, after removal of Ca^{2+} and a simple gel filtration of the extracts, carbonic anhydrase activity could be measured. Further purification steps led to the isolation of two isoenzymes, each having an apparent molecular size of about 28,000 daltons. Carbonic anhydrase from the soluble matrix catalyzes the reversible reaction of $CO_2 + HOH \rightleftarrows H^+ + HCO_3^-$. This is probably the primary action of the matrix enzyme, although esterase activity has also been recorded. Among other substances, sulfonamides inhibit the enzyme activity (11,28,30). These observations corroborated earlier reports demonstrating the existence of carbonic anhydrase in the biocrystalline layer of the eggshell based on histochemical methods (44,45). Other authors were not able to find carbonic anhydrase activity in eggshell powder (8-10). However, general agreement exists on the presence of carbonic anhydrase in uterine mucosal cells not bound to subcellular structures (3,10). High carbonic anhydrase activity in the shell gland of laying hens is correlated with high egg production, but during moult a reduced carbonic anhydrase activity in the eggshell gland and a constant carbonic anhydrase activity in erythrocytes has been observed. It has been assumed that the structure of this enzyme in the uterine cells depends on the action of progesterone (48). A dependence of progesterone action on carbonic anhydrase activity in the mucosal cells of the shell gland of chicken has been found very recently (41). Obviously, carbonic anhydrase activity in the shell gland of birds is associated with the formation of the biocrystalline layer of the shell since inhibitors of this enzyme, e.g., acetazolamide and other sulfonamides prevent the complete formation of eggshells. In some cases, no shell at all was formed although the calcium level of the blood was normal. However, the mechanisms of action of sulfonamides, e.g., the renal loss of bicarbonate, have also been discussed (39).

The functions of that shell enzyme probably include the provision of bicarbonate for the calcification process and the localized dissolution of the shell at the site of mammillary cores, a phenomenon which has been frequently observed in hatched eggs (12,26,49,52,56). Calcium also may be made available from the biocrystalline layer of the shell by changing the pH in a microenvironment. Calcium could be absorbed for the construction of the embryonic skeleton.

Other soluble organic macromolecules from the biocrystalline layer of the eggshell have been demonstrated as being able to bind Ca^{2+}. Several Ca-binding macromolecules have been isolated, and most of them have been purified to homogeneity (15,18). Some of them have been partly characterized and found to be composed of isoprenoid hydrocarbons, carbohydrates, and proteins. An open question is still the precise binding site for Ca^{2+}, although γ-carboxyglutamic acid is a possible candidate for this position (34). Ca-binding peptides have also been detected in dinosaur eggshell (36). Soluble components of the organic matrix exhibit the phenomenon of self-organization to insoluble products binding calcium, when HCO_3^- is then provided, $CaCO_3$ crystals are formed on top of the insoluble material. This system could serve as a model for biomineralization processes during eggshell formation (37), thus confirming the matrix hypothesis on a molecular basis. Ca-binding systems from mucosal cells and uterine fluid resemble each other. Their amino acid sequences are very similar. They correspond with that of a comparable Ca-binding polypeptide of the eggshell. This demonstrates that a segment of the comparable polypeptide produced by mucosal cells of the shell gland appears in the eggshell (Krampitz, Hotta, and Samata, in preparation).

Insoluble Matrix

The insoluble matrix of avian eggshells is primarily constructed of proteins and polysaccharides with low contents of water and lipids (50). Systematic studies of the amino acid composition of the insoluble matrix of various species of birds have been carried out (19,29,32,40), although the precise chemical structure is still unknown. In equilibrium dialysis experiments the organic matrix of eggshells exhibits remarkable Ca-binding properties (Krampitz, Hotta, and Drolshagen, in preparation). Additional effort is necessary to elucidate chemical structure and biological functions of the insoluble matrix of avian eggshells.

Studies on the structure of organic matrix components have revealed several interesting facts, but more experiments have to be done to provide answers to the many questions still open.

REFERENCES

(1) Balch, D.A., and Cooke, R.A. 1970. A study of the composition of hen's egg shell membrane. Ann. Biol. An. 10: 13-18.

(2) Beedham, G.E., and Owen, G. 1965. The mantle and shell of Solemya parkinsoni. Proc. Zool. Soc. London 145: 405-430.

(3) Bernstein, R.S.; Nevalainen, J.; Schraer, F.; and Schraer, H. 1968. Intracellular distribution and role of carbonic anhydrase in the avian (Gallus domesticus) shell gland mucosa. Biochem. Biophys. Acta 159: 367-372.

(4) Bronsch, K., and Diamantstein, T. 1964. Beiträge zur Biochemie der Schalenbildung. 2. Mitteilung: Über Mucopolysaccharide der Eischale. Zentralb. Vet. Med. 2: 323-330.

(5) Crenshaw, M.A. 1972. The soluble matrix from Mercenaria mercenaria shell. Biomineralisation 6: 6-11.

(6) Crenshaw, M.A. 1977. Sonotropic nucleation by molluscan shell matrix. In The 3rd International Symposium on Mechanisms of Biomineralization in the Invertebrates and Plants. Kashikojima, Japan, Abstract No. 16.

(7) Crenshaw, M.A., and Ristedt, H. 1976. The histochemical localization of reactive groups in septal nacre from Nautilus pompilius L. In The Mechanisms of Mineralization in the Invertebrates and Plants, eds. N. Watabe and K.M. Wilbur, pp. 353-367. Columbia, SC: University of South Carolina Press.

(8) Diamantstein, T. 1966. Über die lokale Rolle der Carboanhydrase im Hinblick auf die Eischalenverkalkung. Arch. f. Geflügelkde. 20: 309-312.

(9) Diamantstein, T.; Bronsch, K.; and Schlüns, F. 1964. Carbonic anhydrase in the mammillae of the hen's egg shell. Nature 203: 88-90.

(10) Diamantstein, T., and Schlüns, F. 1964. Lokalisation und Bedeutung der Carboanhydrase im Uterus von Legehennen. Acta histochem. 19: 269-303.

(11) Engels, J.; Heindl, I.; and Krampitz, G. 1976. Methodisches zur Isolierung von Carboanhydratase aus Eischalen und Erythrocyten des Haushuhns. Arch. Geflügelkde. 40: 1-7.

(12) Erben, H.K. 1970. Ultrastrukturen und Mineralisation rezenter und fossiler Eischalen bei Vögeln und Reptilien. Biomineralization 1: 1-54.

(13) Gordon, J., and Carriker, M.R. 1980. Sclerotized protein in the shell matrix of a bivalve mollusc. Mar. Biol. 57: 251-260.

(14) Grégoire, C. 1972. Structure of the molluscan shell. In Chemical Zoology, eds. M. Florkin and B.T. Sheer, vol. 7, Mollusca, pp. 45-102. New York: Academic Press.

(15) Hamm, M. 1976. Molekularmechanismen der Biomineralisation: Vorkommen und Darstellung Calcium-bindender Systeme. Dissertation, University of Bonn.

(16) Hardebeck, H.; Sülz, M.; and Krampitz, G. 1974. Die Toxizität von Cadmium beim Huhn. Arch. Geflügelkde. 38: 86.90.

(17) Hatjipanagiotou, A. 1977. Die Wirkungen von Berylliumverbindungen auf das Haushuhn. Dissertation, University of Bonn.

(18) Hausmanns, G. 1977. Molekularmechanismen der Biomineralisation: Charakterisierung und Biosynthese Calcium-bindender Systeme. Dissertation, University of Bonn.

(19) Hedding, M. 1979. Aminosäuren-Kompositionen rezenter, subfossiler und fossiler Eierschalen von Struthionidae und Aepyornithidae. Biochemische und Phylogenetische Beziehungen. Dissertation, University of Bonn.

(20) Hotta, S. 1969. Infra-red spectra and conformation of protein constituting the nacreous layer of molluscan shell. Earth Sci. Tokyo 23: 133-136.

(21) Hunt, S. 1970. Polysaccharide-protein Complexes in Invertebrates. London: Academic Press.

(22) Krampitz, G. 1980. Calcium binding proteins in mollusk shells. Haliotis 10: 82.

(23) Krampitz, G. 1981. Die Eischale. In Handbuch der Geflügelphysiologie, eds. A. Mehner and W. Hartfiel. Jena: VEB Gustav Fischer, in press.

(24) Krampitz, G.; Boehme, W.; Kriesten, K.; and Hardebeck, H. 1974. Die Aminosäurenzusammensetzung von Reptilieneischalen in biochemischer und evolutiver Sicht. Z. zool. Systematik u. Evolutionsforschung. 12: 1-12.

(25) Krampitz, G., and Engels, J. 1975. Beiträge zur Chromatographie von wasserlöslichen Biopolymeren aus der Eischalenmatrix. Biomineralization 8: 21-31.

(26) Krampitz, G., and Engels, J. 1975. Water-soluble proteins from egg shell matrices. In Proteins and Related Subjects, ed. H. Peeters, vol. 22, pp. 327-331. Oxford: Pergamon Press.

(27) Krampitz, G.; Engels, J.; and Cazaux, C. 1976. Biochemical studies on water-soluble proteins and related components of gastropod shells. In The Mechanisms of Mineralization in the Invertebrates and Plants, eds. N. Watabe and K.M. Wilbur, pp. 355-367. Columbia, SC: University of South Carolina Press.

(28) Krampitz, G.; Engels, J.; and Heindl, I. 1975. Definition und biologische Bedeutung von Carboanhydratase. Arch. Geflügelkde. 39: 189-293.

(29) Krampitz, G.; Engels, J.; Heindl, I.; Heinrich. A.; Hamm, M.; and Faust, R. 1974. Biochemische Untersuchungen an Eischalen. Arch. Geflügelkde. 38: 197-210.

(30) Krampitz, G.; Engels, J.; and Helfgen, I. 1975. Über das Vorkommen von Carboanhydratase in der Eischale des Huhns. Experientia 30: 228.

(31) Krampitz, G.; Kriesten, K.; and Boehme, W. 1973. Untersuchungen über Ultrastruktur und Aminosäurenzusammensetzung der Eischale von Natrix natrix. Experientia 29: 416-417.

(32) Krampitz, G.; Kriesten, K.; and Faust, R. 1974. Über die Aminosäurenzusammensetzung morphologischer Eischalen-Fraktionen von Ratitae. Biomineralisation 7: 1-9.

(33) Krampitz, G.; Koester, U.; and Fischer, W. 1975. Vergleichende Untersuchungen der Aminosäuren-Kompositionen von Vogeleischalen. Biochemische und taxonomische Beziehungen zwischen Hühner-, Kranich- und Gänsevögeln. Z. zool. Systematik u. Evolutionsforsch. 13: 125-152.

(34) Krampitz, G.; Meisel, H.; and Witt-Krause, W. 1980. Identification of γ-carboxyglutamic acid in ovocalcin. Naturwissenschaften 67: 38.

(35) Krampitz, G.; Sülz, M.; and Hardebeck, H. 1974. Die Verteilung von Cadmium im Organismus. Arch. Geflügelkde. 38: 100-104.

(36) Krampitz, G.; Weise, K.; Potz, A.; Engels, J.; Samata, T.; Becker, K.; Hedding, M.; and Flajs, G. 1977. Calcium-binding peptide in dinosaur egg shells. Naturwissenschaften 64: 583.

(37) Krampitz, G., and Witt, W. 1979. Biochemical aspects of biomineralization. Topics Curr. Chem. 78: 57-144.

(38) Meenakshi, V.R.; Hare, P.E.; and Silbur, K.M. 1971. Amino acids of the organic matrix of neogastropod shells. Comp. Biochem. Physiol. 40B: 1037-1043.

(39) Mongin, P. 1970. Etude de l'anhydrase carbonique dans les oiseaux. Ann. Biol. An. 10: 119-125.

(40) Paeffgen, D. 1979. Aminosäuren-Kompositionen der Eierschalen von Papageien, Tauben, Spechten und Kukucken. Biochemische und Phylogenetische Beziehungen. Dissertation, University of Bonn.

(41) Potz, A., and Krampitz, G. 1981. Untersuchungen zur Frage des molekularen Mechanismus der Eischalenbildung. 8. Mitt.: Hormonelle Beeinflussung von Ca-Bindung und Carboanhydratase-Aktivität in Mucosa-Zellen der Schalendrüse des Huhns. Arch. Geflügelkde. 45: 9-12.

(42) Price, T.J.; Thayer, G.W.; La Croix, M.W.; and Montgomery, G.P. 1976. The organic content of shells and soft tissues of selected estuarine gastropods and pelecypods. Proc. Natl. Shellfish Assoc. 65: 26-31.

(43) Ravindranatz, M.H., and Ravindranatz, M.H.R. 1974. The Chemical nature of the shell of molluscs: I. Prismatic and nacreous layers of a bivalve Lamellidans marginalis (Unionidae). Acta histochem. 48: 26-41.

(44) Robinson, D.S. 1970. The structure of the mammillary layer of the domestic hens' egg shell. Ann. Biol. An. 10: 27-35.

(45) Robinson, D.S., and King, N.R. 1963. Carbonic anhydrase and formation of the hen's egg shell. Nature 199: 497-499.

(46) Samata, T., and Krampitz, G. 1982. Ca^{2+}-binding polypeptides in Oyster shells. Malacologia 122: 225-233.

(47) Samata, T.; Sanguansri, P.; Cazaux, C.; Hamm. M.; Engels, J.; and Krampitz, G. 1980. Biochemical studies on components of mollusc shell. In The Mechanisms of Biomineralization in Animals and Plants, eds. M. Omori and N. Watabe, pp. 37-47. Tokyo: Tokai University Press.

(48) Schlüns, F., and Diamantstein, T. 1966. Behaviour of carbonic anhydrase in the uterus of hens during moult. Nature 209: 304-305.

(49) Schmidt, W.J. 1965. Morphologie der Kalkresorption an der ausgebrüteten Vogelschale. Z. Zellforsch. 68: 874-882.

(50) Simkiss, K., and Tyler, C. 1967. A histochemical study of the organic matrix of the hen's egg shell. Q. J. Mic. Sci. 98: 12-17.

(51) Sülz, M.; Hardebeck, H.; and Krampitz, G. 1974. Langzeiteinfluß von Cadmiumgaben auf Futterverzehr, Gewichtszunahme, Legeleistung und Eischalenqualität bei Legehennen. Arch. Geflügelkde. 38: 150-154.

(52) Terepka, A.R. 1963. Organic-inorganic interrelationships in avian egg shell. Exp. Cell. Res. 30: 183-189.

(53) Tyler, C., and Simkiss, K. 1959. Some changes in the shell during incubation. J. Sci. Food Agric. 10: 611-616.

(54) Watabe, N. 1965. Studies on shell formation. XI. Crystal-matrix relationships in the inner layers of mollusc shells. J. Ultrastruct. Res. 12: 351-370.

(55) Weiner, S. 1979. Aspartic acid rich proteins: Major components of the soluble organic matrix of mollusc shells. Calc. Tiss. Int. 29: 163-167.

(56) Weiner, S., and Hood, L. 1975. Soluble protein of the organic matrix of mollusk shell: a potential template for shell formation. Science 190: 987-989.

(57) Weiner, S.; Lowenstam, H.A.; and Hood, L. 1976. Characterization of 80-million year old mollusc shell proteins. Proc. Natl. Acad. Sci. (USA) 73: 2541-2545.

(58) Weiner, S.; Lowenstam, H.A.; and Hood, L. 1977. Discrete molecular weight components of the organic matrices of mollusc shells. J. Exp. Mar. B. 30: 45-51.

(59) Weiner, S., and Traub, W. 1980. X-ray diffraction study of the insoluble organic matrix of mollusk shells. FEBS Lett. 111: 311-316.

(60) Wilbur, K.M. 1976. Recent studies in invertebrate mineralization. In The Mechanisms of Mineralization in the Invertebrates and Plants, ed. N. Watabe and K.M. Wilbur, pp. 79-108. Columbia, SC: University of South Carolina Press.

(61) Wilbur, K.M., and Manyak, D.M. 1981. Biochemical aspects of shell mineralization. ONR Symposium on Marine Biodeterioration, in press.

(62) Wilbur, K.M., and Simkiss, K. 1968. Calcified shells. In Comprehensive Biochemistry 26 A, eds. M. Florkin and E.H. Stotz, pp. 229-295. New York: Elsevier.

(63) Wilbur, K.H., and Watabe, N. 1963. Experimental studies on calcification in molluscs and the alga Coccolithus huxleyi. Ann. NY Acad. Sci. 109: 82-112.

REVIEW ARTICLES

(1) Degens, E.T. 1976. Molecular mechanisms on carbonate, phosphate and silica deposition in the living cell. Topics Curr. Chem. 64: 1.

(2) Hunt, S. 1970. Polysaccharide-Protein Complexes in Invertebrates. London, New York: Academic Press.

(3) Krampitz, G., and Witt, W. 1979. Biochemical aspects of biomineralization. Topics Curr. Chem. 78: 58-144.

(4) Omori, M., and Watabe, N., eds. 1980. The Mechanisms of Biomineralization in Animals and Plants. Tokyo: Tokai University Press.

(5) Simkiss, K., ed. 1981. Biomineralization Newsletter, nr. 4.

(6) Watabe, N., and Wilbur, K.M., eds. 1976. The Mechanisms of Mineralization in the Invertebrates and Plants. The Belle W. Baruch Library in Marine Science, Number 5. Columbia: University of South Carolina Press.

(7) Wilbur, K.M., and Manyak, D.M. 1981. Biochemical aspects of shell mineralization. ONR Symposium on Marine Biodeterioration, in press.

Biological Mineralization and Demineralization, ed. G.H. Nancollas, pp. 233-241.
Dahlem Konferenzen 1982. Berlin, Heidelberg, New York: Springer-Verlag.

Mechanisms of Normal Mineralization in Bone and Cartilage

H. Fleisch
Pathophysiologisches Institut der Universität Bern
3010 Bern, Switzerland

Abstract. Despite intensive investigations, normal and abnormal calcification mechanisms are still poorly understood. The great number of theories which have been and are still being proposed reflect the complexity of the subject and the uncertainty of our knowledge.

MORPHOLOGICAL ASPECTS

Calcium phosphate can be deposited initially both within the cells and extracellularly. Intracellularly, mainly in the mitochondria of osteoblasts, osteocytes, osteoclasts, and epiphyseal cartilage cells, it is present in the form of amorphous granules with a maximum diameter of 100 nm. In the epiphyseal cartilage, the calcium deposits increase in number from the resting to the hypertrophic zone and then disappear at the onset of extracellular calcification. This leads to the suggestion that the ions forming the mineral are transported in some way from the cell to the site of extracellular mineralization.

Extracellularly, precipitation starts at two main locations: in the matrix vesicles and along the collagen fibrils. The matrix vesicles are round bodies limited by a trilaminar membrane, probably originating from cellular processes. They were first discovered in the longitudinal septa of epiphyseal cartilage and later found in other calcifying tissues such as woven and lamellar bone and in dentine. The initial precipitation of mineral seems to be associated with the membrane. Later the first vesicular deposits grow into large clusters, suggesting that once deposition has started crystallization proceeds readily. The clusters then tend to coalesce into a single mass of heavily calcified extracellular tissue, within and outside the collagen fibrils.

Certain investigators believe that the precipitation may also start on collagen fibrils. The first mineral deposits seem to appear at specific locations corresponding to the so-called "hole" regions, as round particles of 1-5 nm with a low molar Ca/P ratio of 1.2 to 1.5. With progressing mineralization the periodicity of collagen is first accentuated, suggesting that growth follows a specific pattern within the fibrils. Later the periodicity becomes blurred, probably because of mineral deposition within and between the fibrils.

Presently it is still being disputed as to which of the two types of calcification is more important. Possibly the first type predominates in woven bone and cartilage, while collagen-induced mineralization is more important in lamellar bone. This would explain why the crystals are more ordered in the latter case.

Whatever the location, about 2/3 of the total mineral deposition in the bone occurs rapidly, i.e., within a few hours, by an event which has been called "primary mineralization" and which is characterized by an increase in the number of crystals. The remaining event, called "secondary mineralization," proceeds much more slowly over months. As the mineralization takes place, water and non-collagenous proteins are displaced from the calcification site.

CHEMISTRY OF CALCIFICATION

While there is general agreement that the end stage in the calcification process is the formation of an "impure" hydroxyapatite, some uncertainty still exists as to the structure of the initial and intermediate phases. It appears that the first phase is not hydroxyapatite, but a precursor calcium phosphate of yet undetermined structure, possibly octacalcium phosphate or even brushite (see Nancollas, this volume).

In vitro during de novo precipitation the first phase formed, at least at high supersaturation, is often an <u>amorphous calcium phosphate</u>. Whether this is also true in the presence of preformed crystals or whether the new phase is then directly hydroxyapatite or some other phase is not known. This amorphous precipitate is thought to be composed of small clusters of about 1 nm diameter of a composition corresponding to $Ca_9(PO_4)_6$, closely packed into large spherical particles of 30-100 nm diameter with water held in the interstices. This phase is unstable and transforms in vitro spontaneously into a crystalline material with a hydroxyapatite diffraction pattern. The transformation is inhibited by various substances, among others magnesium, carbonate, citrate, pyrophosphate, adenosine triphosphate, and bisphosphonates. These substances may act by inhibiting the dissolution of the precursor phase and/or the crystallization of the subsequent phase.

The question as to whether an amorphous phase is formed initially in vivo is not yet settled, although various data suggest that this may be the case. Once formed, however, the amorphous phase in normal bone is rapidly transformed into crystalline apatite, except at specific locations such as inside the mitochondria. It is also possible that an amorphous coat, stabilized by inhibitors, can be maintained at the crystal surface. If true, this could provide an explanation for the relatively high solubility of bone mineral, compared to that of apatite.

Uncertainty also exists as to the type of crystalline phase which is formed initially in vivo. This phase strongly depends on the in situ conditions present, especially the state of supersaturation and the pH. In vitro <u>brushite</u> ($CaHPO_4$ $2H_2O$) is formed preferentially at low pH and high supersaturation. In vivo this salt is thought to exist in urine and by some researchers also in bone and dental calculus. In vitro octacalcium phosphate can also be formed. At physiological conditions of supersaturation and pH, it can transform in a second step into hydroxyapatite. The formation of an octacalcium phosphate phase in vivo is supported by its presence in mineralized tissues. The solid-solid transformation of octacalcium phosphate into hydroxyapatite would also explain the plate-like habit of the bone crystals which is unusual for a salt-like apatite but is consistent with the structure of octacalcium phosphate.

THEORIES OF CALCIFICATION

It has been realized for over 50 years that blood ultrafiltrate is supersaturated in calcium and phosphate with respect to hydroxyapatite, i.e., its calcium and phosphate content is above the solubility of this salt. However, it is also undersaturated with respect to the formation of calcium phosphate de novo. This difference is due to the fact that a higher energy is needed to form a solid phase de novo than to have already formed crystals continue to grow (Garside, this volume). In the calcium phosphate system the gap is especially large, since the first phase formed is not hydroxyapatite, but a precursor phase with a higher solubility. The undersaturation with respect to formation raises the question as to how precipitation is induced in vivo under these circumstances. Various theories have been proposed.

Booster Theory

The oldest theory, termed the booster theory, states that the calcium x phosphate ion product is to be increased locally up to the level of precipitation of the first solid calcium phosphate phase. Various ways have been proposed to obtain this rise in ion concentration. One possibility is that a phosphate ester could become hydrolyzed by alkaline phosphatase, thus increasing the local inorganic phosphate concentration. This hypothesis would provide an explanation for the well-known fact that alkaline phosphatase is always present at sites of calcification. However, alkaline phosphatase is present in many tissues other than those which mineralize. Furthermore, the attempt to find an intermediate phosphate compound with a concentration high enough to lead to a sufficient increase in inorganic phosphate (Pi) has been without success. Finally, data obtained from micropuncture studies have shown that, at least in calcifying cartilage, the calcium x phosphate product is not elevated. Thus, if such a booster mechanism exists, it would have to operate in a microvolume of space not analyzable by current techniques. Possibly the matrix vesicles could fulfil this role, since they are rich in alkaline phosphatase.

Another possibility is that the calcium x phosphate product is increased locally by an active cellular process. Examples are the mitochondria, which are indeed known to transport calcium actively through the membrane into their interiors. Other candidates are the matrix vesicles, although the suggestion that they transport calcium actively is unlikely since, with the vesicles originating from the cell membrane, such a transport would actually drive calcium from inside the vesicles to the outside.

Nucleation Theory

A second theory proposes that the organic matrix might facilitate precipitation through heterogenous nucleation. After an initial interest in glycosaminoglycans, attention was directed mainly towards collagen. This protein has indeed been shown to induce calcium phosphate precipitation in vitro. The groups responsible for such an induction are not yet known. One possibility could be a phosphate group covalently linked to serine as O-phospho-serine. It has also been suggested that calcium could be initially bound. Whichever the case, the bound ions should be favorably oriented with respect to the lattice parameters of calcium phosphate to be efficient.

Other candidates serving as possible nucleators are lipids. The interest in the possible role of lipids in the calcification process originates from histological findings that lipids concentrate at the calcification front, but are not found to do so in vitamin D rickets. The lipids involved are thought to be mainly acidic phospholipids, especially phosphatidyl serine and phosphatidyl inositol. Indeed, these are present in various calcified tissues, often in the form of a calcium-phospholipid-phosphate complex, and are able to induce calcium phosphate precipitation in vitro from solutions metastable with respect to this salt. The lipids present in the membranes of the matrix vesicles might also explain the nucleating activity of these structures.

Other proteins, such as a phosphoprotein extracted from dentine, the protein containing the amino acid γ-carboxyglutamic acid, and the newly described osteonectin when bound to collagen, have been also invoked as nucleators.

Another facet of the nucleator theory is that crystals of one salt can induce the precipitation of salts of another kind. This mechanism is thought to play a role in the formation of urinary stones, explaining why stones are often a mixture of different salts.

Inhibition Theory

The fact that some of the nucleators, such as collagen, are also present in non-mineralizing tissues raises the question as to why more tissues do not calcify in vivo. One reason for this phenomenon may be found in the fact that collagen molecules are more closely packed in soft tissues than in bone, thus impeding the diffusion of phosphate ions to the intrafibrillar nucleation sites. Another possibility is that soft tissues are protected in vivo by circulating inhibitors of crystallization, whereas at sites of calcification these inhibitors are destroyed. Indeed plasma, urine, and other biological fluids contain powerful crystallization inhibitors.

The first inhibitor to be recognized and investigated was inorganic pyrophosphate, which, like other compounds characterized by a P-O-P bond, inhibits precipitation at very low concentrations and is present in blood, urine, saliva, and joint fluid at concentrations effective for blocking the growth of hydoxyapatite. Since alkaline phosphatase is known to have pyrophosphate activity, this enzyme might act by destroying locally the circulating inhibitor, thus allowing mineralization to take place. The matrix vesicles possess strong pyrophosphatase and ATPase activities, so that one of the modes of inducing calcification could be through hydrolysis of such inhibitors as pyrophosphate and ATP. However, since many tissues which do not calcify contain alkaline phosphatase, the pyrophosphate mechanism cannot be considered as the only operating inhibitory mechanism.

Plasma and urine contain several other inhibitors besides pyrophosphate, such as citrate and magnesium. Albumin is another inhibitor in plasma. However, at least in plasma and urine, these known inhibitors do not account for the whole inhibitory activity by any means. Unfortunately, the relative importance of each one is difficult to assess since the results obtained in vitro greatly depend on the technique used. Saliva also contains one or several macromolecular inhibitors. One of them, called statherin, has recently been characterized as a polypeptide with 43 amino acids.

In addition to the circulating crystal growth inhibitors, local, non-circulating inhibitors, are also likely to be of importance. An inhibitor of unknown structure with a molecular weight of about 100,000 daltons has been extracted from tendon. Recently, interest has been revived as to the possible role of proteoglycans in calcification. Epiphyseal cartilage fluid contains a high molecular weight inhibitory fraction consisting of proteoglycans in the aggregated form. When this fraction is disaggregated (possibly by lysozyme) into subunits, it loses its inhibitory activity. Therefore, part of the regulatory process, at least in the epiphyseal cartilage, might consist of a change in the aggregation of proteoglycans, transforming a non-mineralizing milieu into a calcifying one. In bone, Gla-protein as well as other extractable non-collagenous proteins are active as inhibitors. Since the non-collagenous organic fraction in bone matrix diminishes during mineralization, it is possible that it does play a role in the calcification process.

Besides inhibitors of crystal growth, inhibitors of the aggregation process of calcium phosphate crystals also have to be considered. Both plasma and urine contain inhibitors of calcium oxalate and calcium phosphate aggregation. Glycosaminoglycans, pyrophosphate, and citrate are all powerful inhibitors of crystal aggregation, although here, too, their relative importance cannot be estimated. It is thought that the aggregation process is essential in the formation of certain calcifications, especially in urolithiasis, where it might be one of the mechanisms distinguishing normal people who simply form crystals in their urine which can be eliminated, from stone formers forming large aggregates which they are unable to eliminate.

CONCLUSIONS

At present, none of the theories can be singled out as being more important than the others. It could well be that more than one, if not all, are partially correct, the calcification process being controlled by more than one mechanism.

BACKGROUND REVIEWS

(1) Anderson, H.C. 1976. Matrix vesicles of cartilage and bone. In The Biochemistry and Physiology of Bone, ed. G.H. Bourne, vol. 4, pp. 135-157. New York, London: Academic Press.

(2) Boskey, A.L. 1981. Current concepts of the physiology and biochemistry of calcification. Clin. Orthop. 157: 225-257.

(3) Fleisch, H., and Russell, R.G.G. 1977. Experimental and clinical studies with pyrophosphate and diphosphonates. In Calcium Metabolism in Renal Failure and Nephrolithiasis, ed. D.S. David, Ch. 9, pp. 293-336. New York: John Wiley and Sons, Inc.

(4) Glimcher, M.J. 1976. Composition, structure, and organization of bone and other mineralized tissues and the mechanism of calcification. In Handbook of Physiology - Endocrinology VII, pp. 25-115. Washington: American Physiological Society.

(5) Neuman, W.F. 1980. Bone material and calcification mechanisms. In Fundamental and Clinical Bone Physiology, ed. M.R. Urist, pp. 83-107. Philadelphia, Toronto: J.B. Lippincott Co.

(6) Urist, M.R. 1976. Biochemistry of calcification. In The Biochemistry and Physiology of Bone, ed. G.H. Bourne, vol. 4, pp. 2-54. New York, London: Academic Press.

Biological Mineralization and Demineralization, ed. G.H. Nancollas, pp. 243-257.
Dahlem Konferenzen 1982. Berlin, Heidelberg, New York: Springer-Verlag.

Mechanisms of Normal Biological Mineralization of Calcium Carbonates*

M. A. Crenshaw
Dental Research Center and Dept. of Pedodontics
University of North Carolina, Chapel Hill, NC 27514, USA

Abstract. Normal biogenic calcium carbonate is deposited as calcite or aragonite with preferred crystal orientation. The site of mineralization is isolated, and the parent medium may have a unique composition. In most cases, the mineral is deposited on an organic matrix. The insoluble matrical fraction is composed of varying quantities of different classes of macromolecules, and its surface may be hydrophobic. The soluble matrical fraction from bivalve molluscs is primarily a sulfated, high molecular weight glycoprotein that selectively binds calcium. This glycoprotein is assigned a primary role in crystal nucleation in the ionotropic nucleation hypothesis. Postnucleation growth may be controlled in varying degrees by the bathing medium and by the insoluble matrix.

INTRODUCTION

Calcium carbonate is deposited by almost all animal phyla and by some groups of algae in specific crystal patterns that are characteristic of each organism. Most of the structures deposited in normal mineralization are calcite or aragonite, and the majority are in the form of polycrystalline aggregates with preferred orientation, usually with the c-axes in a common direction. Some structures contain a solid solution of calcium

---

* This paper is dedicated to Professor Karl M. Wilbur on his retirement.

carbonate and magnesium carbonate as high magnesium calcite. Amorphous calcite has also been reported. With the possible exceptions of the high magnesium calcites of sponges and echinoderms, an organic matrix is associated with the mineral (24).

Any general theory of calcium carbonate mineralization must account for: (a) the nucleation of the mineral, (b) the transformation of any unstable mineral that may be deposited initially to a more stable form, (c) the determination of the calcium carbonate polymorph deposited, and (d) the control of crystal orientation and crystal growth. Evidence for the mechanisms by which these processes are accomplished is incomplete. It is possible that all the parameters are determined by the nucleation process and are expressed by postnucleation growth.

Mineralization in molluscs illustrates processes generally important to mineralization systems. This review will give information of the molluscan system from this viewpoint. The shell lies outside the animal's body and is formed by a single epithelial layer of the mantle. The intervening extrapallial space may be separated from the environment by the attachment of the mantle to the shell, by the periostracum which is a multilayered, tanned protein that extends from the outer edge of the mantle over the outside of the shell, or by the close approximation of the mantle to the shell. The molluscan shell is composed of two or more layers, each of which has a distinctive microstructure. All layers in one shell may contain only one calcium carbonate polymorph, or aragonite and calcite may be in different layers. In the latter case, each polymorph appears to be associated with a different region of the mantle. Sometimes amorphous calcium carbonate is observed in shells.

The organic matrix is considered to play a principal role in the normal mineralization of calcium carbonate. The primary hypothesis of the mechanisms by which the deposition of calcium carbonate is controlled is the matrix hypothesis, which assumes that the organic matrix is deposited as a substrate that nucleates crystals by a process similar to epitaxy (24,31).

MINERALIZATION MEDIUM

Almost without exception, the sites where calcium carbonate is deposited are isolated from the environment and from the general body fluids of the organism. These sites range from the Golgi vesicles of coccolithoporids to the extrapallial fluid of molluscs. This isolation of the mineralization sites may allow the formation of a medium with a distinct composition in which mineralization takes place. Knowledge of the composition of this medium during mineralization would allow the determination of the chemical driving force for calcium carbonate mineralization and would indicate the effects of the medium on the calcium carbonate polymorph deposited and its growth rate.

It is difficult to obtain samples of the mineralization medium that are large enough for analyses. Some initial analyses of the molluscan extrapallial fluid have been made (4,22). In marine bivalves, the concentrations of major chemical constituents of the extrapallial fluids were very similar to seawater. Various organic materials that may form complexes with calcium were also found. Therefore, the composition of the extrapallial fluid was significantly different from seawater. The calcium concentration in fresh water animals was lower than in marine animals, but the apparent molar product of calcium and carbonate was large because of a higher carbonate concentration in the extrapallial fluid of fresh water molluscs. Magnesium, sodium, potassium, and sulfate concentrations were lower in the fresh water animals. The inorganic composition of extrapallial fluids also fluctuated with season and physiological condition of the animal. Changes in organic composition are poorly studied.

The pH of extrapallial fluids has been monitored with electrodes implanted through the shell (4,30) and by sequential sampling (19). The pH in one animal may fluctuate between 7.0 and 7.8. The lower values were obtained when the valves were closed, and the higher values when the valves were open and the animals were ventilating. Therefore, the pH of the extrapallial fluid appears to reflect the alternating of anaerobic and aerobic periods of the animal (4). Shell growth may be greater during aerobic periods when the extrapallial fluid pH is higher (30).

The analyses of the molluscan extrapallial fluid that have been reported must be regarded as preliminary, for they are incomplete, especially with respect to the organic materials present. The nature of the organic materials and their concentrations in the extrapallial fluid may exert significant control over the calcium carbonate polymorph deposited (12). The rate of crystal growth may be determined by the concentration of crystal poisons such as phosphates present (18) and by the activities of enzymes such as carbonic anhydrase.

Most considerations of the physical chemistry of biogenic calcium carbonate mineralization have been made in terms of the solubility product of the mineral deposited. Recent advances in the chemical modeling of calcium carbonate precipitation indicate that, although thermodynamic considerations provide the driving force for precipitation, precipitation may be determined to a greater extent by kinetic factors. Transport processes are particularly important. For example, in crystal growth and dissolution experiments, the pCO_2 on the crystal surface may be much greater and the pH much less than in the bulk fluid (17). Consequently, the extrapallial fluid may be significantly supersaturated with respect to the mineral of the shell, but crystal growth may not take place because of the slow rate of diffusion to and from the shell surface. Of course, the unstirred layer is a general kinetic problem not confined to crystallization.

In many molluscs the inner, growing surface of the shell is covered by one or more organic layers (15,23,25). These layers probably are permeable to materials diffusing to the shell surface, but they reduce the rate of diffusion to the shell. Both effects would reduce the exchange of materials between the extrapallial fluid and the growing shell surface.

The physical chemistry of the mineralization medium is a very difficult problem, but it must be resolved so that the role of this medium in controlling mineralization can be understood. The molluscan extrapallial fluid seems to be the best experimental model.

ORGANIC MATRIX

Experimentally, the matrix is seen as the organic material into which the mineral is embedded or as the residue left after the mineral is dissolved. There is no a priori reason to assume that all components of the matrix have a function in the mineralization process or that mineralization is the only function of the matrix. For example, the organic matrix appears to have a role in determining the mechanical properties of the shell of molluscs (1).

Most of the studies on the organic matrices of calcium carbonate mineralization have used the molluscan shell matrix. It is readily separated from extraneous tissue and is available in large quantities. The organic content of the external molluscan shell varies from 0.01% to 5% (33). Proteins, glycoproteins, and glycosaminoglycans have been identified in shell matrices which have both water soluble and insoluble components (9).

Insoluble Matrix

The insoluble matrix is the classical conchiolin. As much as 70% of the total nitrogen in this fraction is present as amino nitrogen. Glycine and alanine are the most abundant amino acids and amount to 30-60% of the total amino acid residues (9).

A large number of the observations made on conchiolin have been enumerations of the amino acids present. Any differences found in the amino acid composition of the conchiolin associated with a particular microstructure or calcium carbonate polymorph are less than the variation in the amino acid composition of that particular conchiolin within a taxonomic group (9). This is not surprising because conchiolin contains varying amounts of several proteins and polysaccharides.

An example of the complexity of conchiolin is furnished by the serial extraction of nacreous conchiolin from cephalopods, bivalves, and archeogastropods (9). Nacre is aragonitic and, in its simplest form, in vertical section appears as a brick wall; the crystalline tablets being the bricks and matrix the mortar. When the mineral is dissolved, the vertical intercrystalline matrix collapses upon the horizontal interlamellar matrix to form outlines of the positions of the original polygonal crystals. The interlamellar conchiolin forms lacelike networks with pore patterns that are characteristic of the molluscan class (9). When it is extracted with borate, nacreous conchiolin retains its lacelike structure. The extract contains about 24% of the total nitrogen, and 30% of the amino acids are glycine and alanine. A second extraction with 5% sodium hydroxide removes 71% of the total nitrogen. Glycine and alanine amount to approximately 55% of the total amino acid residues in this extract. The residue is fibrous and contains 6% of the total nitrogen. As much as 35% of this residue is chitin. Over 80% of the amino acids are glycine and alanine. While the significance of the observations in terms of calcium carbonate mineralization is not known, they confirm earlier indications that conchiolin is a refractory mixture and show that total amino acid analyses of conchiolin may be of limited value in the determination of any role conchiolins may have in mineralization.

Additional studies of the structure of conchiolin have used X-ray diffraction to determine any repeating order within this matrix that may serve as a template for calcium carbonate nucleation (28). The demineralized and dried conchiolin adopted an antiparallel β-sheet configuration. The presence of chitin was confirmed. A coincidence of matrix axis with crystallographic axis was found only in the case of the c-axis of aragonite in one cephalopod (17). Such a coincidence was observed earlier and was considered to be due to the growth direction and movement of the mantle (9). Therefore, the alignment of the two axes may not be directly related, and the orientation of the c-axis of the crystal may not be determined by any orientation in the matrix. However, the X-ray diffraction studies are incomplete, and they may yield important insight into the possible role of conchiolin in calcium carbonate mineralization and crystal orientation.

The formation of initial calcium carbonate crystals on the surface of the organic matrix has been observed in many organisms (24). However, the characteristics of this surface have not been investigated. The striking feature from the amino acid analyses of conchiolin (9) is the preponderance of amino acids having hydrophobic side chains (20). The hydrophobicity of conchiolin may be greater than indicated by the amino acid composition because the hydrophilic effect of the peptide bond may be nullified by the formation of interpeptide bonds formed in its β-sheet structure (20). Conchiolin stains with Sudan Black, a typical lipid stain (3,7). However, exhaustive extraction of matrix with chloroform-methanol, acidified chloroform-methanol, or with diethyl ether did not reduce this staining, nor could any lipids be detected in the extracts (Crenshaw, unpublished). Therefore, the Sudan Black staining, which depends upon partition rather than a chemical reaction, appears to provide additional evidence that conchiolin is hydrophobic. The hydrophobic nature of coralline aragonite suspensions (13) may also be due to a hydrophobic matrix which may, in addition, affect the transformation of aragonite to calcite. The possible hydrophobic nature of the insoluble matrix is not directly documented. However, this characterization of this surface is of paramount importance in deducing any possible role that it may have in calcium carbonate mineralization.

Soluble Matrix

From 20 to 50% of the total non-dialyzable, organic matrix from fresh molluscan shell is soluble in water. The soluble matrices from bivalves and cephalopods are similar in composition and distinctly different from those of gastropods. The soluble matrices from bivalves and cephalopods are 40 to 80% protein. From 60 to 70% of the amino acids are aspartic acid, serine, and glycine. Aspartic acid amounts to about 30% of the total (5,26). The protein core may be a polymer of $(Asp\text{-}Y)_n$, where Y is predominately glycine or serine (26). Aspartic acid, however, appears to be present in the native protein as its amide, asparagine (6,32). Therefore, the polymer may be $(Asn\text{-}Y)_n$.

The soluble matrix from the clam, Mercenaria mercenaria, (aragonite) consists of one primary component that is a highly sulfated glycoprotein of 160,000 daltons (5). The soluble matrix from the oyster, Crassostrea virginica, (calcite) appears to be similar with a molecular weight of 170,000 daltons (29). These molecular weights are approximate because the glycoprotein is fibrillar. The carbohydrate content also prevents their staining in electrophoretic gels with Coomassie blue, the stain most commonly used to detect proteins. The soluble fraction may also contain a number of smaller molecular weight components collectively representing up to 10% of the total protein (27).

The soluble matrix from Mercenaria mercenaria and Crassostrea virginica selectively bind calcium (5,29). Binding studies on the protein of M. mercenaria by flow rate dialysis indicate that this material has 0.65 µmoles of high affinity calcium binding sites per mg protein with a dissociation constant of 1×10^{-5} M and 1.3 µmoles of lower affinity sites per mg protein with a dissociation constant of 3×10^{-4} M. The binding parameters were not significantly different over a pH range of 3.2 to 7.4 (Crenshaw, unpublished). These values are remarkably similar to those of the soluble, calcium-binding polysaccharide isolated from coccoliths (11). The soluble matrix of C. virginica has a much higher binding capacity of 20-25 μmoles calcium/mg protein (29).

Gastropod shell matrix tends to have a larger soluble fraction (33). This fraction contains a variety of components, none of which is larger than 50,000 daltons (14). The amino acid compositions of the components are variable, but serine, glycine, and lysine are predominant. Surprisingly, some water-soluble fractions are reported to contain isoprenoid hydrocarbons. Certain fractions of gastropod soluble matrix bind calcium (14), but it is not certain that the binding is selective because no competing cation appears to have been included in the assay medium.

The soluble matrix of nacre may be localized in the pores of the interlamellar matrix (10). When nacre was demineralized with agents that dissolve the soluble matrix, the pore patterns typical of the molluscan class were evident in the interlamellar matrix. The interlamellar matrix was solid, with no pores evident, when the demineralization was done with agents that fixed the soluble matrix as it was exposed.

Calcium-binding, sulfated, polysaccharidic sites were found in the interlamellar matrix when cephalopod nacre was demineralized with low ionic strength solutions containing a quaternary ammonium salt (8). When the quaternary ammonium salt was omitted or solutions of high ionic strength were used, the calcium-binding sites were not detected and the typical cephalopod pore pattern was seen instead. The solubility characteristics of the quaternary ammonium complex, with respect to ionic strength, led to the conclusion that the calcium-binding sites were the soluble matrix. The location of the calcium-binding sites was correlated with the site of initial crystal deposition.

A recent study of the initial deposition of mineral onto glass coverslips inserted between the mantle and shell of bivalves yielded results that are consistent with the hypothesis that the soluble matrix is at the site of initial crystal formation (21). Microprobe analyses showed that only those organic granules with a high sulfur content mineralized. Histochemical staining indicated that the specific calcifying granules contained protein, very acidic glycosaminoglycans, and ester sulfate.

Several roles for the soluble matrix have been suggested. In solution, the soluble matrix may reduce the rate or delay the onset of the precipitation of calcium carbonate from solutions sufficiently supersaturated to precipitate spontaneously (29). However, the action of the soluble matrix may be just opposite when it is fixed in its normal position in the soluble matrix on the shell surface (see below). The soluble matrix may serve as a template for calcium carbonate nucleation. It has been suggested that the aspartic acid residues bind calcium at distances corresponding to some multiple of the spacing of calcium ions in the crystal plan to provide a nucleation surface. Where the interaspartyl distances do not match the crystal, the insoluble matrix may inhibit crystal growth (28). If the native protein contains asparagine rather than aspartic acid (6,32), this mechanism is not possible.

Another proposed function envisages the soluble matrix to act, alternatively, as an inhibitor and as a nucleator in the following manner. The glycoprotein is selectively absorbed onto the growing centers of a crystal because the centers have the highest fixed concentration of free calcium, and it inhibits further growth of that crystal. The free portions of the absorbed glycoproteins selectively concentrate calcium, and secondarily, carbonate to nucleate a second crystal by the process of ionotropy (8). Consequently, the soluble matrix may have opposite functions, depending upon whether it is in solution or absorbed onto a surface.

CONCLUSION

The weight of available evidence indicates that calcium carbonate mineralization occurs on the surface of the organic matrix and that the matrix may have a dominant role in mineralization. Ionotropy is proposed as the nucleation mechanism. Postnucleation crystal growth may be controlled by the matrix and the bathing fluid.

The following description of the ionotropic hypothesis is made in terms of molluscan nacre, although it is generally applicable. The insoluble matrix has hydrophobic surfaces exposed to the aqueous environment. These surfaces cannot nucleate mineral, but they may exert some control over postnucleation growth. The soluble, calcium-binding matrix is fitted into the pores of the soluble matrix. The total assembled matrix presents a hydrophobic field interrupted by specifically placed hydrophilic, calcium-binding sites. These sites bind calcium and concentrate additional calcium from the parent fluid by the Donnan effect (2). A secondary layer, specifically enriched in carbonate, is formed around the calcium-enriched sites. A local concentration change, such as an increase in pH or a decrease in carbon dioxide, occurs, and a crystal seed is formed at the site. Some of the seeds grow and others dissolve. The survival of a seed may be determined by local transport processes or by the original, local concentration of calcium-binding matrix. The initial, rapid growth of the crystal is along the vertical c-axis.

Growth along the c-axis is stopped by the next episodic deposition of a complete matrix assembly. The calcium-binding sites of this matrix assembly register on the growth centers of the crystal, such as surface or screw dislocations, because these growth centers have the highest localized concentration of free and loosely-bound calcium. The horizontal growth of the crystal continues, albeit at a slow rate, until the space between adjacent growing crystals is filled.

The specific concentration of carbonate in the secondary layer around the calcium-binding sites was predicted from calculations using the EQUIL and SEAWAT computerized chemical models of seawater (16). In seawater, calcium forms complexes, primarily with sulfate, bicarbonate, and carbonate. The equilibrium constants for the formation of each of these complexes were calculated from the molal concentrations and activity coefficients of each ion and each complex predicted by the chemical models. The equilibrium constant for the formation of the calcium-carbonate complex was one to two orders of magnitude greater than for either of the other two complexes. These data cannot be applied directly to the molluscan system because the extrapallial fluid has a composition different from seawater and the data are equilibrium data rather than kinetic data. However, the equilibrium constants of formation do indicate the relative stability of the complexes.

The ionotropic nucleation hypothesis proposed here is tentative for it is based on some speculation, as well as data. However, this hypothesis does focus attention to some of the important questions about the mechanisms of calcium carbonate mineralization.

Acknowledgement. We thank N. Watabe and S. Weiner for providing unpublished manuscripts, K. Wilbur for reviewing the manuscript, and Y. Kitano for his critical review. Supported, in part, by NIH Grant DE 02668.

REFERENCES

(1) Abell, A.K.; Crenshaw, M.A.; and Turner, D.T. 1981. Limiting hardness of polymer-ceramic composites. In Biomedical and Dental Applications of Polymers, ed. F.F. Koblitz, pp. 347-355. Oxford: Pergamon Press.

(2) Adamson, A.W. 1976. Physical Chemistry of Surfaces. New York: John Wiley and Sons.

(3) Beedham, G.E. 1954. Properties of the non-calcareous material in the shell of Anodonta cygnea. Nature (London) 174: 750.

(4) Crenshaw, M.A. 1972. The inorganic composition of molluscan extrapallial fluid. Biol. Bull. 143: 506-512.

(5) Crenshaw, M.A. 1972. The soluble matrix from Mercenaria mercenaria shell. Biomin. Res. Rep. 6: 6-11.

(6) Crenshaw, M.A. 1980. Mechanisms of shell formation and dissolution. In Skeletal Growth of Aquatic Organisms, eds. D.C. Rhoads and R.A. Lutz, pp. 115-132. New York: Plenum Press.

(7) Crenshaw, M.A., and Heely, J.D. 1967. Sudanophilia at sites of mineralization in molluscs. J. Dent. Res. 49B: 65.

(8) Crenshaw, M.A., and Ristedt, H. 1976. The histochemical localization of reactive groups in the septal nacre from Nautilus pompilius L. In Mechanisms of Mineralization in the Invertebrates and Plants, eds. N. Watabe and K.M. Wilbur, pp. 335-367. Columbia: University of South Carolina Press.

(9) Grégoire, C. 1972. Structure of the molluscan shell. In Chemical Zoology, eds. M. Florkin and B. Scheer, vol. 7, pp. 45-102. New York: Academic Press.

(10) Iwata, K. 1975. Ultrastructure of the conchiolin matrices in molluscan nacreous layer. J. Fac. Sci. Hokkaido Univ. Ser. 4 17: 173-229.

(11) de Jong, L.W.; Dam, W.; Westbroek, P.; and Crenshaw, M.A. 1976. Aspects of calcification in Emiliania huxley. In Mechanisms of Mineralization in the Invertebrates and Plants, eds. N. Watabe and K.M. Wilbur, pp. 135-153. Columbia: University of South Carolina Press.

(12) Kitano, Y.; Kanamori, N.; and Yoshioka, S. 1976. Influence of chemical on the crystal type of calcium carbonate. In Mechanisms of Mineralization in the Invertebrates and Plants, eds. N. Watabe and K.M. Wilbur, pp. 191-202. Columbia: University of South Carolina Press.

(13) Kitano, Y.; Kanamori, N.; and Yoshioka, S. 1980. Aragonite to calcite transformation in corals in aquatic environment. In The Mechanisms of Biomineralization in Animals and Plants, eds. M. Omori and N. Watabe, pp. 269-278. Tokyo: Tokai University Press.

(14) Krampitz, G.; Engels, J.; and Cazaux, C. 1976. Biochemical studies on water-soluble proteins and related compounds of gastropod shells. In Mechanisms of Mineralization in the Invertebrates and Plants, eds. N. Watabe and K.M. Wilbur, pp. 155-193. Columbia: University of South Carolina Press.

(15) Nakahara, H. 1979. An electron microscope study of the growing surface of nacre in two gastropod species, Turbo cornutus and Tegula pfeifferi. Venus (Jap. J. Malac.) 38: 205-211.

(16) Nordstrom, K.K.; Plummer, L.N.; Wigley, T.M.L.; Wolery, T.J.; Ball, J.W.; Jenne, E.A.; Bassett, R.L.; Crerar, D.A.; Florence, T.M.; Fritz, B.; Hoffman, M.; Holdern, G.R.; Lafon, G.M.; Mattigod, S.V.; McDuff, R.E.; Morel, F.; Reddy, M.M.; Sposito, G.; and Thrailkill, J. 1979. A comparison of computerized chemical models for equilibrium calculations in aqueous systems. In Chemical Modeling in Aqueous Solutions, ed. E.A. Jenne, pp. 857-892. Washington: American Chemical Society.

(17) Plummer, L.N.; Parkhurst, D.L.; and Wigley, T.M.L. 1979. Critical review of the kinetics of calcite dissolution and precipitation. In Chemical Modeling in Aqueous Solutions, ed. E.A. Jenne, pp. 537-573. Washington: American Chemical Society.

(18) Reddy, M.M. 1977. Crystallization of calcium carbonate in the presence of trace concentrations of phosphorus-containing anions. J. Cryst. Growth 41: 287-295.

(19) Richardson, C.A.; Crisp, D.J.; and Runham, N.W. 1981. Factors influencing shell deposition during a tidal cycle in the intertidal bivalve Cerastoderma edule. J. Mar. Bioil. Ass. U.K. 61: 465-476.

(20) Tanford, C. 1980. The Hydrophobic Effect: Formation of Micelles and Biological Membranes. New York: John Wiley and Sons.

(21) Wada, K. 1980. Initiation of mineralization in bivalve molluscs. In The Mechanisms of Biomineralization in Animals and Plants, eds. M. Omori and N. Watabe, pp. 79-92. Tokyo: Tokai University Press.

(22) Wada, K., and Fujinuki, T. 1976. Biomineralization in bivalve molluscs with emphasis on the chemical composition of the extrapallial fluid. In Mechanisms of Mineralization in the Invertebrates and Plants, eds. N. Watabe and K.M. Wilbur, pp. 175-190. Columbia: University of South Carolina Press.

(23) Waller, T.R. 1980. Scanning electron microscopy of shell and mantle in the order Arcoida (Mollusca: Bivalvia). Smithsonian Contributions to Zoology (Washington) 313.

(24) Watabe, N. 1981. Crystal growth of calcium carbonate in invertebrates. In Progress in Crystal Growth and Characterization, ed. B. Ramplin, vol. 4. Oxford: Pergamon Press.

(25) Watabe, N., and Erben, H.K. 1974. Crystal formation and growth in bivalve nacre. Nature (London) 248: 128-130.

(26) Weiner, S., and Hood, L. 1975. Soluble protein of the organic matrix of mollusk shells: A potential template for shell formation. Science 190: 987-989.

(27) Weiner, S.; Lowenstam, H.A.; and Hood, L. 1977. Discrete molecular weight components of the organic matrices of mollusc shells. J. Exp. Mar. Biol. Ecol. 30: 45-51.

(28) Weiner, S., and Traub, W. 1981. Organic-matrix-mineral relationships in mollusk-shell nacreous layers. In Structural Aspects of Recognition and Assembly in Biological Macromolecules, eds. M. Balaban, J. Sussman, A. Yonat, and W. Traub, pp. 467-482. Glenside: International Sciences Services.

(29) Wheeler, A.P.; George, J.W.; and Evans, C.A. 1981. Control of calcium carbonate nucleation and crystal growth by soluble matrix of oyster shell. Science 212: 1397-1398.

(30) Wijsman, T.C.M. 1975. pH fluctuations in Mytilus edulis, L. in relation to shell movements under aerobic and anaerobic conditions. In The Biochemistry, Physiology and Behavior of Marine Organisms in Relation to Their Ecology, ed. H. Barnes, pp. 139-149. Aberdeen: University of Aberdeen Press.

(31) Wilbur, K.M. 1964. Shell formation and regeneration. In Physiology of Mollusca, eds. K.M. Wilbur and C.M. Yonge, vol. 1, pp. 243-282. New York: Academic Press.

(32) Wilbur, K.M. 1976. Recent studies of invertebrate mineralization. In Mechanisms of Mineralization in the Invertebrates and Plants, eds. N. Watabe and K.M. Wilbur, pp. 79-108. Columbia: University of South Carolina Press.

(33) Wilbur, K.M., and Simkiss, K. 1968. Calcified shells. In Comprehensive Biochemistry, eds. M. Florkin and E.H. Stotz, vol. 26A, pp. 229-295. Amsterdam: Elsevier.

Biological Mineralization and Demineralization, ed. G.H. Nancollas, pp. 259-270.
Dahlem Konferenzen 1982. Berlin, Heidelberg, New York: Springer-Verlag.

Abnormal Mineralization

L. H. Smith
Mayo Clinic and Foundation
Rochester MN 55905, USA

Abstract. Abnormal mineralization as it occurs in some of its forms in nature is reviewed. Where possible, etiologic factors are identified, and the mechanisms that produce the abnormalities are defined.

In this brief review of abnormal mineralization, one can only identify some of the examples present in nature along with general mechanisms that may be responsible for the crystal formation. In many situations that involve abnormal mineralization, knowledge of the etiology and mechanisms is incomplete, and at times, speculation will be used in the presentation in order to precipitate discussion. Abnormal mineralization includes a spectrum of disorders from a by-product of normal healing to a primary disease process involving cell death and tissue destruction.

Mineralization, as it occurs in the living organism, is composed of at least two parts: the precipitating crystalline phase(s) and a supporting structure usually termed matrix. This is true whether the mineralization is normal as in shells, bones, and teeth, or abnormal as with stones. Identifiable similarities are present in the setting of normal and abnormal mineralization. In addition, there may be features of certain types of abnormal mineralization that set it apart.

Considering that biological fluids are supersaturated for specific crystal systems from time to time, it is surprising that abnormal mineralization is not more common in living creatures.

Etiologic factors that may influence the occurrence of abnormal mineralization include the state of saturation; heterogenous nucleation, including epitaxy, inhibitors of crystal growth, and aggregation; and matrix that may be involved in the promotion of crystal formation (2). As will be discussed, these categories may not be mutually exclusive. Other questions that may be important in specific types of abnormal mineralization include the mechanisms of initial crystal retention, the effect of crystals on tissue, crystal habit, and phase transformation. A detailed review of these etiologic factors is presented by Finlayson (this volume).

Supersaturation of the precipitating phase must always be present, although its presence by no means insures that abnormal mineralization will occur. In complex biological solutions it is dependent not only on the concentration of the precipitating ions, but also on the ionic strength, complexation, and pH of the solution as they affect the free ion activity. The trigger mechanism that induces crystal nucleation in a supersaturated environment is poorly understood in most types of abnormal mineralization, although some form of heterogenous nucleation seems most likely in biological systems. In body fluids such as saliva or urine, the presence of supersaturation and even crystals does not insure or necessarily predict that abnormal mineralization will occur (5,21).

Inhibitors of crystal growth and/or aggregation have been described for both calcium phosphate and calcium oxalate crystal systems (4,15-17). Some of these inhibitors are known to be present in biological fluids and cells. The site of their action as inhibitors is thought to be at the crystal surface, perhaps at specific growth sites (14). These substances typically exert their effect at micromolar concentrations. Once adsorbed to the crystal surface they may alter, or even inhibit, crystal dissolution. Listed in Table 1 are inhibitors that have been identified as being present in biological organisms.

TABLE 1 - Inhibitors of Crystal Growth and/or Aggregation

	Source
I. Calcium Phosphate Inhibitors	
Statherin	Saliva
Proline-rich proteins	
PRP1	Saliva
PRP3	Saliva
Magnesium	Cells & fluids
Citrate	Cells & fluids
Pyrophosphate	Cells & fluids
*Phosphocitrate	?
*Citrate-metal complexes	?
ATP	Cells
II. Calcium Oxalate Inhibitors	
Citrate (minor inhibitor)	Urine
Pyrophosphate	Urine
Glycosaminoglycans	Urine
RNA-like fragments	Urine
Acidic glycopeptide	Urine
*Phosphocitrate	?
*Citrate-metal complexes	?

*Not demonstrated to be present in biological systems yet.

A decreased concentration or inactivation of these inhibitors theoretically could permit and even promote abnormal mineralization. Certainly, at least one suggested mechanism for normal bone mineralization involves enzymatic destruction of pyrophosphate by pyrophosphatase or ATP by ATPase with release of calcium and phosphate from the mitochondria (1,4,7,13). Yet, where inhibitor deficiencies have been sought, primarily in urine and saliva, the reports have varied. This may in part be due to differences and/or deficiencies in methods whereby inhibition or specific inhibitors are measured.

Important considerations such as the pK's of the inhibitor and the pH of the biological solution have often been ignored (3). Intact biological fluids are difficult to work with since they may contain a gross excess of inhibitors. On the other hand, diluting biological fluids may distort the interrelationships between specific inhibitors, the solution, and the crystal. Finally, at least in the situation of dental calculus formation, normal amounts of inhibitors may be present initially, but then, these are at least in part destroyed by the oral bacterial flora allowing the calculus formation.

Matrix is always present in abnormal mineralization, yet its specific role is not known. It could act as a nucleator of crystal formation, a promoter of crystal aggregation and growth, a skeleton for crystal organization, or simply as a passive substance adsorbed to the area of abnormal mineralization. Work to date would suggest that matrix is an integral part of the process of abnormal mineralization, being involved at the earliest stages (2). At least in urine it is suggested that matrix exists in a soluble form of relatively small size. In the setting of abnormal mineralization it condenses or polymerizes by some mechanism to form the large insoluble form of matrix. This condensation may be stimulated by the presence of crystals or may precede the crystal formation.

As already mentioned, heterogenous nucleation best fits most situations of abnormal biological mineralization. Perhaps some stage or form of matrix is involved in this process. It has long been known that certain types of stone matrix tightly bind calcium (2). More recently a calcium binding amino acid, γ-carboxy glutamic acid, has been found in stone matrix (8). This, or other similar compounds, may be involved in the early stages of crystal formation. A special kind of heterogenous nucleation called epitaxy has been suggested as a possible means of crystal nucleation (9,11). This would infer that a crystal of one type could induce the nucleation of a crystal of another type if the states of saturation were favorable. Mixed crystalline composition that is so commonly found in urinary stones might support the presence of such a mechanism, and theoretical and in vitro studies indicate that there are favorable epitaxial relationships between at least some of the crystal forms found in urine and urinary calculi. In spite of the apparent logic of such a mechanism, it has not been possible to demonstrate that epitaxial nucleation has a role in the formation of urinary calculi.

Listed in Table 2 is a working outline of abnormal mineralization. From this list several specific mechanisms involved in the abnormal crystal formation can be defined, recognizing that detailed knowledge in regard to the total mechanism are often missing or poorly understood. It is the purpose of the remainder of this review to identify some of these mechanisms with illustrations of how they may apply to specific types of abnormal mineralization.

TABLE 2 - Abnormal Mineralization

I. Abnormal tissue mineralization
   A. Tissue healing following cell death and necrosis
   B. Ectopic (metastatic) mineralization
   C. Crystal membranolysis

II. Calculus formation
   A. Sites
      1. Oral
         a. Dental
         b. Salivary glands - sialolithiasis
      2. Gastrointestinal
         a. Gallbladder - cholelithiasis
         b. Pancreas
      3. Urinary
         a. Kidneys and bladder - urolithiasis
         b. Prostate
   B. Etiologic factors
      1. Supersaturation
      2. Inhibitor deficiency
      3. Matrix
      4. Crystal retention

Abnormal mineralization within tissues may be divided into at least three types. The first type, normal tissue healing, may be associated with calcification when cell death with necrosis has occurred. At least two mechanisms may be involved in the mineralization process. First, the integrity of the cell membrane is destroyed, allowing calcium to enter the cells from the surrounding fluids. Second, with cell death there is a release of lysosomal enzymes including pyrophosphatases and ATPases that destroy compounds such as pyrophosphate and ATP that may be acting as surface inhibitors to "package" calcium and phosphate within the mitochondria (1,4). Both situations allow gross supersaturation of the injured area in terms of calcium phosphate with subsequent calcification. Other factors that may also be involved in this process include heterogenous nucleation of the calcium phosphate by cell debris and local changes in the pH and ionic strength within the dead tissue. This process may occur in living organisms wherever there is localized cell death with necrosis surrounded by living tissue. It occurs locally without systemic hypercalcemia and may be seen in any tissue within the body.

Ectopic mineralization is most often due to excessive supersaturation for the precipitating crystalline phase. It may occur locally as with nephrocalcinosis in the kidney due to hypercalcemia or hyperoxaluria, or in the condition termed tumoral calcinosis with calcium phosphate deposits around joints. It may be a generalized process with crystal deposits throughout the organism as seen in oxalosis and extreme hypercalcemia, hyperphosphatemia, or both. Although other etiologic factors may be involved, the principle disturbance is supersaturation of the precipitating crystalline phase.

The process of crystal membranolysis is thought to be fundamental in the various types of crystal deposit diseases such as gout, pseudogout, and asbestosis. Calcium pyrophosphate dihydrate (CPPD) crystal deposition disease (pseudogout) is representative of this group of diseases (12). CPPD crystals are normally present as perilacunar deposits and tophus-like masses in cartilage and synovium. Crystal shedding into the joint fluid may occur with changes in concentration of the crystal ions, trauma, enzyme activity, infection, or perhaps even with metabolically-induced changes in the ground substance of cartilage. Once the CPPD crystal is in the synovial space, it is probably coated with IgG globulin which promotes phagocytosis by surrounding inflammatory cells (polymorphonuclear leukocytes and monocytes). After phagocytosis of the CPPD crystals, there is release of a low molecular weight chemotactic factor and lysosomal breakdown with enzyme release, cell destruction, and release of crystals back into the synovial space. All crystals do not have the capacity to induce membranolysis. For instance, in the inherited disorder, cystinosis, cystine crystals that are deposited ectopically do not induce an inflammatory reaction. Studies of single crystal structure suggest that the irregular surface that is characteristic of crystals that induce membranolysis is necessary for the crystals to bind to lysosomal membranes and induce membranolysis and the inflammatory process characteristic of crystal deposition diseases (10).

The formation of crystals that may aggregate and grow to produce calculus may occur in biological fluids that are supersaturated at least intermittently for the precipitating crystalline phase(s). Examples of calculus formation are given in Table 3. In the case of calculus formation within the urinary tract where the crystalline component of the calculus may be composed of different crystal types, it is important to indicate the relative frequency that specific crystal types may occur. Listed in Table 4 is the frequency distribution as a percentage of crystal types which occur as major crystalline components in urinary calculi (19).

Although Table 4 suggests that stones formed within the urinary tract are commonly of a single crystal composition, this is rarely the case. Most urinary calculi have a major crystal form but also other minor crystal habits. Also, matrix is always present.

TABLE 3 - Calculus Formation

Site	Crystal Types
A. Oral	
Dental	Calcium phosphate
Salivary glands	Calcium phosphate
B. Gastrointestinal	
Pancreas	Calcium carbonate
Gallbladder	Calcium bilirubinate
	Calcium apatite
	Cholesterol
C. Urinary	
Prostate	Calcium phosphate
Kidneys, ureters, bladder	Calcium oxalate
	Monohydrate (Whewellite)
	Dihydrate (Weddellite)
	Calcium phosphate
	Hydroxyapatite
	Calcium hydrogen phosphate dihydrate (Brushite)
	Carbonate apatite
	Tricalcium phosphate (Whitlockite)
	Magnesium ammonium phosphate
	Hexahydrate (struvite)
	Uric acid
	Sodium acid urate
	Ammonium acid urate
	Cystine
	Xanthine
	2,8-hydroxyadenine
	Drugs and drug metabolites

TABLE 4 - Composition of Urinary Calculi - Major Crystalline Component

	Percent
Calcium oxalate	58.8
Mixed calcium oxalate - calcium phosphate	11.4
Calcium phosphate	8.9
Struvite	9.3
Uric acid/urates	10.1
Cystine	0.7
Others	0.8

Saliva and urine have been studied most extensively in terms of ion composition, pH, ionic strength, complexation, and the states of saturation for pertinent crystal habits (5,18). From both of these biological fluids, normally occurring compounds that inhibit the growth and aggregation of specific crystal systems have been isolated (see Table 1) (4,15,17). Similar compounds may also be present at the sites of other calculus formation, but this has not been examined adequately, primarily due to the difficulty of obtaining fluid from sites such as the gallbladder, pancreas, and prostate. The mechanisms of calculus formation within the urinary tract will be reviewed in some detail as an example of calculus formation. This is not to infer increased importance of this type of calculus formation, but simply that more information is available in part due to the accessibility of the media (urine) and the product (stones).

The first element of calculus formation that must always be present, at least intermittently, is supersaturation for the precipitating crystal phase(s). At the distal nephron and collecting tubule, urine is concentrated, often becoming supersaturated, even in normal subjects. Crystals may form but are usually freely excreted in the urine without difficulty. The crystal type that forms may vary from time to time through the day and is due, at least in part, to diet and fluid intake. Crystalluria only becomes a problem if the crystals are retained and grow, so that an important question in regard to calculus formation is the mechanism and site of crystal retention. If one considers the minute size of crystals (<5 microns) in relation to the relatively large size of the conduit (>100 microns) and the rapid flow of urine from glomerulus to bladder (<5 minutes), some mechanisms for retention are needed. Two alternate theories have been suggested, although they are not mutually exclusive. The first would suggest that gross supersaturation exceeding the formation product for the precipitating phase occurs with massive precipitation and sludging of the involved passages (20). Both aggregation and crystal growth occur to further increase the crystalline mass. This is followed by a metastable state for the same crystalline phase. This has been termed the free particle theory of calculus formation. The second theory, the fixed particle hypothesis, would suggest that crystals are bound to membranes of the collecting system and possibly even adsorbed into cells, resulting in crystal retention, aggregation, and growth (6). This mechanism could involve a similar set of events that are thought to occur with crystal membranolysis as described in crystal deposit diseases. Certainly in some animal models that are designed to produce crystal formation within the urinary tract, cell death may occur where crystals are present within the nephron of the kidney. Other factors that may be involved in crystal retention are matrix and inhibitors for crystal formation. On the other hand, matrix could promote crystal aggregation and even heterogenous nucleation, and it could be involved in the attachment of crystal aggregates (microstones) to the cellular membranes of the collecting system. Inhibitors could act in a beneficial manner to prevent crystal growth and aggregation, thus preventing accumulation of crystal mass.

Physiologic and metabolic factors can affect at least the states of saturation for potential crystal types, and they may also influence inhibitors and matrix both qualitatively and quantitatively. These factors, along with diet and normal body metabolism, determine urine volume, pH, ionic strength, and solute load. They also influence the major ionic species present in urine that are important in ion complexation.

In summary, an attempt has been made to identify major forms of abnormal mineralization that may occur. Some specific mechanisms responsible for this abnormal mineralization have been reviewed with the hope that this may provide a basis for in-depth review and discussion of the multiple forms of abnormal mineralization.

Acknowledgement. This investigation was supported in part by research grant AM 20605 from the National Institutes of Health, Public Health Service.

REFERENCES

(1) Betts, F.; Blumenthal, N.C.; and Posner, A.S. 1981. Bone mineralization. J. Cryst. Growth (Amsterdam) 53: 63-73.

(2) Boyce, W.H. 1968. Organic matrix of human urinary concretions. Am. J. Med. (New York) 45: 673-683.

(3) Curreri, P.A.; Onoda, G.; and Finlayson, B. 1981. A comparative appraisal of adsorption of citrate on whewellite seed crystals. J. Cryst. Growth (Amsterdam) 53: 209-214.

(4) Fleisch, H. 1981. Inhibitors of calcium phosphate precipitation and their role in biological mineralization. J. Cryst. Growth (Amsterdam) 53: 120-134.

(5) Grøn, P. 1973. Saturation of human saliva with calcium phosphate. Arch. Oral Biol. (Elmsford, NY) 18: 1385-1392.

(6) Khan, S.R.; Finlayson, B.; and Hackett, R.L. 1979. Scanning in electron microscopy of calcium oxalate crystal formation in experimental nephrolithiasis. Lab. Invest. (Baltimore) 41: 504-510.

(7) Lehninger, A.L. 1970. Mitochondria and calcium ion transport. Biochem. J. (Colchester, England) 119: 129-138.

(8) Lian, J.B.; Prien, E.L., Jr.; Glimcher, M.J.; and Gallup, P.M. 1977. The presence of protein-bound γ-carboxyglutamic acid in calcium-containing renal stones. J. Clin. Invest. (New York) 59: 1151-1157.

(9) Lonsdale, K. 1968. The solid state. Epitaxy as a growth factor in urinary calculi and gallstones. Nature Basingstoke, Hants, England) 217: 56-58.

(10) Mandel, N.S. 1976. The structural basis of crystal-induced membranolysis. Arthrit. Rheum. (New York) 19: 439-445.

(11) Mandel, N.S., and Mandel, G.S. 1981. Epitaxis between stone-forming crystals. In Urolithiasis: Clinical and Basic Research, eds. L.H. Smith, W.G. Robertson, and B. Finlayson, pp. 469-480. New York: Plenum Press.

(12) McCarty, D.J. 1976. Calcium pyrophosphate dihydrate crystal deposition disease - 1975. Arthrit. Rheum. (New York) 19: 275-285.

(13) Meyer, J.L.; McCall, J.T.; and Smith, L.H. 1974. Inhibition of calcium phosphate crystallization by nucleoside phosphate. Calc. Tiss. Res. (New York) 15: 287-293.

(14) Nancollas, G.H. 1976. The kinetics of crystal growth and renal stone formation. In Urolithiasis Research, eds. H. Fleisch, W.G. Robertson, L.H. Smith, and W. Vahlensieck, pp. 5-23. London: Plenum Press.

(15) Oppenheim, F.G.; Hay, D.I.; and Franzblou, C. 1971. Proline-rich proteins from human parotid saliva. I. Isolation and partial characterization. Biochem. (Columbus) 10: 4233-4238.

(16) Robertson, W.G.; Scurr, D.S.; and Bridge, C.M. 1981. Factors influencing the crystallization of calcium oxalate in urine - critique. J. Cryst. Growth (Amsterdam) 53: 182-194.

(17) Schlesinger, D.H., and Hay, D.I. 1977. Complete covalent structure of statherin, a tyrosine-rich acidic peptide which inhibits calcium phosphate precipitation from human parotid saliva. J. Biol. Chem. (Baltimore) 252: 1689-1695.

(18) Smith, L.H. 1976. Application of physical, chemical, and metabolic factors to the management of urolithiasis. In Urolithiasis Research, eds. H. Fleisch, W.G. Robertson, L.H. Smith, and W. Vahlensieck, pp. 199-211. London: Plenum Press.

(19) Smith, L.H. 1979. Urolithiasis. In Strauss and Welt's Diseases of the Kidney, 3rd ed., eds. C.W. Gottschalk and L.E. Earley, pp. 893-932. Boston: Little, Brown and Co.

(20) Vermeulen, C.W., and Lyon, E.S. 1968. Mechanisms of genesis and growth of calculi. Am. J. Med. (New York) 45: 684-692.

(21) Werness, P.G.; Bergert, J.H.; and Smith, L.H. 1981. Crystalluria. J. Cryst. Growth (Amsterdam) 53: 166-181.

Biological Mineralization and Demineralization, ed. G.H. Nancollas, pp. 271-285.
Dahlem Konferenzen 1982. Berlin, Heidelberg, New York: Springer-Verlag.

Pathologic Mineralization, Nucleation, Growth, and Retention

B. Finlayson
Division of Urology, Dept. of Surgery, University of Florida
Gainesville, FL 32607, USA

Abstract. Many different materials have been identified as being involved in mineralizations in mammals. Although solubility properties of these materials are known, their kinetics of formation in vivo are largely unknown. A general paradigm for the study of in vivo mineralization has yet to be developed. The whole subject area is rife with speculation and half-baked ideas, some of which are my own. Some kinetic and mechanistic data about urinary stones are discussed.

INTRODUCTION

A variety of materials precipitate in mammals. Some of these precipitations may have catastrophic effects. While many details about the various precipitations are unknown, it is useful to make some generalizations. Questions about abnormal precipitations in mammals divide into two categories: (a) How does it happen (mechanism)? and (b) What effect does it have on the host (pathology)? We will deal as best we can with mechanisms.

GENERAL COMMENTS ABOUT MECHANISM

In most cases, mass transport mechanisms are poorly understood. Precipitates in animals are either luminal (joint spaces or organ ducts), or nonluminal (in cells or between cells). Mineralizations have been reported in every tissue and in every lumen. The position of a nonluminal precipitate remains fixed unless the host or neighboring cell(s) move it (e.g., slough). Thus mass accretion and heat dissipation of nonluminal precipitates occur by diffusion along a path well beyond the hydrodynamic shear plane. In the case of extracellular nonluminal precipitates, at some distance from the precipitate, arterial, venous, and lymphatic convective flows provide a more or less steady state boundary condition. Nonluminal precipitates grow within a matrix. Although coalescence of neighboring precipitates can occur, mass acquisition by aggregation is not possible. Nonluminal precipitates lose mass only by dissolution, by scavenging, by macrophages, and by extrusion into a lumen. In the case of luminal precipitates, because of normally occuring shear fields, mass and heat transport may be convective up to the hydrodynamic shear plane. The shear field also increases the chance for aggregation, and convective flow gives the possibility of the solids being swept away.

Often a source of the supersaturation can be identified; e.g., the calcium pump in mitochondria, or water absorption by the renal tubule, gut, or gall bladder. Sodium acid urate urinary stones in the Dalmation coach hound are due to a sink failure (readsorption of uric acid by the renal tubule). Another example of sink failure is calcium phosphate stones in hyperparathyroidism (the renal tubular reabsorption of calcium is overloaded). In other cases (e.g., calcified phleboliths in the true pelvis and protein crystalloids of Reinke in the testes), there is no clue as to how the concentration develops.

The mechanism of nucleation of all pathologic in vivo precipitation is obscure. A body of evidence is growing that urinary and gallstones (luminal precipitates) nucleate near or on a membrane or polymeric structure. Nucleation of all nonluminal precipitates probably occurs at a catalytic interface. Most biological fluids contain so much submicron-sized debris, it seems unlikely that any in vivo precipitation would have homogeneous nucleation. Considerations of momentum and particle number density suggest that secondary or contact nucleation is not an important phenomenon in biology.

Mechanisms of growth of most abnormal biological precipitates are unknown. Often some relevant thermodynamic features are known, such as solubility and its temperature dependence. However, mechanism is inferred from kinetics, and there have been few studies seeking the time dependence of the course of mineralization in vivo. This failure, in part, is due to the difficulty of quantifying a chondrocalcinosis, or a calcifying tendinitis, let alone obtaining an adequate description of the pertinent conditions. The failure is also in part due to a general lack of a kinetic point of view. In our laboratory, we have investigated the kinetics of brushite and struvite foreign body stone formation in rats (4) and the kinetics of calcium oxalate nephrocalcinosis and its disappearance in rats (9). Biologic variation required expenditure of many animals to gain small amounts of kinetic data.

In vitro testing of mineralization also has problems. It requires simulation of incompletely defined biological fluids. It is difficult to simulate in vitro the arterial, venous, and lymphatic contributions to mass transport. Lack of symmetry makes three-dimensional theoretical modeling of such systems formidable. A recently described system with solutions flowing past both sides of thin polymer matrix (2) seems a good experimental first approximation of the nonluminal biological conditions. The relevance of in vitro simulation of luminal precipitations depends in large part on appreciating the effects of the biologically extant shear field. Shear associated with flows in salivary ducts and inertia mixing in the gallbladder is not of the same degree or duration as shear in a well-mixed in vitro reactor. However, shear in the renal tubules is reasonably approximated by a well-mixed reactor. Considerations of shear are important, because mixing caused by shear minimizes the diffusion path for mass transport, and shear permits orthokinetic aggregation.

Well conducted in vitro studies provide upper limits on in vivo growth and aggregation rates. Crystal growth shown to be rate-limited by surface reactions in well-mixed systems is likely to experience a significant diffusion impediment to growth in many in vivo situations. It seems reasonable to expect that many in vivo nonluminal growth rates will have a concentration dependence closer to first order (diffusion controlled) than to second order (surface controlled).

The importance of aggregation depends on location. Nonluminal precipitates large enough to see by light microscopy do not experience sufficient translational motion to aggregate. Fusion of these particles can occur by interparticulate mass transport. Luminal particles can aggregate by classical perikinetic or orthokinetic mechanisms or by polymer flocculation, or a particle attached to a lumen wall can act as a collector of convectively moving particles. Except for flocculation by polymers, aggregation occurs by the particles falling into a well-defined primary or secondary potential well. DLVO (Derjaguin, Landau, Verweg, and Overbeek) theory describes these interactions (1). The biological importance of flocculation by polymers is not known; however, this principle has been used to simulate calcium carbonate biliary lithiasis. With particles bigger than a few micrometers, the secondary potential may well be much deeper than 10 kT, which means that aggregation occurs even though the closest distance of approach is > 1.0 nm (1).

Extrapolating from experience with whewellite and hydroxyapatite (11), all precipitates in vivo probably have approximately a monolayer or more of polymers adsorbed to their surfaces. The composition of the adsorbed layer is moot. Nonluminal crystals will have proteins and glycosaminoglycans competing for precipitate surface. Luminal precipitates may have RNA fragments and polymers of bacterial origin as well in their adsorbed polymer layer. The competition of polymers for precipitate surface will also be perturbed by low molecular weight adsorbents, such as citrate and pyrophosphate. The quantitative features and consequences of competitive adsorption by in vivo precipitates have not been studied in detail. Work with urinary stone crystals suggests that many adsorbates (e.g., citrate, pyrophosphate, glycosaminoglycans, RNA, and acid polypeptides) can be expected to modify rates of crystal growth and aggregation and possibly nucleation. Questions about heterogeneity of adsorption sites and the relation between adsorption site and growth sites have not been studied in detail.

MECHANISMS IN URINARY STONES

The most detailed available information about the mechanisms and kinetics of pathologic crystallization deals with urolithiasis. Of the five common kinds of stones, calcium oxalate, calcium phosphate, struvite, uric acid, and cystine, the most is known about calcium oxalate. The following is a précis of kinetic and mechanistic information about stone formation, especially calcium oxalate urinary stones.

A conventional discussion of the mechanisms and kinetics of the formation of an object presupposes detailed knowledge of its structure. It may seem strange, but to this day, no one can say for certain what holds ordinary urinary stones together. One view is that the crystals are held together by matrix. Another is that the crystals are held together by primary crystalline unions. Another is that matrix holds the crystals together until crystalline unions develop. Measurements have not been made of factors such as density, internal surface area, porosity, connectedness, diffusion transport, and crystal size distribution. Without basic measurements like these, it is difficult to make an informed approach to the problem. Nevertheless, we will try. The principal component in urinary stones may be any one of at least 20 different substances. The list is so varied that mechanisms of urinary stone formation must be quite general. We arbitrarily define a urinary stone to be a precipitate in the lumen of the urinary passages.

We define stone disease to be the result of a urinary stone passing through the urinary system at a rate slower than the convective rate of urine through the same passage. It is the possibility of disease that evokes our interest. We all occasionally make stones in the form of crystalluria, but only about 10% of Westernized populations experience stone disease.

The cause of urinary stones is urinary supersaturation. Correcting the supersaturation stops the stone formation. Normal urine is periodically undersaturated with respect to the stone material, with the exception of calcium oxalate. Because of intermittent ingestion of food and fluid, calcium phosphate and uric acid concentrations in normal urine may fluctuate between undersaturation and supersaturation several times a day, but the periods of undersaturation are sufficiently prolonged to remove or dissolve any particles retained in the urinary tract. With calcium oxalate, normal urine is almost constantly supersaturated with superimposed postprandial fluctuations of supersaturation above and below the formation concentration. Reports from two laboratories have shown that examining urine for calcium oxalate supersaturation and crystallization inhibitor activity provides a means for predicting with ~ 90% confidence who is a calcium oxalate stone-former (13,14). This implies that in the case of calcium oxalate stone disease, kinetics is as important as thermodynamics. There are four kinetic factors - nucleation, crystal growth, aggregation, and particle retention. The first three considerations, and possibly the fourth, are modulated by the inhibitor activity of urine.

Retention

Retention is an important consideration, because order of magnitude calculations indicate that without retention, stones would be washed out before becoming large enough to cause disease. Particle retention is a reflection of precipitate-host interaction and has not yet been quantified. It is likely that in Robertson's scattergraph of urinary inhibitor activity and calcium oxalate concentration (14), the position of the discriminating zone is a function of particle retention. For a given urinary inhibitor activity and calcium oxalate concentration, the retentive urinary tract would favor developing stone disease. Our observations on calcium oxalate precipitation in rat kidneys show that the calcium oxalate particles stick to the walls of the renal tubules and are expelled from the kidney in first order kinetics (halftime = 28 hr) (9). In principle, a test could be devised to determine whether the urinary tracts of persons with urinary stone disease are more retentive than normal. Acid mucopolysaccharides (a major class of urinary stone inhibitors) have been shown to reduce the particle retentiveness of rat urothelium.

Nucleation

Plots of $Ln(N)$ vs $(Ln(RS))^{-2}$ and $\bar{V}$ vs RS, in which N is the concentration of precipitate particles, RS is relative supersaturation, and $\bar{V}$ is mean precipitate particle volume, suggest that the homogeneous nucleation of calcium oxalate does not predominate until RS > 70 (3). Since the calcium oxalate RS of mammalian voided urine does not exceed 30, all nucleation in urine is probably heterogeneous. Calcium phosphate has been suggested as the heterogeneous substrate. Lattice matches are favorable. It is ubiquitous in urine, and calcium phosphate is often found at the center of calcium oxalate stones. Also, seeding experiments indicate that hydroxyapatite is a suitable substrate for whewellite growth (10). The initial nucleating phase of calcium oxalate in urine is moot. Calcium oxalate trihydrate (COT) has been suggested as the initial phase. The transformation halftime for COT to whewellite (COM) (mediated by dissolution) is ~ 4.5 h at 37°C. Calcium oxalate crystalluria generally occurs within 3 h postprandial, and COT rarely occurs in these crystallurias. Either COT is uncommon in urine, or urine contains unappreciated accelerators of the transformation. There is also no morphologic evidence in calcium oxalate stones of COT capture by a urolith. In our laboratory, we have been unable to identify

COT with X-ray diffraction in fresh precipitates formed with various calcium and oxalate concentrations. Frey-Wyssling, one of the early describers of COT, now considers the "COT" to be weddellite with zeolitic water. If COT is excluded from consideration in urinary lithiasis, the mechanistic picture is simplified.

Gill has presented evidence that the 30 min calcium oxalate formation product is 1.25 times greater in the rat urinary bladder than in a glass vessel (7). He suggests that nucleation in a glass vessel occurs predominantly on the glass wall. A critical test of this hypothesis would be to demonstrate a saturation type of concentration dependence of nucleation rate. The concentration dependence of total particle production suggests that nucleation in the heterogeneous region does not show a saturation type of concentration dependence. Observations on rat kidneys suggest that the anatomic site of nucleation depends on the mode of causing supersaturation. After single parenteral injections of oxalate (causing brief large concentrations in serum), nucleation appears to be confined to the lumen of the proximal tubule. Chronic ethylene glycol poisoning causes intracellular nucleation and transport of the precipitates to the tubular lumen. With continuous infusions of small amounts of oxalate, nucleation appears to occur between the distal tubular cells and their basement membranes. In all instances, the onset of calcium oxalate precipitation is accompanied by cellular necrosis. Once in the lumen, calcium oxalate precipitates are covered with a single layer osmophilic membrane easily seen by dissolving the crystals. The membrane envelope often reveals an unsuspected defective structure in the crystals.

Considerable attention has been paid to lattice matching between various urinary crystals with a view to predicting heterogeneous nucleating efficiency. A great many good matches exist, but thus far accurate predictions have been limited.

Uromucoid and insoluble stone matrix appear to be nucleation promoters. The mechanism is not known.

It is of interest to note that the time course of the initiation of growth on brushite and struvite foreign body stones in rats was exponential with a halftime of 29 days (relative supersaturation unknown) (4). Any relationship between the initial growth of foreign body stones and nucleation is arguable, but initiation requires nucleation and at least one "sticking" event on the foreign body.

Urinary Inhibitors of Crystallization

It is common to measure the effect of diluted urine (1/20th to 1/100th) on crystal growth or aggregation. Workers have observed inhibition by the diluted urine on the order of 60% by normal urine. At these dilutions, stone-forming urine usually has approximately half the inhibitory activity of normal urine. The inhibitory effect appears to have a Langmuir adsorption concentration dependence. The well recognized inhibitors are citrate, pyrophosphate, chondroitin, sulfate, ribonucleic acid, and acidic polypeptides.

Although the results of inhibitor assays of dilute urine are substantially different (~50%) between normal subjects and stone-forming patients, Langmuir adsorption isotherm theory predicts that expected differences in the inhibitory activity of whole urine are much smaller (~ 4%). Although there are indisputable observations that crystals in normal urine are smaller and less aggregated than those in stone-forming urine, it is reasonable to question the validity of using Langmuir adsorption theory to extrapolate the behavior of whole urine, albeit the explanation for these facts may be less straightforward than has been supposed. For example, the Stern potential of the seed crystal-solution interface may be different in whole urine than in the test solution. If Langmuir adsorption theory has validity in this case, inhibition of crystal growth or aggregation is probably not the reason that the inhibitor assays are useful in distinguishing between stone-forming and non-stone-forming urine. We must seek a phenomenon involving an interface with a lower affinity for the inhibitory substances than that of mature calcium oxalate surfaces. One possible candidate for such a phenomenon is the coating of injured mucosa to reduce adhesion and retention times of particles. Gill (7) has shown this phenomenon, but has not measured its dependence on inhibitor concentration.

Another lower affinity phenomenon could have to do with an early event in calcium oxalate precipitation, possibly nucleation. The inhibitors could be poisoning the heterogeneously nucleating surfaces, or the inhibitors could have a lowered affinity for the surface of fresh calcium oxalate nuclei. These possibilities are suggested because formation product ratios of calcium oxalate in whole urine coupled with supersaturation, discriminate stone-forming urine from non-stone-forming urine with the same resolution as do the dilute urine inhibitory assays when coupled with supersaturation. Either or both kinds of low affinity phenomena could be operating.

In the active search for the significant inhibitors in urine, the general strategy has been to fractionate urine in various ways and then look at the inhibitory activity of various fractions. I suggest that the most straightforward separation strategy would be to chromatograph urine on a calcium oxalate column and examine the most retarded species for inhibitory activity. The product of retention time and area under the elution curve probably predicts the relative effectiveness of a given fraction as an inhibitor.

Calcium Oxalate Crystal Growth

The growth rates of weddellite have not been measured as a function of supersaturation. In a continuous crystallizer with an ~10 min mean retention time that receives an artificial urine-like solution, the linear growth rate is ~ 0.1 μm/min. In a seeded crystallizer at RS ≈ 18 (well into stone-former range), the linear growth rate of whewellite is ~ 0.1 µm/min. With linear growth rates of this magnitude, speculative calculations indicate that calcium oxalate particles cannot accrete mass rapidly enough by crystal growth to be retained by virtue of their size. A retention mechanism(s) was invoked. Subsequently, considerable morphological evidence has shown that particles adhere to renal tubular mucosa and injured urothelium. The speculative calculations ignored the possible effect of aggregation on the linear growth rate. Particles that have precipitated in a body lumen are exposed to hydrodynamic shear (G). We briefly explore the implications for calcium oxalate. Hydrodynamic shear can increase mass accretion rates by minimizing diffusion distances and by promoting orthokinetic aggregation. Shear in a renal tubule can be estimated as follows: in the absence of turbulence, if V is velocity, r is distance from tubule center, and r_o is tubule inner radius, then V is approximated with (6):

$$V = V_{max}(r_o{}^n - r^n)/r^n,$$

i.e., in the limit as $n \rightarrow \infty$, the flow changes from laminar to plug flow. The assumption of no turbulence is plausible since estimated Reynold's numbers for flow in the renal tubules are < 0.3 (5). The average absolute value of shear $\langle G\rangle$ is calculated as:

$$\langle G\rangle = V_{av}(n+2)/nr_o,$$

with an average velocity, V_{av}, of 0.1 cm/sec, $\langle G\rangle$ in the distal renal tubule (r_o = 0.01 cm) varies from 20 sec^{-1} for n=2, to 10 sec^{-1} for $n=\infty$ ($G_{r=r_o}$ ranges from 40 sec^{-1} at n=2 to 120 at n=10). A 200 ml system stirred at 400 rpm with a paddle having 1 cm blades (2 cm^2 paddleface), using Ive's equation, has an estimated $\langle G\rangle$ of 20 sec^{-1}. With this shear, in vitro whewellite seeded systems are growth rate limited by surface processes and aggregation is orthokinetic. We expect similar behavior in the renal tubule.

Aggregation of whewellite is inevitable in most in vitro systems. DLVO calculations show that aggregation may occur in either a primary energy minimum (separation < 1 nm), or in a secondary energy minimum with separations of up to 16.0 nm (1). At urine-like ionic strengths, the electrical double layer is so compressed that a deep secondary energy minimum may occur not far from a primary minimum. This tendency to aggregate has given an apparent simplicity to studies of seeded crystal growth. Unpublished studies using a Coulter Counter show that the increase of surface area resulting from crystal growth is approximately balanced by the loss of surface area by aggregation. Apparently, constant surface area has simplified theoretical modeling with the various growth models. The apparent average net rate of linear growth (aggregation + crystal growth) measured with a Coulter Counter ranged from 0.6 μm/min at RS = 11 to 0.1 μm/min at RS ≈ 4. With no logical reason for doing so, we note that in vivo linear growth rates for brushite and struvite foreign body stones in rats was ~ 0.1 μm/min (supersaturation unknown).

Urinary Stone Matrix

The role of urinary stone matrix in urolithiasis is still controversial. Tawashi et al. (8) reported that ethylene diamine (0.5N quanternary ammonium hydroxide in Toluene) causes calcium oxalate stone matrix to dissolve and the stone to disintegrate sufficiently for particle sizing by Coulter Counting. Tawashi carefully refrained from making inferences about matrix, but a simple explanation would be that matrix was holding the stone together. Our laboratories have been unable to confirm either feature of this report. The matrix content of stones is ordinarily 2-3% by weight, however, individual calcium oxalate stones have been reported to have as much as 10% matrix. It seems reasonable to expect the matrix content to correlate with its structural significance. Tawashi did not report the matrix content of the stones he treated. Even if successful extraction of matrix failed to cause the stone to disintegrate; matrix could have functioned as a bonding substance in the formative stages of a layer until mineral coarsening or cementing stabilized the structure.

Typical urinary stone matrix is a mixture of acidic glycoproteins with low sialic acid content and no hydroxyproline. There is some variation in the amino acid composition of matrix extracted from pools of stones of various crystal types. The significance of these differences is not known. Amino acid composition of matrix varies appreciably from stone to stone, although within a given stone, the amino acid composition is fairly uniform. Matrix is about 40% soluble in mild aqueous solutions. The insolubility of the remaining 60% is unexplained. EGTA, 7M urea, and β-mercapto ethanol will not dissolve it.

Deposition of matrix in urinary stones is, to a large extent, probably by adsorption. However, there is insufficient structural information about a stone to estimate internal surface and predict matrix content caused by monolayer adsorption. If stones had an average crystal size of 5 μm, then ≥ 40% of the matrix could be explained by monolayer adsorption (9) at ordinary urine mucoprotein concentrations.

One gets the impression that the matrix content of a stone is largely dependent on the amount of protein in the solution that grows the stone. Furthermore, the importance of matrix to a stone is related to the amount of matrix in a stone. Urine has relatively small amounts of dissolved protein (compared to saliva or bile). Most urinary stones have scanty amounts of matrix, and it is probable that matrix is not a major factor in urolithiasis. On the other hand, gallstones appear to have a greater matrix content than urinary stones. Strong indirect evidence indicates that matrix is crucially involved in gallstone formation in prairie dogs (12). In this mammal, acetylsalycilic acid therapy inhibits mucus formation and condensation in the gallbladder and is completely effective in stopping gallstone formation induced by a lithogenic diet.

FINAL COMMENT

In closing and for the sake of argument, one may ask which of the four mechanisms, nucleation, crystal growth, aggregation, and retention, is most important to the development of stone disease. Nucleation seems an unlikely candidate because crystalluria is so common and stone disease is so relatively infrequent. Crystal growth and aggregation are also not likely candidates inasmuch as demonstrated uninhibited linear growth rates resulting from crystal growth and aggregation are quite leisurely, less than 1 μm/min. By a process of exclusion, we conclude that the most important mechanism(s) for developing urolithiasis is the mechanism(s) for causing the retention of a particle until it has grown so big that it can pass only with difficulty, or not at all.

Acknowledgement. Supported by NIH Grant No. AM 20586-04.

REFERENCES

(1) Adair, J.H. 1981. Coagulation of calcium oxalate monohydrate suspensions. Ph.D. Thesis, University of Florida, USA.

(2) De Jong, A.S.H.; Hak, T.J.; and Van Duijn, P. 1980. The dynamics of calcium phosphate precipitation studied with a new polyacrylamide steady state matrix-model: Influence of pyrophosphate, collagen and chondroitin sulfate. Connective Tissue Res. 7: 73-79.

(3) Finlayson, B. 1978. Physicochemical aspects of urolithiasis. Kidney Int. 13: 344-360.

(4) Finlayson, B., and Miller, G.H. 1970. Observations on the kinetics of experimental foreign body lithiasis. J. Urol. 104: 169-175.

(5) Finlayson, B., and Reid, F. 1978. The expectation of free and fixed particles in urinary stone disease. Invest. Urol. 15: 442-448.

(6) Giles, R.V. 1962. Fluid Mechanics and Hydraulics, 2nd ed. New York: Schaum Publishing Co.

(7) Gill, W.B.; Ruggiero, K.J.; and Fromes, M.C. 1980. Elevation of the metastable limits and absence of container surface nucleation for calcium oxalate crystallization in a urothelial-lined system as compared to glass containers. Invest. Urol. 18: 158-161.

(8) Ismail, S.I., and Tawashi, R. 1980. Size distribution characteristics of mineral phase in renal stones. J. Pharm. Sci. 69: 829-831.

(9) Jordan, W.R.; Finlayson, B.; and Luxenburg, M. 1978. Kinetics of early time calcium oxalate nephrolithiasis. Invest. Urol. 15: 465-468.

(10) Koutsoukos, P.G.; Sheehan, M.E.; and Nancollas, G.H. 1981. Epitaxial considerations in urinary stone formation. II. The oxalate-phosphate system. Invest. Urol. 18: 358-363.

(11) Leal, J., and Finlayson, B. 1977. Adsorption of naturally occurring polymers onto calcium oxalate surfaces. Invest. Urol. 14: 278-283.

(12) Lee, S.P.; Carey, M.C.; and LaMont, J.T. 1981. Aspirin prevention of cholesterol gallstone formation in prairie dogs. Science 211: 1429-1431.

(13) Pak, C.Y.C., and Galasy, R.A. 1980. Propensity for spontaneous nucleation of calcium oxalate. Quantitative assessment by urinary FRR-APR discriminant score. Am. J. Med. 69: 681-689.

(14) Robertson, W.G. 1976. Saturation-inhibition index as a measure of the risk of calcium oxalate stone formation in the urinary tract. N. Eng. J. Med. 294: 249-252.

Biological Mineralization and Demineralization, ed. G.H. Nancollas, pp. 287-301.
Dahlem Konferenzen 1982. Berlin, Heidelberg, New York: Springer-Verlag.

Mechanisms and Regulation of Normal and Pathologic Demineralization

L. G. Raisz
Dept. of Medicine, University of Connecticut Health Center
Farmington, CT 06032, USA

Abstract. Organisms use mineralization for structure and storage. Physiologic demineralization may occur because the stored minerals are needed to serve as buffers, to supply soft tissues, or to remodel the structure. These processes are largely carried out by cells which are under the control of systemic hormones or local factors. Pathologic demineralization can occur as the consequence of excess of such hormones or factors, abnormal demands on mineral stores, as in calcium or phosphate deficiency or acidosis, proliferation of abnormal demineralizing cells or exogenous agents such as bacteria. Mineralized tissues also contain an organic matrix, and this may have to be removed to provide access to the mineral. Although many factors which affect demineralization have been identified, and the cells involved have been described in detail, the precise biochemical mechanisms are largely unknown.

INTRODUCTION

Many eukaryotes have mineralized structures which serve as reservoirs of ions, protectors of soft tissue, and providers of a mechanical system for movement. Some system of demineralization is required not only to recover the stored ions, but to remodel the structure for growth and repair. Pathologic demineralization can occur because of abnormalities in regulation, unusual demands on mineral stores, proliferation of abnormal cells which act on mineralized structures, or the presence of some exogenous factor which acts on the structure. This review will summarize the major forms of demineralization and describe the cells involved and the factors which are known to regulate their function. The precise biochemical mechanisms of demineralization are not known, but there are a number of theories which will be mentioned. Much of the available information is in reviews and texts (4,8,19,35,45), and only key articles will be cited.

TYPES OF DEMINERALIZATION

Just as there are many forms of mineralization in eukaryotes (25), there are many forms of demineralization. Invertebrates most often deposit mineral in exoskeletons, but may store mineral in cells, for example, in the hepatopancreas (19). Growth often requires shedding of the exoskeleton, and many invertebrates have the capacity to remove mineral prior to shedding and to use it to remineralize the next exoskeleton formed. Although they have not been as carefully studied, the processes involved in this mineral transport may be similar to those in the internal translocation of mineral during shell formation in egg-laying vertebrates. Another mechanism which invertebrates share with vertebrates is the removal of mineral by organic acids, which serves to buffer these acids (7). It is possible that further study of invertebrate mineral transport will uncover novel mechanisms of demineralization and shed light on processes in higher forms.

In the vertebrates bone resorption is by far the most important form of demineralization. In addition, egg-laying vertebrates can deposit mineral in the shell and yolk, and amphibia and reptiles can deposit mineral in paravertebral lymph sacs. Demineralization of eggshells has been the most extensively studied (50) and appears to involve both acid secretion and active calcium transport by the chorioallantonic membrane. The calcium in the yolk is probably bound to phosphate-rich proteins (phosvitin) which can be taken up by the developing embryo (8). Fish scales represent another potential source of mineral in vertebrates, which can be resorbed under certain circumstances and are also a sink for mineral, since removal of scales results in large amounts of mineral being mobilized from bone (52).

Certainly any discussion of demineralization must center on the mechanisms of resorption of bone and calcified cartilage. These processes are responsible for maintaining soft tissue supply when exogenous sources are limited and the structural functions of the skeleton during growth, remodeling, and repair. Cellular mechanisms appear to be strikingly similar among all the vertebrates.

CELLS INVOLVED IN DEMINERALIZATION

Mechanisms which do not involve direct contact between resorbing cells and mineralized tissue are probably relatively unimportant in physiologic demineralization. However, the mobilization of mineral by acids may occur at some distance from the acid producing cells (7,12,21,44,52). The skeleton is an important source of buffer because demineralization can release carbonate and phosphate ions which bind hydrogen ions at normal body pH. Among the cellular mechanisms for demineralization, resorption of bone, dentin, and calcified cartilage by osteoclasts, odontoclasts, and chondroclasts is the most important (20,28). These cells have a specialized functional unit consisting of a ruffled border and a surrounding clear zone which appears to be the active site for bone resorption.

The role of other cell types in bone resorption is less clear. The concept of osteocytic osteolysis has been presented based largely on increases in the size of periosteocytic lacunae in pathologic states (2). However, it is not possible to distinguish whether an enlarged lacunae was formed by resorption or due to less deposition of mineral and matrix by the enclosed osteocyte. In any case, it is unlikely that this process plays a major role in demineralization (34).

Osteoclasts and osteoblasts are clearly derived from separate lineages of precursor cells (20). The osteoclast precursors appear to resemble the monocyte-macrophage series, but it has not been possible to show a direct transformation of macrophages into morphologically typical osteoclasts. Mononuclear macrophages and multinucleated giant cells derived from macrophages may play an important part in both physiologic and pathologic bone resorption. Macrophages can break down uncalcified connective tissue (43), can attack dead bone (29), and could be responsible for removal of ectopic calcification. They may also act on bones in conjunction with osteoclasts, stimulating resorption by their production of prostaglandins and helping to remove matrix by their production of collagenase (11,18,22).

There is a frequent association between vascular invasion and the initiation of demineralization in the skeleton. Capillary endothelial cells are often seen in contact with osteoclasts (47), and one of the products of blood vessels, prostacyclin, may be a potent stimulator of osteoclastic bone resorption (40).

These processes generally involve direct contact between the mineralized tissue and the cell. However, in some situations tissues could be demineralized by an acid-secreting and calcium transporting system such as the chorio-allantoic membrane which is not in direct contact with the mineralized tissue. For example, in experimental models, supernatant fluids from monocytes or malignant cells can demineralize bone and degrade its matrix by the action of soluble factors which they release (13).

CHEMICAL MECHANISMS OF DEMINERALIZATION

The precise mechanism by which mineral is removed by bone-resorbing cells remains unknown. The extracellular calcium and phosphate concentrations are supersaturated with respect to bone apatite (32), and calcium and phosphate exchange rapidly between bone and blood. Hence, demineralization must involve work against an overall physicochemical concentration gradient. However, this may not be true directly at the bone surface. Cells or matrix components cover much of the bone surface. At sites of demineralization, a local change in concentration gradient could be produced. This might be the result of active transport with an extrusion of calcium by bone-resorbing cells, release of an agent which lowers the calcium and phosphate ion activities on the bone surface by chelation or crystal dissolution due to acid production. Hydrogen ions have been shown to be important in several different kinds of demineralization and are likely to be directly involved in crystal dissolution in bone. Whatever the process for releasing calcium, there must also be removal of any organic materials which prevent access to the mineral. Thus the separation of covering cell layers and the digestion of non-mineral components of the matrix may be necessary prerequisites for demineralization.

STRUCTURE AND FUNCTION OF OSTEOBLASTS

The morphology of osteoclasts has been studied in great detail, and responses to factors which influence osteoclast function have been analyzed quantitatively. Biochemical studies of osteoclast function are much less complete. The osteoclast is a large multinucleated cell. It has a characteristic resorbing apparatus which consists of a ruffled border surrounded by a clear zone. The cell contains abundant mitochondria and lysosomes and relatively little endoplasmic reticulum. Bone resorption is evident in the ruffled border area by the disruption of the bone surface with the appearance of fragments of mineral and collagen between the cell processes. The clear zone contains actin filaments and probably is responsible for holding the cell on the bone surface. Since it is closely adherent to the bone and surrounds the ruffled border area, the extracellular fluid at the site of active resorption could be sequestered and quite different in composition from extracellular fluid generally. Thus, high concentrations of enzymes, hydrogen ions, and chelating agents could be present. The cytoskeleton of the osteoclast is clearly important to its function since agents such as colchicine, which disrupt microtubules, block osteoclastic resorption. There is considerable indirect evidence that osteoclasts secrete lysosomal enzymes. When bone resorption is stimulated in organ cultures, lysosomal enzymes are released into the medium (14), and inhibitors of lysosomal enzymes can block bone resorption (9). In contrast, there is less consistent correlation between the amount of collagenase in the culture medium and resorptive response (24,41). This is particularly true for the inhibitory effects of calcitonin and colchicine in which changes in osteoclast morphology and calcium and lysosomal enzyme release are parallel, but collagenase is not inhibited. This paradoxical result was one of the bases for the hypothesis that nonosteoclastic cells may be responsible for collagen degradation in resorbing bone (18). Carbonic anhydrase has been localized in osteoclasts and eggshells and could play a role in acid production (16). While hydrogen ions are probably essential for an attack on mineral crystals, chelating agents could also be central to calcium removal. The only potential chelator of calcium thus far identified in resorbing tissues has been hyaluronic acid, production of which is increased when bone resorption is stimulated (26). The possibility that osteoclasts remove mineral from bone by a mechanism similar to transepithelial transport of calcium in intestine and other tissues has not been ruled out, but seems unlikely since parathyroid hormone, a potent stimulator of bone resorption, appears to increase cell permeability to calcium and decrease membrane potential (49).

PHYSIOLOGIC REGULATION OF DEMINERALIZATION

Resorption of mineralized tissues is necessary both for maintenance of ion supply and for structural remodeling of the skeleton. It is not surprising that the control of demineralization is complex and that data on hormone effects often conflict. Parathyroid hormone, 1,25-dihydroxyvitamin D (calcitriol) and calcitonin are the major determinants of the rate of movement of mineral from bone to blood in response to metabolic need.

Both parathyroid hormone and calcitriol can stimulate osteoclastic bone resorption, and it is not clear which of these hormones is the more important in physiologic regulation (48). The difference between the two hormones lies in their effects on other organ systems. Calcitriol is a major stimulator of intestinal absorption of calcium and phosphate, while parathyroid hormone acts on the kidney to increase the net tubular reabsorption of calcium and decrease the tubular reabsorption of phosphate. As a result, parathyroid hormone tends to increase serum calcium concentration and decrease serum phosphate concentration, while calcitriol can increase the concentration of both ions. Because of its powerful effect on gut absorption calcitriol can actually decrease bone resorption in vivo by increasing exogenous calcium supply and decreasing the need for demineralization to maintain calcium levels in blood and soft tissue.

Calcitonin is a direct inhibitor of osteoclastic bone resorption which is probably important physiologically in preventing hypercalcemia during periods of rapid intestinal absorption of calcium (1). This effect is likely to have its greatest importance in rapidly growing animals with high rates of bone resorption. Calcitonin does not appear to be as important in calcium regulation in adult mammals, but it may be more important in lower vertebrates, particularly fish, in which it may control calcium transport by the gills (27).

The synthesis and secretion of these three hormones is regulated by calcium and phosphate concentration, and all three hormones can influence demineralization of bone throughout the skeleton. Additional mechanisms are required for local control of demineralization. Local removal of bone is important in marrow development and in the many structural adaptations which the skeleton undergoes during its growth and development. Three factors have been identified which are produced by normal cells and which could be important in these responses: a) prostaglandins, b) osteoclast activating factor, and c) epidermal growth factor. Prostaglandins can be produced by bone in response to stress (46) and possibly also by invading blood vessels (40). Although many different prostaglandins can be shown to stimulate osteoclastic bone resorption, prostaglandin E_2 appears to be the most potent. Osteoclast activating factor can be produced by normal as well as neoplastic lymphoid cells (30) and could be the mediator of bone resorption involved in the development of a hematopoietic marrow. The role of epidermal growth factor is unknown. It can stimulate bone resorption directly (39) or by increasing prostaglandin production by bone (51). It is present in sufficient concentrations to act on the skeleton, but thus far there is no evidence which might suggest a physiologic role.

There are a number of other hormones which can influence calcium movement and may affect bone demineralization. Thyroid hormone is a direct stimulator of bone resorption which is required for normal bone turnover (31). It probably has its major role in growth rather than calcium mobilization. Glucocorticoids can inhibit bone resorption and have been demonstrated to have a direct effect at physiologic concentrations in vitro (36), although in vivo their inhibitory effects on gut absorption of calcium and bone formation usually predominate. Estrogens can clearly slow the rate of demineralization of the skeleton in postmenopausal women, but have not been shown to inhibit bone resorption directly at the physiologic concentrations in vivo (5).

There are a number of factors which may be important in calcium regulation in lower vertebrates but which have not yet been clearly identified. In fish the corpuscles of Stannius appear to secrete an important antihypercalcemic hormone. It is not known whether this hormone influences demineralization or has its major effect on calcium transport in the gill (6). Pituitary hormones such as prolactin can influence calcium regulation in fish, and further studies are needed to determine whether or not they influence movement of mineral from bone to blood (33).

It is likely that there are as yet undiscovered hormones and local factors which regulate bone mineralization and demineralization. For example, bone mineral is mobilized during pregnancy and lactation in animals who are deprived of both parathyroid hormone and the active forms of vitamin D (17). Certain tumors can produce a factor which mobilizes bone mineral and causes hypercalcemia. There are probably local factors released during bone remodeling which influence mineralization and demineralization. These factors could be responsible for the close coupling of bone resorption and formation which is observed in the remodeling skeleton (3). Not only is osteoclastic bone resorption usually followed by new bone formation at the same site, but the factors which stimulate bone resorption also have important effects on the bone-forming cells or osteoblasts. The possibility that osteoblasts release a bone-resorbing factor has been suggested, and this factor might be a prostaglandin.

PATHOLOGIC DEMINERALIZATION

Excessive synthesis or secretion of the physiologic stimulators of bone resorption or deficiency of inhibitors could produce pathologic demineralization. The most obvious examples are primary hyperparathyroidism and intoxication with vitamin D in which bone is demineralized, and the resultant hypercalcemia produces toxic effects in soft tissues. There is evidence that prostaglandins and osteoclast activating factor may be involved in osteolytic lesions in malignancy and inflammatory disease (23,42). However, most of this evidence is indirect because it is so difficult to analyze for the presence of these materials in tissue. Moreover, there is no convincing evidence that the administration of prostaglandins or osteoclast activating factor can produce pathologic demineralization in vivo.

Demineralization can occur either as the direct or indirect consequence of change in ions (37). For example, calcium deficiency and phosphate excess can stimulate parathyroid hormone secretion and hence increase bone resorption. Hydrogen ion excess can cause bone dissolution (12). Magnesium deficiency can produce hypocalcemia in some animals, while it produces hypercalcemia in others. These changes appear to be largely due to effects on parathyroid hormone secretion.

In vitro these ions have quite different effects. Reduction in calcium can cause passive dissolution of bone mineral, but does not enhance osteoclastic bone resorption, while increased phosphate concentration diminishes both dissolution and cell-mediated resorption of bone. Low magnesium concentrations appear to have some inhibitory effect on bone resorption.

The most common disorder of demineralization is age-related osteoporosis. This disease, which is particularly prevalent in postmenopausal women, is probably due in part to accelerated bone resorption, although diminished bone formation also plays a role. In some forms of osteoporosis, demineralization may be even more important, such as that which occurs in association with immobilization, while in still others, diminished bone formation may be the primary factor, for example, in patients with glucocorticoid excess, malnutrition, and alcoholism.

Exogenous agents can have important effects on bone demineralization. Bacterial products have been shown to stimulate osteoclastic bone resorption in vitro and may be responsible for alveolar bone resorption as well as for the loss of mineral from teeth in dental disease (38).

There is great interest in the use of therapeutic agents which inhibit demineralization. These include not only calcitonin but cytotoxic agents, lysosomal enzyme inhibitors, and membrane stabilizing drugs which can act directly on osteoclasts (10) as well as inhibitors of crystal dissolution. Among the most potent inhibitors of demineralization are the diphosphonates. Their mechanism of action was originally thought to be by an effect on dissolution of bone mineral, but it seems likely that the most effective diphosphonates have some selective toxic effect on osteoclasts (15).

UNSOLVED PROBLEMS AND OPEN QUESTIONS

This brief summary should serve to point to some important questions for future research: a) Are there mechanisms of demineralization in invertebrates which can shed light on this process in the vertebrates? b) What produces the physicochemical disequilibrium between bone and blood, and what role does this mechanism play in demineralization? c) How do osteoclasts remove mineral from bone? d) What is the relative importance of each of the calcium regulating hormones in demineralization? e) What factors control demineralization in structural remodeling of the skeleton? f) What factors are responsible for pathologic demineralization in osteoporosis, malignancy, and inflammation? Answers to these questions would not only help us to understand the processes involved in demineralization, but will lead to a better understanding of the causes and treatment of many diseases of bone and other mineralized tissues.

REFERENCES

(1) Austin, L.A., and Heath, H. III. 1981. Calcitonin: Physiology and pathophysiology. New Eng. J. Med. 304: 269-278.

(2) Baud, C.A., and Boivin, G. 1978. Effects of hormones on osteocyte function and perilacunar wall structure. Clin. Orthop. Rel. Res. 136: 270-281.

(3) Baylink, D.J., and Liu, C.C. 1979. The regulation of endosteal bone volume. J. Periodontol. 50: 43-49.

(4) Bentley, P.J. 1976. Comparative Vertebrate Endocrinology. Cambridge: Cambridge University Press.

(5) Caputo, C.B.; Meadows, D.; and Raisz, L.G. 1976. Failure of estrogens and androgens to inhibit bone resorption in tissue culture. Endocrinol. 98: 1065-1068.

(6) Copp, D.H., and Ma, S.W.Y. 1981. Teleocalcin and calcium regulation in boney fish. In Hormonal Control of Calcium Metabolism, eds. D.V. Cohn, R.V. Talmage, and J.L. Matthews, pp. 298-306. Amsterdam: Excerpta Medica.

(7) Crenshaw, M.A., and Jeff, J.M. 1969. Decalcification of the mantle-shell interfact in molluscs. Am. Zool. 9: 881-885.

(8) Dacke, C.G. 1979. Calcium Regulation in Submammalian Vertebrates. New York: Academic Press.

(9) Delaissse, J.-M.; Eeckhout, Y.; and Vaes, G. 1980. Inhibition of bone resorption in culture by inhibitors of thiol proteinases. Biochem. J. 192: 365-368.

(10) Dietrich, J.W.; Mundy, G.R.; and Raisz, L.G. 1979. Inhibition of bone resorption in tissue culture by membrane-stabilizing drugs. Endocrinol. 104: 1644-1648.

(11) Dominguez, J.H., and Mundy, G.R. 1980. Monocytes mediate osteoclastic bone resorption by prostaglandin production. Calcif. Tissue Intl. 31: 29-34.

(12) Dominguez, J.H.D., and Raisz, L.G. 1979. Effects of changing hydrogen ion, carbonic acid and bicarbonate concentrations on bone resorption in vitro. Calcif. Tissue Intl. 29: 7-13.

(13) Eilon, G., and Mundy, G.R. 1978. Direct resorption of bone by human breast cancer cells in vitro. Nature 287: 726-728.

(14) Eilon, G., and Raisz, L.G. 1978. Comparison of the effects of stimulators and inhibitors of resorption on release of lysosomal enzymes and radioactive calcium from fetal bone in organ culture. Endocrinol. 103: 1969-1975.

(15) Fleisch, H., and Felix, R. 1979. Diphosphonates. Calcif. Tissue Intl. 27: 91-94.

(16) Gay, C.V., and Mueller, W.J. 1974. Carbonic anhydrase and osteoclasts: Localization by labeled inhibitor autoradiography. Science 183: 432-434.

(17) Halloran, B.P., and DeLuca, H.F. 1980. Skeletal changes during pregnancy and lactation: The role of vitamin D. Endocrinol. 107: 1923-1929.

(18) Heersche, J.N.M. 1978. Mechanism of osteoclastic bone resorption: A new hypothesis. Calcif. Tissue Res. 26: 81-84.

(19) Highnam, K.C., and Hill, L. 1969. Comparative Endocrinology of the Invertebrates. New York: Elsevier.

(20) Holtrop, M.E., and King, G.J. 1977. The ultrastructure of the osteoclast and its functional implication. Clin. Orthop. Rel. Res. 123: 177-196.

(21) Jenkins, N.K., and Simkiss, K. 1968. The calcium and phosphate metabolism of reproducing reptiles with particular reference to the adder vipera berus. Comp. Biochem. Physiol. 26: 865-876.

(22) Jones, S.J.; Hogg, N.M.; Shapiro, I.M.; Slusarenko, M.; and Boyde, A. 1981. Cells with Fc receptors in the cell layer next to osteoblasts and osteoclasts on bone. Metab. Bone Dis. Rel. Res. 2: 357-362.

(23) Josse, R.G.; Murray, T.M.; Mundy, G.R.; Jez, D.; and Heersche, J.N.M. 1981. Observations on the mechanism of bone resorption induced by multiple myeloma. Science 211: 1126-1131.

(24) Lenaers-Claeys, C., and Vaes, G. 1979. Collagenase, procollagenase and bone resorption effects of heparin parathyroid hormone and calcitonin. Biochim. Biophys. Acta 584: 375-388.

(25) Lowenstam, H.A. 1981. Minerals formed by organisms. Science 211: 1126-1131.

(26) Luben, R.; Goggins, J.F.; and Raisz, L.G. 1974. Stimulation by parathyroid hormone of bone hyaluronate synthesis in organ culture. Endocrinol. 94: 737-745.

(27) Milhaud, G.; Rankin, J.C.; Bolis, L.; and Benson, A.A. 1977. Calcitonin - Its hormonal action on gill. Proc. Natl. Acad. Sci. 74: 4693-4696.

(28) Miller, S.C. 1977. Osteoclast cell-surface changes during the egg-laying cycle in Japanese Quail. J. Cell Biol. 75: 104-118.

(29) Mundy, G.R.; Altman, A.J.; Gondek, M.D.; and Bandelin, J.G. 1977. Direct resorption of bone by human monocytes. Science 196: 1109-1111.

(30) Mundy, G.R.; Eilon, G.; Orr, W.; Spiro, T.P.; and Yoneda, T. 1980. Osteoclast activating factor: Its role in myeloma and other types of hypercalcemia of malignancy. Metab. Bone Dis. Rel. Res. 2: 173-176.

(31) Mundy, G.R., and Raisz, L.G. 1979. Thyrotoxicosis and calcium metabolism. Miner. Electrolyte Metab. 2: 285-292.

(32) Neuman, W.F.; Diamond, A.G.; and Neuman, M.W. 1980. Blood/bone disequilibrium. IV. Reciprocal effects of calcium and phosphate concentrations on ion fluxes. Calcif. Tissue Intl. 32: 229-236.

(33) Olivereau, M., and Olivereau, J. 1978. Prolactin, hypercalcemia, and corpuscles of stannius in seawater eels. Cell Tissue Res. 186: 81-96.

(34) Parfitt, A.M., and Chir, B. 1977. Cellular basis of bone turnover and bone loss - rebuttal of osteocytic resorption-bone flow theory. Clin. Orthop. Rel. Res. 127: 236-247.

(35) Raisz, L.G. 1972. Mechanisms of bone resorption. In Handbook of Physiology; Section on Endocrinology, eds. E. Astwood and R. Greep, vol. 7, pp. 117-136. Baltimore: Williams and Wilkins.

(36) Raisz, L.G. 1980. Effects of corticosteroids on calcium metabolism. Prog. Biochem. Pharmacol. (Switzerland) 17: 212-219.

(37) Raisz, L.G., and Niemann, I. 1969. Effect of phosphate, calcium, and magnesium on bone resorption and hormonal responses in tissue culture. Endocrinol. 85: 446-552.

(38) Raisz, L.G.; Nuki, K.; Alander, C.; and Craig, R.G. 1981. Interactions between bacterial endotoxin and other stimulators of bone resorption in organ culture. J. Periodont. Res. 16: 1-7.

(39) Raisz, L.G.; Simmons, H.A.; Sandberg, A.L.; and Canalis, E. 1980. Direct stimulation of bone resorption by epidermal growth factor. Endocrinol. 107: 270-273.

(40) Raisz, L.G.; Vanderhoek, Y.J.; Simmons, H.A.; Kream, B.A.; and Nicolaou, K.C. 1979. Prostaglandin synthesis by fetal rat bone in vitro. Evidence for a role of prostacyclin. Prostaglandins 17: 905-914.

(41) Sellers, A.; Meikle, M.C.; Reynolds, J.J. 1980. Collagenase and collagenase inhibitor levels following changes in bone resorption in vitro. Calcif. Tissue Intl. 31: 35-43.

(42) Seyberth, H.W.; Raisz, L.G.; and Oates, J.A. 1978. Prostaglandins and hypercalcemic states. Ann. Rev. Med. 29: 23-29.

(43) Silvestrini, G.; Ricordi, M.E.; and Bonucci, E. 1979. Resorption of uncalcified cartilage in the diaphysis of the chick embryo tibia. Cell Tissue Res. 196: 221-235.

(44) Simkiss, K. 1968. Calcium and carbonate metabolism in the frog during respiratory acidosis. Am. J. Physiol. 214: 627-634.

(45) Simmons, D.J. 1971. Calcium and skeletal tissue physiology in teleost fishes. Clin. Orthop. Rel. Res. 76: 244.

(46) Somjen, D.; Binderman, I.; Berger, E.; and Harell, A. 1980. Bone remodeling induced by physical stress is prostaglandin (PGE_2) mediated. Biochim. Biophys. Acta 627: 91-100.

(47) Soskolne, W.A. 1979. The osteoclast-endothelium interface during bone resorption in the femurs of young rabbits. Cell Tissue Res. 203: 487-492.

(48) Stern, P.H. 1980. The D vitamins and bone. Pharmacol. Rev. 32: 47-80.

(49) Trumbore, D.C.; Heideger, W.J.; and Beach, K.W. 1980. Electrical potential difference across bone membrane. Calcif. Tissue Intl. 32: 159-168.

(50) Tuan, R.S. 1980. Calcium transport and related functions in the chorioallantoic membrane of cultured shell-less chick embryos. Dev. Biol. 74: 196-204.

(51) Voelkel, E.F.; Tashjian, A.H., Jr.; and Levine, L. 1980. Cyclooxygenase products of arachidonic acid metabolism by mouse bone in organ culture. Biochim. Biophys. Acta 620: 418-428.

(52) Weiss, R.E., and Watabe, N. 1979. Studies on the biology of fish bone. III. Ultrastructure of osteogenesis and resorption in osteocytic (cellular) and anosteocytic (acellular) bones. Calcif. Tissue Intl. 28: 43-56.

Biological Mineralization and Demineralization, ed. G.H. Nancollas, pp. 303-324.
Dahlem Konferenzen 1982. Berlin, Heidelberg, New York: Springer-Verlag.

Mechanism of Dental Caries

J. Arends
Laboratory for Materia Technica, University of Groningen
9713 AV Groningen, The Netherlands

Abstract. Dental caries is caused by acids generated by the dental plaque adhering to the tooth surface. This paper gives the definition of dental caries, its importance, and introductory remarks on enamel structure and composition. Four different approaches on the mechanisms of the acid interaction with dental enamel are discussed: a) the semipermeable membrane mechanism, in which a pH drop in the enamel is thought to be due to large positively charged mucine cations and small penetrable anions in the enamel. In this mechanism the acid is acting indirectly through salivary mucins on the enamel. The practical importance is most likely minor; b) dissolution mechanism, based on a direct attack of H^+ and undissociated acid (HA) on the enamel. This mechanism, giving a reasonable empirical description of experimentally obtained data, assumes that caries is mainly determined by the kinetics of the acid-enamel interaction and that second phases (DCPD) make the enamel crystallites partly insoluble; c) dissolution-precipitation mechanism(s). In these mechanism(s) the surface layer covering the lesion is crucial. During lesion formation (assuming quasi equilibrium), the rate of dissolution and rate of precipitation of the surface layer are equal (or nearly equal). Lesion depth progresses proportional to the square root of time; the apparent diffusion constant $\approx 10^{-12}$ $cm^2.sec^{-1}$; and d) inhibitor controlled mechanism. In this approach, caries is considered (as in b) a kinetic phenomenon in which mainly the inhibitor penetration determines the process. Lesion depth progresses with $t^{1/3}$ and the apparent diffusion constants are $\approx 10^{-5}$ $cm^2.sec^{-1}$.

Finally, the enamel crystallite dissolution - preferential axial dissolution along the c-axis most likely due to the presence of "hollow core" dislocations - is considered.

INTRODUCTION

Dental caries is the localized progressive decay of teeth, initiated by demineralization of the outer tooth surface due to organic acids produced locally by bacteria in dental plaque. Carious lesions develop under dense bacterial masses (termed dental plaque) which adhere strongly to the tooth surface. The bacteria are separated from the enamel mineral by a thin organic film of salivary origin, the acquired pellicle. Streptococci and gram-positive filamentous bacteria are the most prominent organisms on human coronal tooth surfaces; they number about 10^{10} per gram wet plaque. The cariogenic bacteria, especially streptococcus mutant, ferment dietary carbohydrates in situ principally to lactic acid. These cariogenic bacteria metabolize sucrose in such a way that they produce extracellularly an adhesive polysaccharide (dextran) and lactic acid. Furthermore, they store intracellar polysaccharides during periods of carbohydrate abundance and utilize it with the formation of lactic acid in periods of carbohydrate deficiency. In human dental plaque the pH drops from 7 to 5.5 (or lower) for a period of 20 min after a 2 min rinse with glucose. Subsequently, the pH in the plaque rises slowly to neutral in the following 40 min. In practice this happens many times a day. When the acid decreases the pH of the plaque to 5.5 or less, the underlying enamel starts to lose calcium and phosphate ions. For caries to occur, the acid formed during breakdown of carbohydrates by the bacteria in the plaque must be capable of attacking the enamel before flowing saliva can remove the acid. Two plaque properties are most important: a) the high concentrations of bacteria are capable of producing large quantities of acids in a short period, and b) the rate of acid formation is faster than the acid diffusion from the plaque because the plaque matrix makes diffusion slow. Saliva helps in caries reduction because it washes away acids, bacteria, etc.; furthermore, it acts as a chemical buffer (HCO_3^-/H_2CO_3).

Dental caries is the most prevalent disease in modern society other than the common cold. Nearly everybody experiences caries before adulthood. It is the leading cause of tooth loss before the age of 35. To repair the damage caused by dental caries in the USA, more than $10 billion would be necessary annually. Dental caries can involve the hard tissues enamel, dentine, or cementum. Because enamel covering the crown of our teeth is always attacked first by the caries process, we will limit ourselves to enamel caries.

DENTAL ENAMEL

Enamel is a hard calcified material that during genesis is produced by cells: the ameloblasts. The cellular formation of the enamel is reflected in the structure, especially in the enamel prisms (sometimes called enamel rods). These prisms are the basic units if we consider enamel at a scale of about 10-100 µm. The individual enamel prisms have a diameter on the order of 4-5 µm, are in general "keyhole" shaped, and run from the dentine in a slightly curved pattern to the anatomical enamel surface. Between prisms we have interprismatic areas with an average width of about 0.1 µm. In the caries process the interprismatic area is the easiest pathway for acids.

Although the prisms are important structures from a biological viewpoint, they are not the basic units of the enamel structure. Within and between the prisms there are millions of tightly packed minute enamel crystallites. These tiny crystallites are the real mineral units of the enamel and are needle shaped. The length is on the order of 0.5-10 µm, while the diameter is about 40 nm. In Tables 1 and 2 the main enamel components and ultrastructural details are compiled.

ACID DEMINERALIZATION AND PROTEOLYTIC THEORIES

Two completely different theories for dental caries formation have been proposed:

TABLE 1 - The main components of mature human enamel

Component	Details	Amount (wt%)	Amount (vol%)	Density of component ($g\ cm^{-3}$)
Mineral	Non-stoichiometric hydroxyapatite	≈ 95	≈ 87	≈ 3
Organic matrix	Glycoproteins	≈ 1	≈ 1.5	≈ 1.1
Water	"Bound" ¾, "free" ¼	≈ 4	≈ 11.5	1

TABLE 2 - Ultrastructural properties of human mature enamel

Unit	Diameter (µm)	Spaces between units (nm)	length (µm)	Shape in cross section
Prism	4-5	100-500	1000-2000	"Keyhole"
Crystallite	0.04	1-2	≈ 1-10	Flattened hexagonal

a) The acid demineralization theory (34), in which it is assumed that acids produced as described above attack the mineral component of the enamel first.

b) The proteolytic theory, in which the first step in the caries process is the breakdown of the organic matrix of enamel by oral bacteria. Direct evidence for this theory has not yet been found. Modifications of the proteolytic theory have been suggested as well. Proteolysis of the enamel protein may release glutamic or aspartic acid, causing subsequent demineralization of the mineral (8). Schatz et al. (37) proposed that the end products of the proteolysis of the organic matrix may influence solubilization of calcium ions from the mineral (chelation).

Because of the overwhelming amount of evidence in support of the acid demineralization theory and the absence of data supporting the proteolytic theories, we will consider only a). It should be stated that proteolysis of accumulated matrix proteins, particularly in interprismatic areas, can facilitate the basic dissolution process.

DENTAL CARIES GENERAL

At the moment there is consensus in the literature that caries is caused by acids generated by bacteria adhering to the tooth surface. The acid(s) (mainly lactic acid) diffuses in the enamel, reacts with the mineral apatite, and forms soluble calcium and phosphate ions. These ions (or ion complexes) diffuse out of the material and leave a porous subsurface structure.

One special feature of dental caries has been given considerable attention: viz. the formation of a relatively intact surface layer on top of the subsurface lesion. The subsurface layer is (although porous) relatively intact and has a high mineral content. A typical microradiogram of a carious lesion in enamel is given in Fig. 1.

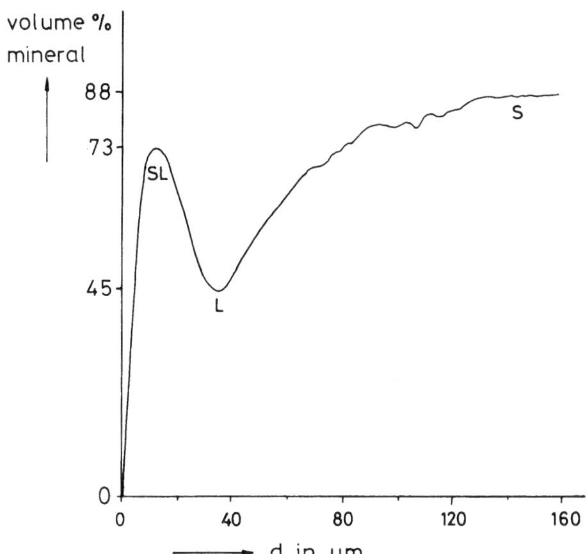

FIG. 1 - Microradiogram of a carious lesion. The plot shows mineral content in vol% versus depth. SL = surface layer, S = sound enamel, L = lesion.

Dental caries can be simulated at the moment very well by means of "artificial caries" solutions. The resulting lesions

(39) show nearly all properties of natural lesions. The use of artificial lesions contributed considerably to the knowledge of the caries process. It made model studies possible on lesion formation, lesion fluoridation, as well as lesion remineralization.

The principal mechanisms proposed for dental caries can be divided into four groups:
a) the semipermeable membrane mechanism,
b) mechanisms of lesion formation involving protective agents rendering the enamel surface insoluble,
c) lesion formation mechanisms assuming equilibria between dissolution and reprecipitation of the calcium phosphate involved, and
d) inhibitor controlled mechanism.

THE SEMIPERMEABLE MEMBRANE MECHANISM

This is a mechanism suggested by von Barteld (9,10) and is based on the well-known fact that enamel is permeable to small ions and molecules and can be considered as a semipermeable membrane. Furthermore, it takes into account that saliva contains considerable amounts of large ionized proteins and protein-like molecules. Saliva contains about 2% mucine-like material. If we denote the sodium compound of the mucine-like material by EOHNa, the reaction with acids generated by the bacterial system can be visualized in the following two steps:

$EOH^- + H^+ \rightarrow HEOH$ and with excess acid, HA,

$$HEOH + HA \rightarrow HE^+ + A^- + H_2O \;.$$

The ions HE^+ will certainly be too large to enter the enamel, but A^- is small enough to be transported <u>in</u> enamel. Von Bartheld suggests that when the second reaction given above takes place, a Donnan-membrane equilibrium is established. A^- ions enter the enamel, while OH^- ions leave the enamel, resulting in a pH <u>decrease</u> causing demineralization. The suggested mechanism has serious disadvantages, in that, a) it cannot explain the typical surface layer formation, and b) subsurface lesions can be formed without the presence of large, positively charged molecules. The membrane mechanism is most likely not a dominant one in caries. On the other hand, this mechanism may play a certain role once the caries process is initiated by one of the following mechanisms (e.g., when the surface layer is already formed). To the author's knowledge no real evidence is available.

DISSOLUTION MECHANISM

A completely different approach was used by Gray et al. (20-22, 28). Schematically the description of the mechanism of the caries process is given in Fig. 2.

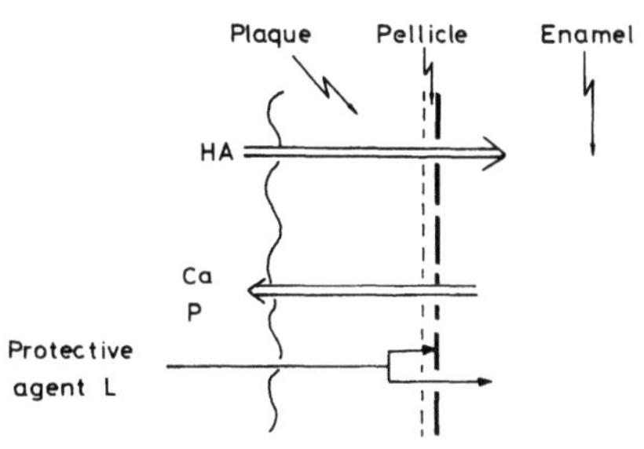

FIG. 2 - Penetration of acid HA (and H^+) in the enamel and the loss of calcium and phosphate.

The ions migrating in the enamel are H^+ and undissociated acid HA, while the ions diffusing out of the enamel are H_2O, A^-, Ca^{2+}, phosphates, and complexes, e.g., CaA^+. Furthermore, the enamel surface is presumed to be insoluble due to the effect of protective agents. Protective agents suggested are: F^-, Sn^{2+}, salivary proteins, diphosphonates, polyphosphonates, tripolyphosphate, etc. In this model the emphasis is on dissolution rate(s), (R) of the enamel dissolving during lesion formation.

The essential process consists of the diffusion of reactants to the reaction sites in the enamel, followed by a heterogeneous reaction and a subsequent diffusion of the reaction products out of the enamel. Two points are important in this mechanism:

a) The overall rate is controlled by the diffusion rate of the reactants to the solid enamel. Because the reactant rate of diffusion is proportional to the concentration (C) of the reactants, the overall rate, R, is proportional to the concentration of the reactants.
b) The data of Gray indicate that the experimental rate of subsurface lesion formation is proportional to C_{HA}, is nonlinear with C_{H^+}, and decreases with $C_{Ca^{2+}}$. Therefore, the dissolution rate can be written as

$$R = \sum_i k_i C_i^n \tag{1}$$

where i is a reactant; the constant n being 1 for diffusion controlled reactions.

Equation 1 can be written as:

$$R = k_O (C_{H^+})^{n_O} + k_1 K_A^{n_1} (C_{H^+})^{n_2} (C_{HA})^{n_3} + k_2 (C_{H^+})^{n_4} (C_{Ca^{2+}})^{n_5}. \tag{2}$$

Here K_A is the dissociation constant of the acid HA. The best fit to the experimental data available for <u>subsurface dissolution</u> of enamel is:

$$R = k_O C_{H^+} + k_1 K_A^{\frac{1}{4}} C_{H^+}^{\frac{1}{4}} \cdot C_{HA} + k_2 f(C_{H^+}, K_A, K_C, C_{A^-}, C_{HA}), \tag{3}$$

with K_C being the complexation constant of CaA_2. A may be lactate because lactic acid is the main acid produced by the dental plaque.

Equation 3 can be considered as reflecting the diffusion controlled contributions of the <u>two</u> hydrogen sources H^+ and HA, while the third term shows the influence of calcium complexation. This mechanism indicates that charges such as those mentioned in the Donnan-membrane model as well as other parameters (e.g., viscosity) do not (or only slightly) influence mineral dissolution.

During carious lesion formation only part of the mineral dissolves (see, e.g., Fig. 1); the lesion is a porous, but still mineral-rich material. In the mechanism proposed here, the fact that only a part of the mineral is dissolved was explained by the precipitation of some of the reaction products in the form of a different phase. These phases (e.g., DCPD in a layer of 2.5 nm covering the enamel crystallites) are supposed to limit further reaction and in an equilibrium situation could prevent access of H^+ ion to the enamel crystallites.

The amount of HPO_4 ions in lesions is considerable; amounts on the order of 15 wt% have been indicated (1). Whether the HPO_4 is present as DCPD, as suggested here, or in other forms (e.g., non-stoichiometric apatite or as surface HPO_4 without formation of a separate phase) is not yet certain.

DISSOLUTION-PRECIPITATION MECHANISMS

A different mechanism was proposed by Moreno et al. (35,36,44). The presented mechanism is schematically given in Fig. 3. This mechanism assumes that acid diffuses to the enamel through the pellicle. Subsequently, in the surface enamel a series of phase transitions occur resulting in DCPD and fluorapatite (FAP). This means that the surface layer of enamel should contain DCPD, FAP, and the original mineral HAP. These phases are presumed to be in quasi-equilibrium with the liquid in the enamel pores. Because the electrochemical potentials μ_{HA} and $\mu_{H_3PO_4}$ are larger in the surface layer than in the inner enamel and $\mu_{Ca(OH)_2}$ is small in the surface layer with respect to the inner enamel, Moreno et al. suggest a flow of basic ingredients from the inner enamel to the surface layer.

In this mechanism the surface layer plays a crucial role. It is supposed to dissolve partly first, then be continuously

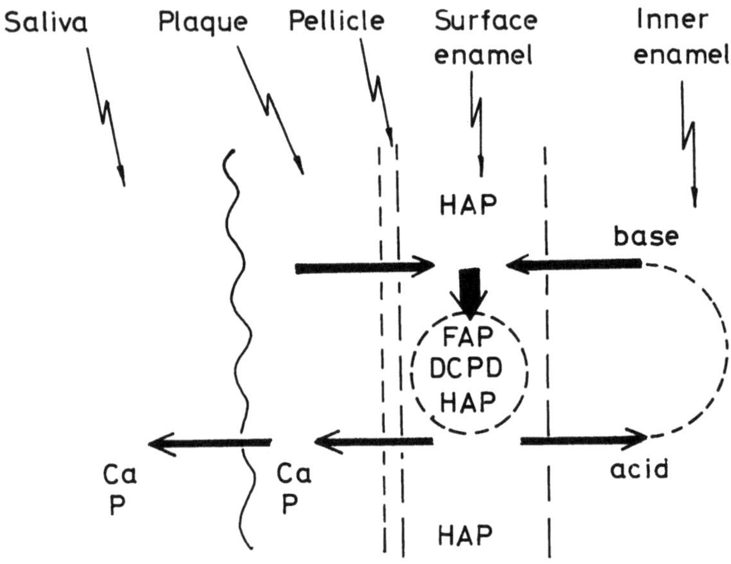

FIG. 3 - Dissolution-precipitation sequence in the caries process.

regenerated. It may even thicken due to slight unbalances between the dissolution rate and the precipitation rate. In this model plaque and pellicle are assumed to be controlling the rate of influx of acid and the outflux of calcium and phosphate. During lesion formation the rate of dissolution (R_d) of the surface layer is in equilibrium with the rate of precipitation (R_p). Thus: $R_d^{SL} = R_p^{SL}$.

As the lesion progresses, R_p slows down due to a longer diffusion path or due to decreased dissolution in the lesion. If $R_d^{SL} > R_p^{SL}$, the surface layer collapses, resulting in cavitation.

An analogous approach was followed by van Dijk et al. (41). They concluded that the surface layer in Fig. 3 may not be caused by the precipitation of second phases like DCPD or FAP, but that the surface layer itself is due to either a porosity gradient, a gradient in the rate constant k of the dissolution reaction, or to a gradient in the solubility product.

A combination of the dissolution mechanism and the dissolution precipitation mechanism has been suggested recently (16-18). Starting with the presumption that initially a protective agent is adsorbed to the outer surface of the enamel, a porous surface is then formed with a loss of about 20 vol% mineral. In inner enamel and surface layer a second phase is deposited (Featherstone suggests monetite: $CaHPO_4$), and R_p^{SL} and the R_s^{SL} determine the process as depicted in Fig. 4.

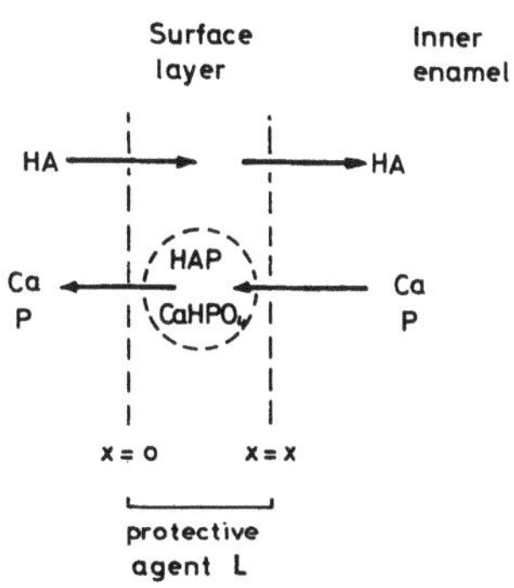

FIG. 4 - Combination of dissolution and dissolution-precipitation mechanisms.

A thickening of the surface layer from 12-36 µm in 5 days was observed (16,17) and suggests an empirical relation:

$$x_{SL}^2 = D_{app} \cdot t , \qquad (4)$$

with x_{SL} the surface layer thickness, and D_{app} being the apparent diffusion coefficient. Featherstone found $D_{app} \approx 10^{-12}$ $cm^2 sec^{-1}$. This value is comparable to $D_{Ca^{2+}}$ and $D_{phosphate}$ observed by Arends and collaborators in SOUND enamel (14,15,19).

INHIBITOR CONTROLLED MECHANISM

Recently a mechanism has been proposed (11) in which the inhibitor penetration determines lesion formation and progress. Basic assumptions are: a) initially tiny holes are formed in the enamel surface by the acid (≈ 0.1-0.5 μm diameter), b) during demineralization acid will penetrate the liquid phase between enamel crystallites causing partial dissolution of crystallites not protected by inhibitor. Inhibitor diffusing into the lesion causes crystallites not to dissolve completely. c) Inhibitors block the apatite dissolution effectively if the inhibitor concentration X_L exceeds the fraction of surface phosphate sites X_L^*. d) A steady state diffusion takes place (for all details see (11)). The inhibitor(s) may come from outside the enamel (e.g., the enamel surface) and/or may <u>originate from the enamel itself</u> (e.g., F^-).

We denote by r_ℓ the lesion depth, by r_0 the initial tiny holes in the enamel surface (see Fig. 5), and by r_1 the depth of the lesion at $t = t_1$. After a transient time t_1, $r_1 \gg r_0$, we have a steady state diffusion process.

The enamel dissolution rate can be written as:

$$J_{HAP}^- = -dn_{HAP}/dt = (A\ \Delta\rho/M)dr_\ell/dt. \tag{5}$$

Here, J_{HAP}^- is the mineral flux from the enamel, n_{HAP} is the amount of apatite in the solid enamel, $\Delta\rho$ is the density difference between sound enamel and the partly demineralized enamel in the lesion, M is the molar mass of hydroxyapatite, A is the area of the demineralizing zone, and φ is the 3-dimensional space angle.

The area A of the demineralizing zone is given by

$$A = \varphi\ r_\ell^2. \tag{6}$$

Inserting Eq. 6 in Eq. 5 and dropping the index ℓ and integrating gives:

$$\int_{r_1}^{r_\ell(t)} r^2 dt = \frac{MJ_{HAP}^-}{\Delta\rho\cdot\varphi} \int_{t_1}^{t} dt \tag{7}$$

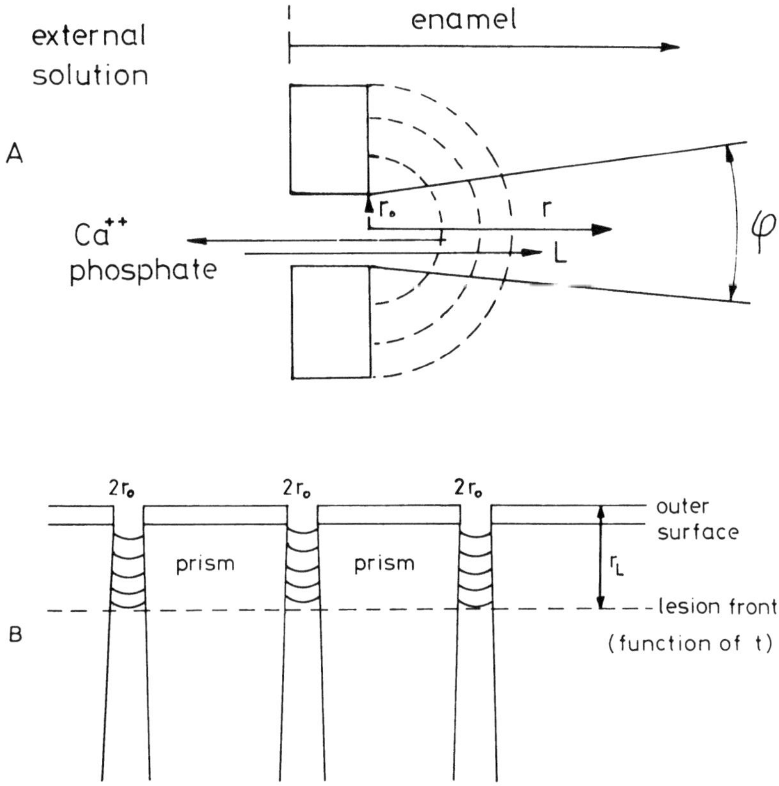

FIG. 5 - Inhibitor controlled mechanism. A) One hole in the enamel in detail; r is distance, r_0 is hole radius, L is the inwards diffusing inhibitor, φ is the space angle. B) Several holes together with the resulting front; r_ℓ is lesion depth, $2r_0$ is the hole diameter.

or

$$r^3 - r_1^3 = \frac{3MJ^-_{HAP}}{\Delta\rho \cdot \varphi}\,(t - t_1). \qquad (8)$$

From which we directly obtain lesion progress as a function of time:

$$r_\ell^3 = \alpha t + q\ , \qquad (9)$$

with $\alpha = 3MJ^-_{HAP}/\Delta\rho\varphi$ and $q = r_1^3 - \alpha t_1$.

Equation 9 shows that from a theoretical calculation we may expect the lesion depth of a lesion to progress proportional to $t^{1/3}$. Figures 6 and 7 show a comparision with experimental data.

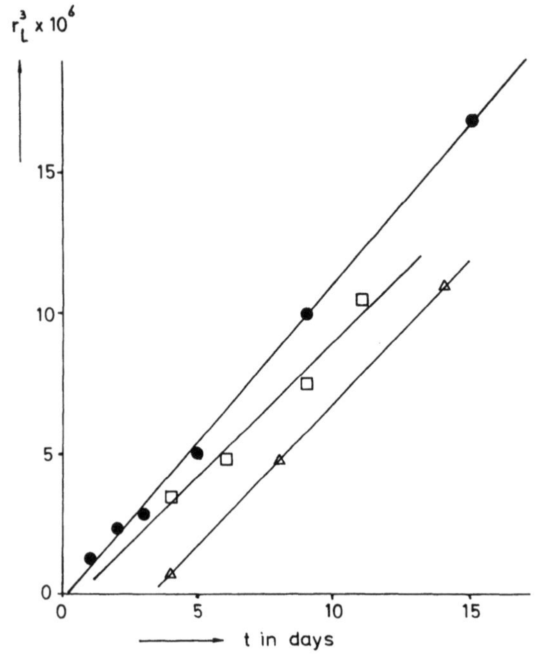

FIG. 6 - Plot of the third power of lesion depth versus time ΔΔΔ from (7), □□ from (6), ••• from (16).

From the slope of the last figures we may calculate the apparent diffusion coefficient of the system (D_{app})

$$\bar{J}_{HAP} = A\, D_{app} (d\, C_{HAP}/dt)_{r=r_O} = \varphi\, D_{app} \cdot r_O \cdot (C_{HAP} - C^E_{HAP}), \quad (10)$$

with C_{HAP} and C^E_{HAP} being the hydroxyapatite concentration in the lesion and in the external solution, respectively.

From Eqs. 5 and 10 and assuming the outer solution to be a sink, it is obvious that

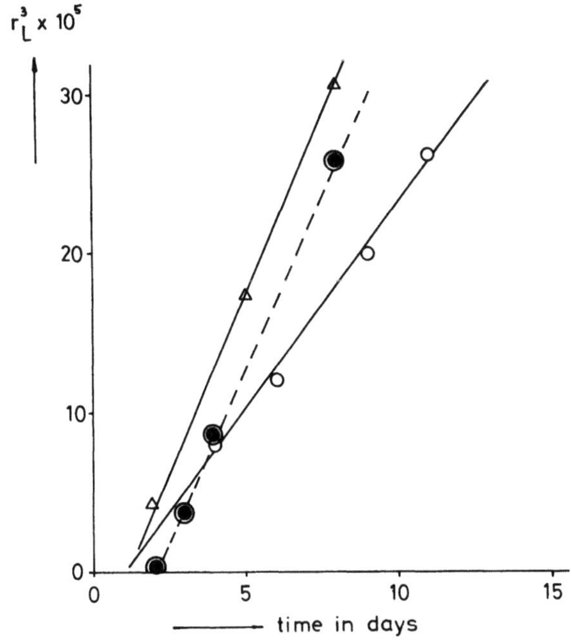

FIG. 7 - Plot of the third power of lesion depth versus time ΔΔΔ Arends (7), human enamel, pH = 4, HEC system. ⊙⊙⊙ Arends (6), bovine enamel, gelatine system, inhibitor F^-. OOO Groeneveld (23), human enamel, pH = 4.5, HEC system.

$$D_{app} = \frac{\Delta\rho}{M} \; \frac{d(r_\ell)^3/dt}{3r_o(C_{HAP} - C^E_{HAP})} \; . \tag{11}$$

$D_{app} \approx 10^{-5}$ cm^2 sec^{-1}, if we employ the chosen data: r_o = 0.25 µm (31), M = 1004 g mol^{-1}, $\Delta\rho \approx 0.5$ g cm^{-3} (13), while the difference $C_{HAP} - C^E_{HAP}$ was chosen as 0.75 C_{sat}. C_{sat} is the saturation value of hydroxyapatite at pH = 4.5 viz. 1.7 x 10^{-3} mol l^{-1} (12).

Considering the magnitude of $D_{app} \approx 10^{-5}$ cm^2 sec^{-1}, it is obvious that it is of the same order of magnitude as $D_{Ca^{2+}}$ and $D_{phosphate}$ in aqueous solution (24,27). The transport of phosphate and calcium out of the lesion through the pores enamel structure has the correct order of magnitude and is comparable to transport through water-filled channels (21).

The demineralization along the tiny holes implies an uncoupled demineralization. Furthermore, it is presumed that in the outer enamel $\varphi\, r_\ell^2$ is smaller than the surface area per hole. If these conditions are not fulfilled, the r_ℓ^3 versus t relation will break down. In this (not observed) case the area will be independent of r_ℓ and one may expect r to be proportional to t.

ENAMEL CRYSTALLITE DISSOLUTION DURING CARIES

During the caries process the tiny enamel crystallites are attacked by the penetrating acid. Two phenomena are important in this respect: a) in a carious lesion the average crystallite diameter is smaller than in sound enamel (32); b) a large percentage of the crystallites shows a preferential dissolution ("pipe formation") parallel to the c-axis of the crystallite. This preferential dissolution has been found by numerous authors (32,40,42,43) and is shown schematically in Fig. 8.

Crystallites in contact with acids lose first the central core; the outer part of the crystallite is much more acid-resistant. This phenomenon is independent of the type of acid employed; it has been found in the pH range 2-5.5.

There is no doubt that the fast acid penetration roughly parallel to the longest axis of the crystallite is the very first noticeable effect of an acid. Central core dissolution has been observed in natural lesions, artificial lesions, as well as in enamel demineralized directly by acids. The experimental diameter of the preferentially removed core is 5-10 nm (29,42).

The origin of this anisotropic dissolution is not fully understood. Most likely it is caused by the presence of one (or a group) of dislocations. Dislocations are linear lattice defects which can be defined with the aid of the so-called Burgers circuit, an atom-to-atom path forming a closed loop made in a single crystal (26). In an ideal crystal with no

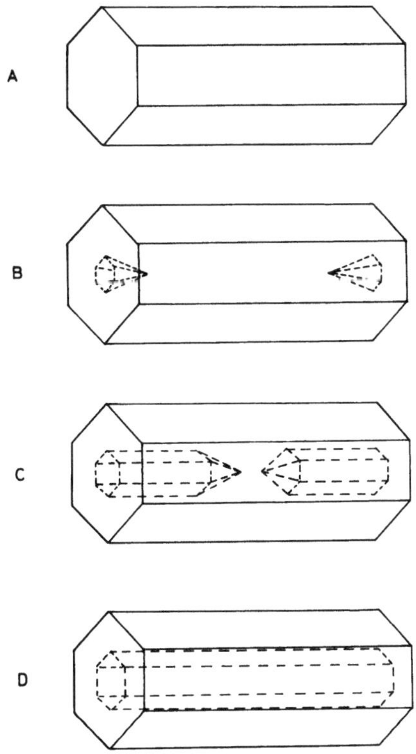

FIG. 8 - Dissolution stages of one crystallite: A) Before acid attack. B) Etch pit formation. C) and D) Progressive stages.

dislocations the vector to complete any circuit is zero. If, however, a dislocation is enclosed in the Burgers circuit, the closure failure is the Burger vector b. Edge dislocations and screw dislocations have a Burgers vector perpendicular or parallel to the dislocation line, respectively.

Around the point where the dislocation line emerges from the crystal surface, several atoms have shifted from minimum energy positions; this is the origin of the preferential dissolution around this special spot.

There is evidence that dislocations are indeed responsible for the preferential dissolution in enamel:

a) Dislocations have been observed in enamel (25,38).
b) In contact with acids, first an "etch pit" (see Fig. 8) is formed and second a central core is dissolved.
c) In contact with acids the destruction follows the crystallographic planes [100] and [010] parallel to the c-axis.

Indirect evidence is based on the following:

a) Energy considerations based on anisotropic elasticity theories indicate that screw dislocations with b parallel to the c-axis have the lowest energy among several other defects (5).
b) Similar anisotropic dissolution phenomena have been observed in synthetic hydroxyapatite (3,30,33).

Recently it has been calculated that dislocations in enamel and apatites are most likely of the "hollow core type" (2,4). This means that around the dislocation a highly strained area exists. In enamel the diameter of the strained core is on the order of 10 nm. Most likely the dislocation (including the highly strained cylinder around it) is caused by carbonate (CO_3) in the crystal lattice. In enamel it may be possible that the highly strained area in the control core of the crystallites is the result of a remaining precursor phase of hydroxyapatite. From the preceding it is obvious that enamel crystallites do not have uniform properties; active sites play a role during dissolution.

The author presumes that inhibitors, e.g., F, diphosphonates, etc., reduce or prevent crystallite dissolution during the caries process. The inhibitors are either adsorbed locally (where the dislocation line meets the surface) or are adsorbed onto the surface. The inhibitors adsorbed to the enamel surface have most likely two effects during a caries attack:

a) They prevent or retard dissolution of the crystal surface by preventing H^+ ions from reaching the surface.
b) They prevent the preferential dissolution discussed above. From model experiments with fluorapatite it is known that

after a treatment with EHDP the number of preferential dissolutions can be reduced 2-3 orders of magnitude (29).

REFERENCES

(1) Arends, J., and Davidson, C.L. 1975. HPO_4^{2-} content in enamel and in artificial carious lesions. Calc. Tiss. Res. 18: 65-79.

(2) Arends, J., and Jongebloed, W.L. 1977. Dislocations and dissolution in apatites. Caries Res. 11: 186-189.

(3) Arends, J., and Jongebloed, W.L. 1979. Ultrastructural studies of synthetic apatite. J. Dent. Res. 58(B): 837-842.

(4) Arends, J., and Jongebloed, W.L. 1981. Apatite single crystals, formation, dissolution and influence of CO_3^{2-} ions. J. Royal Dutch Chem. Soc. 100: 3-9.

(5) Arends, J.; Royce, B.S.H.; Welch, D.O.; and Smoluchowski, R. 1971. Lattice defects and the electrical and transport properties of apatite. In Symposium on Structural Properties of Hydroxyapatite, Gaithersburg, 1969, ed. R.E. Young. New York: Benjamin.

(6) Arends, J., and Schuthof, J. 1980. Microhardness and lesion depth studies of artificial carious lesions: a comparison of gelatin and HEC based systems. J. Biol. Buccale 8: 175-181.

(7) Arends, J.; Schuthof, J.; and Jongebloed, W.L. 1980. Lesion depth and microhardness indentations on artificial white spot lesions. Caries Res. 14: 190-195.

(8) Atkinson, H.F., and Matthews, F. 1949. An investigation into the organic components of the human tooth. Brit. Dent. J. 86: 167-177.

(9) Bartheld, F., von. 1958. Decalcification in initial caries. Tijdschrift voor Tandheelkunde 65: 76-88.

(10) Bartheld, F., von. 1961. Membrane phenomena in carious dissolution of teeth. Archs Oral Biol. (spec. suppl.) 6: 284-303.

(11) Christoffersen, J., and Arends, J. 1982. Progress of artificial carious lesions in enamel. Caries Res., in press.

(12) Christoffersen, J., and Christoffersen, M.R. 1981. Kinetics of dissolution of calcium hydroxyapatite IV. The effect of some biologically important inhibitors. J. Cryst. Growth 53: 42-51.

(13) Davidson, C.L.; Arends, J.; and Hoekstra, I. 1976. Density changes in enamel after decalcification. J. Biomech. 9: 81-85.

(14) De Rooij, J.F. 1981. Diffusion of phosphate and MFP ions in bovine enamel. Thesis, State University of Groningen.

(15) De Rooij, J.F., and Arends, J. 1980. Phosphate diffusion in whole bovine enamel at pH = 7. Caries Res. 14: 393-402.

(16) Featherstone, J.D.B.; Duncan, J.F.; and Cutress, T.W. 1978. Surface layer phenomena in in vitro early caries lesions of human tooth enamel. Archs Oral Biol. 23: 397-404.

(17) Featherstone, J.D.B.; Duncan, J.F.; and Cutress, T.W. 1979. A mechanism for dental caries based on chemical processes and diffusion phenomena during in vitro caries simulation on human tooth enamel. Archs Oral Biol. 24: 101-112.

(18) Featherstone, J.D.B., and Mellberg, J.R. 1981. Relative rates of progress of artificial carious lesions in bovine ovine and human enamel. Caries Res. 15: 109-114.

(19) Flim, G.J., and Arends, J. 1977. Diffusion of ^{45}Ca in bovine enamel. Calc. Tiss. Res. 24: 59-64.

(20) Francis, M.D.; Brines, W.W.; and Gray, J.A. 1973. Chemical agents in the control of calcification processes in biological systems. In Hard Tissue Growth, Repair and Remineralization, CIBA Foundation Symposia, nr. 11, pp. 57-90. Amsterdam: Elsevier.

(21) Gray, J.A. 1977. Chemical events during cariogenesis. In Proceedings of the Symposium on Incipient Caries of Enamel, November 1977, Michigan.

(22) Gray, J.A., and Francis, M.D. 1963. Physical chemistry of enamel dissolution. In Mechanisms of Hard Tissue Destruction, publication nr. 75, pp. 213-260. Washington: American Association for the Advancement of Science.

(23) Groeneveld, A., and Arends, J. 1975. Influence of pH and demineralization period on mineral content, thickness of surface layer and depth of artificial carious lesions. Caries Res. 9: 36-44.

(24) Higuchi, W.I.; Valvani, S.C.; and Hefferen, J.J. 1974. The kinetics and mechanisms of reactions of human tooth enamel. Archs Oral Biol. 19: 737-746.

(25) Hirai, G., and Fearnhead, R.W. 1972. Lattice defects in biological and synthetic apatites. Proceedings 6th International Conference in X-ray Optics and Microanalysis, pp. 863-872. University of Tokyo Press.

(26) Hirth, J.P., and Lothe, J. 1968. Theory of Dislocations. New York: McGraw-Hill.

(27) Hollinghead, E.A., and Gordon, A.R. 1941. The differential diffusion constant of calcium chloride in aqueous solution. J. Chem. Phys. 9: 152-153.

(28) Holly, F.J., and Gray, J.A. 1968. Mechanism for incipient carious lesion growth utilizing a physical model based on diffusion concepts. Archs Oral Biol. 13: 319-334.

(29) Jongebloed, W.L. 1976. An ultrastructural study on the caries process. Thesis, State University of Groningen.

(30) Jongebloed, W.L.; Molenaar, I.; and Arends, J. 1973. Orientation dependent etchpit penetration and dissolution of fluorapatite. Caries Res. 7: 154-165.

(31) Jongebloed, W.L.; Molenaar, I.; and Arends, J. 1975. Effects of MFP in the pathways followed by cariogenic agents. Caries Res. 9: 388-403.

(32) Jongebloed, W.L.; Molenaar, I.; and Arends, J. 1975. Morphology and size distribution of sound and acid treated enamel crystallites. Calc. Tiss. Res. 19: 109-123.

(33) Jongebloed, W.L.; van den Berg, P.J.; and Arends, J. 1974. The dissolution of hydroxyapatite single crystals in acids. Calc. Tiss. Res. 15: 1-9.

(34) Miller, W.D. 1890. The Microorganisms of the Human Mouth. Philadelphia: S.S. White Manufacturing Company.

(35) Moreno, E.C., and Zahradnik, R.T. 1974. Chemistry of enamel subsurface demineralization in vitro. J. Dent. Res. 53: 226-235.

(36) Moreno, E.C., and Zahradnik, R.T. 1979. Demineralization and remineralization of dental enamel. J. Dent. Res. 58 (special issue B): 896-902.

(37) Schatz, A., and Martin, J. 1958. Some perspectives of dental caries research. Ann. Dent. 17: 1-11.

(38) Selvig, K.A. 1970. Periodic lattice images of hydroxyapatite crystals in bone and dental hard tissue. Calc. Tiss. Res. 6: 227-238.

(39) Silverstone, L.M. 1977. Remineralization phenomena. Caries Res. 11 (suppl. 1): 59-84.

(40) Swancar, J.R.; Scott, D.B.; Simmelink, J.W.; and Smith, T.J. 1971. The morphology of enamel crystals. In Tooth Enamel II, eds. R.W. Fearnhead and M.V. Stack, pp. 72-76. Bristol: Wright Ltd.

(41) Van Dijk, J.W.E.; Borggreven, J.M.P.M.; and Driessens, F.C.M. 1979. Chemical and mathematical simulation of caries. Caries Res. 13: 169-180.

(42) Voegel, J.C., and Frank, R.M. 1974. Microscopie électronique de haute résolution du cristal d'apatite d'énamel humain et de sa dissolution carieuse. J. Biol. Buccale 2: 39-50.

(43) Voegel, J.C., and Frank, R.M. 1977. Ultrastructural study of apatite crystal dissolution in human bone and dentine. J. Biol. Buccale 5: 181-194.

(44) Zahradnik, R.T., and Moreno, E.C. 1977. Progressive stages of subsurface demineralization of human tooth enamel. Archs Oral Biol. 22: 585-591.

Group on Structure, Properties, and Function of Mineralized Tissue Components

Standing, left to right: H.J. Arnott, K.P.H. Pritzker, G.P. Krampitz, K.J. Münzenberg, R.K. Schenk, W. Dosch, D.J. Prockop.
Seated: V.C. Hascall, N. Watabe, R.A. Young, J.L. Matthews, W.E. Brown.

Biological Mineralization and Demineralization, ed. G.H. Nancollas, pp. 327-348.
Dahlem Konferenzen 1982. Berlin, Heidelberg, New York: Springer-Verlag.

Structure, Properties, and Function of Mineralized Tissue Components
Group Report

J. L. Matthews, Rapporteur
H. J. Arnott, W. E. Brown, W. Dosch, V. C. Hascall,
R. Hautmann, G. P. Krampitz, K. J. Münzenberg,
K. P. H. Pritzker, D. J. Prockop, R. K. Schenk, N. Watabe,
R. A. Young

INTRODUCTION

A survey of mineralized tissues in nature reveals a variety of crystal forms and mineral components associated with a diverse group of organic matrices, cell types and distribution, bathing fluids, and activities. Mineralized "compartments" are found in association with unicellular plants and animals and in species of most phyla of multicellular plants and animals. These minerals may be found in both intracellular preferred sites and in uniquely characteristic forms and sites in matrices. It is likely that this diversity of form and distribution is necessary to achieve the functions essential for the survival of living systems in differing environments, including conservation of elements, regulation of electrolytes and pH, support, locomotion, etc. The precision of replication of mineral form, type, and distribution in each species is impressive and argues for the existence of finely tuned initiating and regulatory factors, restrictive volumes, and critical surfaces that modulate these mineral forms. Although much is known about the physical dynamics of these minerals from in vitro studies (precipitation phenomena, surface energies, equilibration forms, nucleation requirements, lattice configurations, etc.), we still know little about the essential conditions of nucleation, growth, form, and dynamics of these minerals in biological environments and less about how these minerals interact with the matrix surfaces, the sources of the minerals, fluid flow rates, diffusion, facilitated transport, modifications that are essential to nucleation, etc. To examine these phenomena in an ordered fashion, we elected to select a few examples of different mineralizing systems with the purpose of asking:

1. How is structure of matrix related to the consequences of mineral-matrix interactions?
2. How do bulk and surface structures differ in detail among minerals of intracellular, extracellular, and matrix-free origins?
3. How do physicochemical/biological properties (e.g., dissolution, growth, morphology, space filling, attachment, lysing, etc.) of the biominerals depend on structural detail at both atomic and ultrastructural levels? What are the structures and properties of biomineral aggregates (attachment, heterogeneity, order, strength, etc.)?
4. What methods, new technology, and experimental constraints are necessary to further understanding?

The specific tissues selected for examination are not representative of all of the mineral systems in nature but were chosen as they have received more attention to date by researchers. The systems reviewed were: cartilage, bone, tooth, some aragonite-, calcite-, and vaterite-related invertebrates, and oxalate-forming plants.

STRUCTURE AND FUNCTION IN CARTILAGE AND MINERALIZED CARTILAGE

Epiphyseal cartilage is a useful model for study, as it possesses both temporal and spatial organization. Furthermore, the tissue is avascular in that it lacks a well-defined capillary bed, and the cells are sustained in their matrix lacunae via diffusion of extracellular fluid (ECF) through the interstitial spaces. Cartilage in growth plates is organized such that one region of the plate is principally involved in cell proliferation activities and is followed, in succession, by regions of matrix synthesis, cell hypertrophy, matrix compression, matrix mineralization, and cell death. A steady state exists for the growth period of the animal such that increases in plate height attained by proliferation and interstitial matrix synthesis are matched by matrix mineralization, cell death (partial removal), and bone apposition onto the opposite surface (see Schenk, this volume). Interestingly, longitudinal septa mineralize while transverse septa do not. Changes in the cell volume in the hypertrophic zone when compared to those in the proliferative zone are dramatic, representing approximately a 2.5x increase. The increased cell height, supported by its surrounding matrix (also increased in volume and later mineralized), may account for the growth in length of the plate and represents a good model for physical scientists to study fluid dynamics in connective tissues. Initiation of mineralization occurs in a relatively narrow zone near the metaphysis which traverses the entire plate! Thus both spatial and temporal events converge to achieve the appropriate milieu (compartment - surface and/or saturation) to achieve mineral formation here. It is noted that cartilages other than growth cartilage do not mineralize unless they are invaded by blood vessels or extracted with guanidinium chloride or trypsin (which extracts several factors, including some that act as an anticollagenase and as an anti-endothelial cell proliferation factor).

What distinguishes these cartilages that mineralize from those that do not? What distinguishes the mineralizing zone? The matrices of both mineralizing and non-mineralizing cartilages contain Type II collagen which forms a fibrillar network (see Prockop and Williams, this volume) and proteoglycans (Hascall and Lowther, this volume) that form a condensed phase within this network, where each macromolecule is constrained to occupy much less than its full capacity for expansion observed in ideal solutions of larger volume. Thus the proteoglycan phase exerts a swelling pressure on the collagen network which provides the tissue with a structural organization ideally suited to bear weight with strength and resiliency. Are the major structural components modified at these sites of first mineral formation? Do they remain unchanged in the mineralized form? Are they removed from the site? Do they have special charge or ion binding functions that facilitate nucleation? What role do the cells play in achieving mineralizing conditions? Are vesicles derived from these cells by budding causally related to the later initiation of mineralization? What other organic components may be implicated? These questions were raised, but many were incompletely answered, often due to the limitations of investigative means and models and/or time afforded to the question. A clear recommendation was made for renewed efforts to find means of fixing these materials that achieve retention of location and configuration of organic and cellular elements and ions with minimal translocation in order to study components in an integrated system.

The site of onset of mineralization contains trilaminar membrane vesicles situated in longitudinal matrix septa between cell columns which are produced by the chondrocytes at the level of the proliferating zone. At the level of the hypertrophic cells, they are first observed to contain apatite crystals. As these appear to be the first sites for crystal formation, much investigative work with electron probes has been directed towards assessing at what zone in the interterritorial matrix the vesicles first show increased contents of calcium and phosphate over that of the surrounding fluid. Unfortunately, available data presently differ markedly on this question, ranging from evidence that they contain sufficient Ca^{2+} and HPO_4^{2-} for mineral deposition in ideal circumstances when they first appear, to evidence that the ions increase in concentration in the vesicles as they approach the zone of first crystal formation. Variance in data on matrix (non-vesicular) levels of calcium and phosphate also exists (4). Vesicles have been studied in detail in both intact and in vitro systems and are considered in more detail in the proceedings of the 1981 Vesicle Conference at Spoleto, Italy (1). Although some investigators report that the first mineral crystals form in (or on) this unique "compartment," this view is controversial, as others report finding no mineral in vesicles but rather the deposition in the fibrillar substance. With further crystal growth in the matrix, other questions arise. Is the collagen modified at this site (an important question, as the crystals ultimately surround the fibrils and form within the proteoglycan phase)? Collagen is believed not to change its fibril size from one zone to the next, but an interesting suggestion based upon recent work from several laboratories indicates that collagen types other than Type II $[\alpha 1\ (II)]_3$ are present in the cartilage. Some of these may be present in the mineralizing zone. There is evidence that cells can be made to express different collagen types under other conditions. The possibility of sectioning cartilage into zones (proliferative-hypertrophic) to test the ability of Type II collagen to mineralize was proposed with the goal of comparing it with Type I collagen, which in bone, dentin, and sharpeys fibers, etc., is intimately associated with mineral.

Proteoglycan (PG) configuration and size were compared to other matrix components to evaluate their spatial requirements. In ideal solutions at low solute concentration, growth plate cartilage proteoglycans occupy approximately 50 ml solvent/g, but due to their tissue concentration, they occupy only about 1/5 of their potential volume. This partial compression appears important for effectively fulfilling one of their major functions in cartilage, namely, to provide support and resistance to compressive loading (see Hascall and Lowther, this volume). The absolute physical and chemical state of proteoglycans in situ and their steric relationship with other matrix elements await further elaboration. A study of proteoglycan molecules (i.e., their physical properties at high solute concentration) would be useful. Morphological evidence (Schenk, this volume) indicates close apposition of proteoglycans to both plasma cell membrane (plasmalemma) and vesicle membrane, suggesting specific interactions at these sites. Are these associations mediated by interactions similar to or distinct from those involved in the well-defined hyaluronic acid mediated aggregation of proteoglycans? Would disruption of vesicles associated with (or just prior to) initiation of mineralization account for the decrease in the "super-aggregated" proteoglycan fraction observed by Pita and Howell (6)? Their high ester sulfate and carboxylate residues and their ubiquitous distribution in cartilage raise the question as to the possible role these polyanions may play in facilitating cation (calcium) or impeding anion (phosphate) transport through the matrix and even as transient mineral ion binding sites to facilitate mineral accretion. For example, if the negative phosphate ions are impeded from moving through the proteoglycan phase because of the high fixed negative charge, what mechanisms are available to bring this anion to the mineralizing site (perhaps a neutral precursor form, phospho-lipo-protein, with activation at the site?)? Detailed analysis of the physical chemistry of selective ion binding to the fixed polyelectrolyte phase at high PG concentration and in electrolyte solutions with similar concentrations as those found in the matrix would provide valuable information, since it appears likely that the principal route of appropriate ions or precursors from sources (cells, capillaries) to mineralization sites (vesicles, matrix fibrils) must be through the hydrated PG phase. Even in its condensed macromolecular state, the PG phase is still primarily solvent, providing ample space for facilitated (or impeded) transport and even for crystal formation. (Whether PG is disaggregated, removed, or remains intact during mineralization is still an unresolved question.)

In examining the phenomenon of cell hypertrophy, functional implications of accompanying organellar enlargement were also made. The similarity between the growth afforded by the enlargement of cartilage cells (surrounded by their proteoglycan boundary and ultimately stabilized by mineralization of the longitudinal septa) and the cells of plant tissues (associated by cellulose and later stabilized by lignin) was discussed. Some investigative work on cartilage matrix and cells suggests that both Na^+ and K^+ are increased and that water enters the cells due to a) differences in electrolyte content and b) likely differences in the permeability of the hypertrophic cells compared to other cells in the column. Is this a "programmed" phenomenon? The field of fluid content and fluxes in premineralized tissue matrices has been investigated by only a few workers. Pita and Howell (6) have analyzed some of these fluids and report significant differences from ECF. The presence of fixed and highly charged molecules in matrices were considered with regard to their influence on electrolyte flow, Donnan equilibrium, and the possible presence of significant galvanic potentials. It is clear that a membrane need not separate the fluid phases when charged components (proteoglycans, etc.) are fixed in space, and that much might be learned from additional experiments using available physical methods for measuring net exchanges. These data will be important to determine the accessibility and time dimensions for transport of essential ions such as calcium and phosphate. The vertical orientation of cartilage was equated with the metaphyseal blood vascular source, and it was suggested that the phosphate ion originated from this site as it is closer to the zone of mineralization. The need for more information on modes of phosphate transport was emphasized.

An emerging field of research was discussed relative to the use of mutants for evaluating function of affected molecules. An example of this was discussed from recent work on the cartilage matrix deficient (cmd/cmd) mouse (Barrach and Merker, personal communication). In this mutant, the core protein for the major cartilage proteoglycan is absent, and since this precursor protein is missing, there is no proteoglycan phase in the matrix. The overall cartilage did not show columnar cell organization, but a progression of cell types from resting to very abnormal, pyknotic hypertrophic cells was found. No mineral was found in the matrix, even though normal Type II collagen was present and subjacent bone did mineralize. The absence of PG in the matrix casts doubt on models of cartilage mineralization which assign an obligatory role to PG's in mineralization. This experimental approach of nature was appreciated and elicited interest in finding other useful mutant models. Other available mutants cited were those affecting collagen (Type I), including a patient with Ehlers-Danlos Syndrome with a structural defect in the pro-alpha-2 propeptide which prevents its enzymatic removal (8). One consequence of this defect is a partial loss of the collagen "holes." Interestingly, the patient had no apparent defect in bone structure. Patients with osteogenesis imperfecta, many of whom have genetic defects in the structure or synthesis of collagen, also offer important models for study.

Numerous other noncollagen proteins are present that need further study, including glycoproteins and enzymes. The orderly succession of cell and matrix changes that occur from one zone to the next in cartilage make this an ideal model for examining mineralization in complex organisms.

BONE MORPHOLOGY AND FUNCTION

Bone is usually lined by cells on its free surfaces and has a premineralized (osteoid) and a mineralized phase. Bone consists principally of collagen (Type I), other non-collagenous proteins, and small PG molecules with few branches that are distinctly different from cartilage PG and are impregnated with small hydroxyapatite crystals oriented with respect to the collagen fibers. Consideration was given to other potential structural arrangements of collagen fibrils. Recent analyses indicate that X-ray diffraction patterns of Type I collagen fibers appear to be more consistent with a quasi-hexagonal packing (Hulmes and Miller) than with the earlier pentamer model of Miller et al. (see Prockop, this volume). The quasi-hexagon pattern should now be further studied in terms of observed mineral distribution patterns, and the entire "hole theory" should be reviewed in light of this model. The Ehlers-Danlos patient with deficient hole space (1/2 normal) particularly raised the question as to the precise role these "holes" play in determining mineral distribution.

The fluid of bone per se has been indirectly evaluated by equilibration tests in various media and is thought to be uniquely different from general extracellular fluid (ECF), i.e., high K^+, lower Ca^{2+} and phosphate. Consequently, this partitioning of fluids infers the presence of a functionally limiting membrane, the location of which is not established, although the cells lining the surface have been implicated. These cells, however, do not possess tight junctions, and tests with markers such as horseradish peroxidase and ferritin indicate that relatively large proteins, as well as ions and water, readily flow into the spaces between crystals and deeply buried bone cells. Again, an understanding of the true makeup of the bone fluid and factors that control its exchange is needed.

It is recognized that bone morphology shows considerable variation in form, particularly between woven bone (found in embryonal sites and repair sites) and lamellar bone (a more ordered morphology) that has precise mineralized collagen lamellae demonstrable with polarized light. The latter form was considered in some detail, as it was felt that the higher degree of order might be more readily studied by physical means. In developing lamellar bone, the mineralized surface is separated from the matrix secreting cells by an approximately 10 μm-thick layer of osteoid consisting primarily of collagen fibers and non-collagenous proteins, proteoglycans, and bone fluid. The width of this layer is maintained by simultaneous new osteoid formation and advancement of the mineralizing front (this front being typified by the presence of phosphoproteins and phospholipids). The "lag" in mineral formation presents an interesting experimental challenge: Are the deeper collagen fibers more "mature"? Is there a size dependency? Does further assembly of collagen molecules occur here? (Fibril assembly is observed at the cell.) Does an inhibitor have to be removed? What role do the phosphoproteins play here? If calcium and phosphate come from within the cells, why does mineralization not occur next to the cell? Do other bone-cell-produced molecules such as osteonectin have to diffuse to this site and "condition" the collagen for mineral deposition? (It was reported that osteonectin binds to collagen and that, after binding to collagen, it can nucleate hydroxyapatite solutions (see Termine et al. (11)). This temporal-spatial model was interpreted to represent a "cascade of reactions" worthy of increased attention. The evidence reviewed clearly suggests that special molecular-matrix interactions make this Type I collagen mineralizable and distinguish it from other Type I collagen (skin, tendon) where these bone non-collagenous proteins are not available. The proteoglycan molecule here is very small, occupying much less matrix than proteoglycans of cartilage (Termine and Heinegard, unpublished data). Its function is not apparent nor is that of vitamin K-dependent osteocalcin, α2HS glycoprotein, 24,000 D phosphorylated protein, 62,000 D glycoprotein, or even osteonectin. The significance of large variations in amounts of these proteins in young and old, and in woven and lamellar bone is unknown. It is established that some of these molecules in solution can inhibit crystal formation, while they may have a different effect when bound to some other matrix unit (10). Possibly some configurational change may occur when these proteins interact that is different when isolates are used under in vitro conditions.

Still another useful model is that of rickets, in which the cartilage growth plate widens as does the osteoid seam of bone, thus lacking the ability to mineralize. Osteonectin is reduced in this rachitic bone osteoid (Termine et al., unpublished data). The specificity of antigen-antibody reactions offers still another powerful tool for locating the matrix sites of various components. Highly specific collagen type antibodies are now available and antibodies for other matrix molecules would be useful. Additional technical help is also needed to develop fluorophores that can be specifically attached to different atomic components. A recent review of developing technology for bone research is cited in the reference list (7).

The ultimate experimental goal was felt to be the successful reconstitution of a functioning mineralizing system from isolated components. Lamellar bone offers the unique opportunity to study events in clearly morphologically separable zones.

STRUCTURE AND FUNCTION OF PLANTS

In contrast to cartilage and bone calcification, the formation of mineral (calcium oxalate) deposits in higher plants occurs inside cells, and the relationship between the cell and mineral is unambiguous. Calcium oxalate formation occurs in a compartment(s) inside the vacuole. The process of crystal formation has been studied sufficiently in plants by a number of investigators to allow for a general model of mineralization, even though certain details vary from species to species. Briefly, at some stage in development, the cell forms a vacuole, within which a series of membranes, tubules, and dense areas are formed. These membranes form chambers within which crystal development occurs. Whether one or hundreds of chambers are formed, the crystals develop within the chamber. The chambers are usually associated with tubules or membranes which are found in large numbers within the vacuole. Many authors have identified proteoglycan complexes in developing vacuoles. (These are presented in detail by Arnott, this volume.) Other plant mineral systems are reviewed by Pentecost (5).

Questions that were raised include:

1. What is the nature of the membranes within the vacuole, and how do they relate to the proteoglycans?
2. How is such a system elaborated?
3. Where are the oxalate and calcium ions before crystallization?
4. What is the function of these minerals?
5. How is the calcium and/or oxalate re-utilized by the cell, and how can they be expelled from the cell?
6. How long do plant crystals take to develop?
7. What are the structural details of the proteoglycans, their association with the mineral oxalate, and their ability to regulate nucleation?

The availability of these specimens at various stages offers a magnificent model for study of matrix-mineral relationships. At present, few investigators are using these models, so that few of the questions raised can be approached.

INVERTEBRATE STRUCTURE AND FUNCTION

The variety of these interesting animals is detailed by Crenshaw and Krampitz (both this volume) and Watabe (13). A few representative species were selected to illustrate the variations employed in producing mineral structures which can be identified by their morphology. Most of the calcified structures in the invertebrate are composed of $CaCO_3$ in the form of calcite, aragonite, or vaterite and organic matrix. Vaterite is the least stable and is less common than the other two minerals. These calcified structures show a great diversity in morphology and type of crystals formed by the organisms. They can be categorized in a way dependent on the site of formation:

1. Extracellular sites (outside of the animal body): Examples are some of the mollusc shells which are composed of parallel layers of crystals of aragonite. The organic matrix is found between the layers as interlaminar matrix, between the crystals as intercrystalline matrix, and within the crystals as intracrystalline matrix. The presence of the intracrystalline matrix is controversial. Mantle epithelium covers this crystalline layer and secretes shell substance.
2. Extracellular sites (inside the animal body): Calcareous sponge spicules and sea urchin spines are formed by sclerocytes, but the crystals are extracellular. Both structures are made of high magnesium calcite, and the entire structure of each is a single crystal. No organic matrix has been found to be associated with the crystal.
3. Intracellular sites: Octocoralian spicules are examples of this type of formation. The spicules are formed within vacuoles probably derived from the Golgi complex. First, organic matrix is deposited, followed by calcite crystal formation. The crystals grow and the vacuole expands

the volume. In one species, the crystals may remain within the vacuoles. In the other, the vacuole membranes fuse with the plasma membrane and a breach is formed, resulting in the extrusion of the spicule into the extracellular environment. The spicule is composed of an aggregate of small calcite needles sandwiched by an organic matrix. The morphology of the mature spicule is very regular.

The common features of most invertebrate calcification are the association of organic matrix and the well-regulated, ornate form of the composite end products. What is unknown and important to learn are: a) the factors which regulate the form of the calcified structures; b) the roles of soluble and insoluble components of the organic matrix in nucleation, acceleration, and inhibition of growth of $CaCO_3$ crystals; c) the factors which control the crystal polymorphism, the nature and concentrations of the inorganic ions, and the structural and chemical natures of soluble or insoluble components of the organic matrix; and d) what characterizes the organic matrix at the time of crystal nucleation and after the maturation of the mineral form.

Some efforts to extract organic matrices of oyster shells have suggested a matrix that contains at least ten soluble and some insoluble components. One insoluble component has been called conchiolin and two soluble ones have been studied in detail, both of which have the capacity to bind calcium ions. These preliminary data again indicate that differences exist in the organic components of the premineralized and mineralized structure (3). Invertebrate examples of unique crystal-cell or crystal-matrix interactions call for further attention.

TOOTH STRUCTURE AND FUNCTION

The structural organization of two dental tissues, dentin and enamel, were considered. It is appropriate to consider these two together, for although they have a totally different morphology of crystals and matrix components, their initial embryonal formation is interdependent. It is possible that the mineral phase that always forms in dentin first has some initiating role on the adjacent pre-enamel tissue, which subsequently mineralizes.

Enamel is formed by ameloblasts, columnar-shaped cells that contain an abundance of rough endoplasmic reticulum, a Golgi complex, and secretory vesicles and tubules that accumulate in the dentin-directed distal end of the cells in a process called Tomes process. Enamel development occurs in at least three phases: the forming stage, the maturing stage, and the mature stage. In the forming stage, it consists of partially mineralized tissue with a high concentration of enamel protein, while in the maturing stage it is highly mineralized. The protein matrix of dental enamel changes in content and composition in its maturation phases (9). The protein matrix is originally rich in proline, glutamic acid, leucine, and histidine, but it shows increased composition levels of aspartic acid, serine, glycine, and glutamic acid in the mature (fully mineralized) state. The proteins characteristic of the immature matrix are called amelogenins (9). Interestingly, the amelogenins have a hydrophobic protein index of 70. The amelogenin-crystal mineral complex of enamel has been likened to a gel-crystal system, possibly involving hydrophobicity as an important driving force for mineral growth.

Proteins of the mature enamel are crystal-bound and are called enamelins (9). Two explanations are offered for the protein composition differences (amino acid composition, molecular weight, etc.) in the different phases: a) degradation of a high molecular weight pro-enamel-protein precursor with selective removal of some fragments, and b) preferential degradation and removal of amelogenin proteins leaving crystal-bound enamelin proteins. Thus enamel consists primarily of protein when first secreted from Tomes processes, but this labile gel is rapidly modified and removed as the continuous process of mineralization occurs, ultimately resulting in a tissue that consists of up to 99 percent mineral and a fraction of the original organic material.

Apatite crystals grow into newly secreted matrix within a short morphological and temporal interval (2). The enamel crystals closely resemble mineral hydroxyapatite crystals in their birefringent, X-ray diffraction, and electron diffraction properties. They are imperfect crystals, showing a carbonate content of about 2-4 percent, some molecular water, a hydroxyl group deficiency, and 3-5 percent acid phosphate. Considerable variation likely occurs between species and even within the same tooth. The crystals are long and thin, appearing first as flattened hexagons, but the hexagonal form is often lost with further growth. The crystals are initially ribbon-like rather than needle-shaped. Boyde reports that the number of crystals is established early in the formation phase, and that crystal growth accounts for further maturation. The newly formed crystals are generally formed with their axes approximating a position perpendicular to the surface of the first formed enamel, having an original diameter of 10 nm that grows to 40-60 nm. As the enamel matrix is secreted by the conically shaped Tomes processes, the newly secreted matrix surrounds these processes that create "pits" on the enamel matrix surface. As the crystals are oriented with respect to this irregular surface, discontinuities in enamel crystal organization become apparent at the planes at which crystallite orientation abruptly changes between domains of adjacent process surfaces, producing the prism pattern that typified mature enamel. The prism domains are invested in a prism sheath that also represents residual enamel matrix, but it should be noted that the prism patterns show many variations in discontinuity, and that it might be appropriate to consider enamel as a single continuous phase (2). The grouping of crystals into domains that show crystal orientation provides strong morphological evidence for some regulatory activity of the matrix in which the crystals first nucleate and then grow. X-ray diffraction data on enamel proteins suggest evidence for an organization of the first phase protein molecules into parallel arrays of pleated sheets, separated by approximately 10°A (a width sufficient to contain a unit hydroxyapatite crystal with a basal plane length of

9°A). The proteins are acidic, having proton accepting groups, which can also bind calcium. It thus seems plausible that these proteins may bind crystal "compartments," in which the necessary pH and PO_4^{3-}, Ca^{2+}, and OH^- concentrations necessary for apatite nucleation are achieved (12). The orientation of the protein sheets could then have dual functions of defining nucleation surface boundaries for compartments and influencing crystal orientation! Still to be explained is the large size of the crystals formed here as compared with those in cartilage, bone, and dentin. This system was considered to be a prime model for the convergence of views of physical scientists, chemists, and biologists, as the crystals are large and unique in morphology, and the matrix in which they are formed is apparently limited to a workable number of components. It was noted that enamel formation occurs without any evidence that mineralizing matrix vesicles are involved, and this matrix is also free of proteoglycans, making this a simpler model to evaluate.

Dentin structure resembles lamellar bone structure in several aspects. Dentin-forming cells (odontoblasts, derived from neuro-ectomesenchyme) are initially aligned facing the ameloblasts and move away incrementally, secreting a predentin matrix on the enamel-facing side. This matrix is principally Type I collagen, but it also contains calcium-binding phosphoproteins, acidic glycoproteins, and a small amount of proteoglycans. Phosphoproteins comprise the major non-collagenous matrix components. Some of these contain hexoseamine and sialic acid, but nothing is presently known of the relationships of the carbohydrate moieties to the structure or function of the phosphoproteins, nor of the interactions of these molecules with the collagen and other non-collagenous proteins (10). They deserve further attention as their demonstrated high affinity for apatite and their inhibitory activity upon calcium phosphate crystallization in in vitro systems suggest that they may play a cardinal role in dentin mineralization. It has been suggested by some investigators that mineral-matrix vesicles are produced by the odontoblasts at the earliest formed site of predentin formation and that they are the first site of crystal formation in dentin. Little is known about the properties of these vesicles, and they are not believed to serve a role in subsequent phases of dentin apposition and mineralization. The growing dentin is characterized by lamellae of mineralized dentin, a surface layer of "osteoid-like" predentin (about (about 2-5 μm thick), and the odontoblasts. Long processes of these cells, connected with tight junctions, extend in tubules in the matrix to the dentin-enamel surface. The collagen bounding these tubules is oriented differently, and non-collagenous proteins are concentrated here. Thus, dentin is supplied with living cell processes, but unlike bone, contains no cell bodies in the mineral phase. The fluid within dentin is partitioned from the pulp extracellular fluid by the odontoblasts, probably accounting for the lack of much exchange of Ca^{2+} and HPO_4^{2-} which occurs in bone to maintain appropriate plasma calcium levels. Marker proteins, however, have been shown to traverse this odontoblast layer. The interesting work of Weinstock and Leblond (14) is noteworthy, since they suggest rapid segregation of some organic molecules to the mineralizing front as compared with others (collagen), which are deposited initially at the cell surface. The crystals are bone-like needles of apatite that apparently have a spatial relation to the collagen fibers, as in bone. It is possible to mark mineralization fronts with fluorescent markers such as tetracyclines and to arrest mineralization with high levels of inhibitors such as diphosphonates, making this system available for experimental manipulation and histomorphometric analysis. By using inhibitors, the predentin area can be widened, making the critical zones more accessible. These same techniques are being used in bone and other tissues. Teeth thus represent a powerful model for further study.

MINERAL STRUCTURE AND FUNCTION

The properties of minerals, including biominerals, are explained by their structures. Details of some of the basic crystal structures and properties are presented by Young and Brown (this volume). Logically, analysis of crystal structure, properties, and function is based upon in vitro studies of pure systems, such that the thermodynamics, saturation requirements, kinetics, and influences of impurities and defects can be defined in absolute terms. With this foundation, hope was expressed that physical scientists will further extend their technology to include minerals of biological origin. At least four areas need to be investigated: a) What are the structures of the idealized crystalline forms, and what are the quantities and types of impurities in biominerals? b) What are the types and locations of impurities and defects within the structures of biominerals? c) How are the defects and/or impurities incorporated, i.e., during nucleation and growth, or from precursors? d) What are the effects of defects and impurities on both the atomic structure and ultrastructure, and on their chemical and physical properties?

From a review of the different animal and plant examples, an enormous variety of mineral types and crystal shapes were elaborated. Some of these appear to be produced from nucleation of solutions directly into the final mineral form, while others appear to be associated in some way with matrix or cell surfaces that may even be incorporated into crystal interstices. In reviewing the type of crystals found in nature, it is clear that some crystals that might be considered as precursors to other crystal forms (from in vitro data based on X-ray diffraction) are found in some biological minerals. Octacalcium phosphate, for example, has been reported in fetal bone, dental calculus, and renal stones. In in vitro systems, this crystal form can be readily hydrolyzed to an imperfect apatite. It can be considered as a possible precursor for biological apatite, as can many other calcium phosphate phases. Octacalcium phosphate as a precursor helps explain, and is consistent with, imperfections, impurities, pyrophosphate generation upon heating, solubility properties, morphology, hydration, structure, surface energy, growth mechanisms, calcium phosphate ratio, and model compounds. Dicalcium phosphate dihydrate, for example, occurs naturally and also has strong apatite precursor characteristics, i.e., it requires less total base, has less buffering requirements, its rate of formation is much faster than that of apatite, and its low calcium/phosphate ratio is closer to those of the calcifying fronts of dentin, bone, and tooth enamel (a ratio of approximately 1:3).

As described by Young and Brown (this volume), the roles of precursors may be quite complex. For example, they may participate in nucleation, crystal growth, and hydrolysis. The properties of brushite and amorphous calcium phosphate indicate that they may be the first to form, but they can then dissolve and may reprecipitate as a nearly stoichiometric hydroxyapatite or as octacalcium phosphate, which then hydrolyzes to an impure, defective apatite. It is the last step that appears to have major implications regarding the biochemical properties of apatites in teeth, bones, and other calcium phosphate containing minerals. Possibly several intermediate steps occur, such as the initial formation of dicalcium phosphate dihydrate at physiological pH, transforming octacalcium phosphate which is then hydrolyzed to apatite. Hydrolysis also provides a hydrated layer that may give it more stability. Although these examples refer to calcium phosphate minerals, precursor forms are also known to form in in vitro formation of carbonate and oxalate crystals. Indeed, hydrolyzed oxalate intermediate forms are found naturally in some invertebrates and stones.

From these examples, it appears appropriate to encourage further study of crystal precursors, as they are likely to be of seminal importance in biomineralization.

In many instances (invertebrates, plants, cartilage vesicle inner membranes?), matrix surfaces appear structurally positioned to confer intercrystal nucleating or crystal specificity. One question raised for future consideration is the definition of a "biocrystal" itself (as opposed to a pure synthetic crystal).

In each specific case, is there an inter- and possibly and intra-crystalline matrix? The polymorphic selection exhibited by calcium-carbonate mollusc shells suggests this structural possibility. The potential of specificity of ion binding sites appropriately spaced on "matrix" surfaces was considered for their role as nucleation sites, and consensus was reached that this interesting possibility would be feasible and would indeed serve as an efficient means of equating matrix surface structure, vesicles, membranes, etc., to the ultimate crystal morphology. Little is known about these phenomena to date, as too little information is available for matrix surface "mapping" of mineral ion binding sites. It is conceded that only a few sites in a correct binding configuration could serve to nucleate the entire crystal structure. It might be important in this system for the nucleation site (if present) to be free of strain energy. In evaluating these more complex matrix-biomineral systems, it will, indeed, be difficult to define the nucleation criteria, as we find it difficult to define ion saturation conditions in measurable terms in these living systems. Nevertheless, the complementary structural features of both matrices and crystals impel further investigation.

It was also perceived that not only might the matrix influence crystal formation, but also that the crystal surfaces possess the potential for influencing functional organic molecules. Work in this area is preliminary, but enzyme activation seems a likely possibility. The ability of pleated protein sheets, phosphoproteins, osteonectin, salivary proteins, etc., to block crystal formation in solution argues for unique crystal surface - molecular interaction. Physical chemists are urged to work with this "dirty" material - biominerals.

SUMMARY

The compelling fact emerged that a large variety of mineral forms and matrices exist. Nevertheless, certain common conditions were perceived. Three fundamental mineralizing structure systems are known: a) systems with intracellular compartments,

b) extracellular systems with no apparent matrices, and c) systems with matrix components that bind mineral crystals into aggregates. The interaction between morphologists, chemists, physical scientists, biologists, and physicians was suggested to be essential to foster continued progress in understanding biomineral structure and function in the most appropriate and expeditious way. During the deliberations of this study group, a "wish list" was tabulated in an effort to focus the group's attention on future investigative needs in this field. The length of the list was indicative of the present state of the art, as a wish was made for more structural and functional information on essentially all of the cells, matrix components, and their associated crystals, particularly with regard to their temporal and spatial interrelations. Additional work on developing additional techniques using mutants, chromofluors, fixatives for preservation of tissue and ions in in vivo conditions, and methods for identifying saturation conditions and phenomena of ion transport in tissues was recommended. Elaboration of information on crystal shape, size, surfaces, thermodynamics, nucleation factors, crystal growth and kinetics, surface modifications, etc., coupled with more information on the structural and functional organic matrices and/or cells seems the most sound means of understanding their full structural and functional interactions. It appears appropriate to recommend that investigators converge their thinking upon a few selected models in order to delineate some of the fundamental principles of biomineral systems.

REFERENCES

(1) Ascenzi, A.; Bonucci, E.; and de Bernard, B., eds. 1982. Matrix Vesicles. Proceedings of the 3rd International Conference on Matrix Vesicles, Spoleto, Italy. Milano: Wichtig Editore srl.

(2) Boyde, A. 1976. Amelogenesis and the structure of enamel. In Scientific Foundations in Dentistry, eds. B. Cohen and I.R.H. Kramer, pp. 335-352. London: W. Heineman Med. Books Ltd.

(3) Krampitz, G., and Witt, W. 1978. Biochemical aspects of biomineralization. Topics Curr. Chem. 78: 56-144.

(4) Krefting, E.; Lissner, G.; and Höhling, H.J. 1981. Quantitative Electronenstrahl - Mikroanalyse der Wachstumsfuge. Beitr. Elektronenmikroskop Direktabb. aberfl. 14: 373-384.

(5) Pentecost, A. 1980. Calcification in plants. Int. Rev. Cytol. 62: 1-27.

(6) Pita, J.C., and Howell, D.S. 1978. Microbiochemical studies of cartilage. Joints Synov. Fluid 1: 273-330.

(7) Simmons, D.J., and Kunin, A.S., eds. 1979. Skeletal Research - An Experimental Approach. New York: Academic Press.

(8) Steinmann, B.; Tuderman, L.; Peltonen, L.; Martin, G.R.; McKusick, V.A.; and Prockop, D.J. 1980. Evidence for a structural mutation of procollagen type I in a patient with Ehlers-Danlos syndrome type VII. J. Biol. Chem. 255: 8,887-8,893.

(9) Termine, J.D.; Belcourt, A.B.; Christner, P.J.; Conn, K.M.; and Nylen, M.U. 1980. Tooth matrix proteins I. Principal molecular species in developing bovine enamel. J. Biol. Chem. 255: 9760-9768.

(10) Termine, J.D.; Belcourt, A.B.; Miyamoto, M.S.; and Conn, K.M. 1980. Tooth matrix proteins II. Separation and purification of fetal bovine dentin phosphoprotein. J. Biol. Chem. 255: 9769-9772.

(11) Termine, J.D.; Kleinman, H.K.; Whitson, S.W.; Cohn, K.M.; McGarvey, M.L.; and Martin, G.R. 1981. Osteonectin, a bone-specific protein linking mineral to collagen. Cell 26: 99-105.

(12) Waddell, W.T. 1972. A molecular mechanism for biological calcification. Biochem. Biophys. Res. Comm. 49: 127-132.

(13) Watabe, N. 1981. Crystal growth of calcium carbonate in the invertebrates. Prog. Cryst. Growth Charact. 4: 99-147.

(14) Weinstock, A.; Weinstock, M.; and Leblond, C.P. 1972. Autoradiographic detection of ^{3}H-fucose incorporation into glycoprotein by odontoblasts and its deposition at the site of the calcification front in dentin. Calc. Tiss. Res. 8: 181-189.

Group on Mechanisms of Mineralization (Normal)

Standing, left to right: A. Boyde, J.D. Termine, H.-J. Höhling, C. Anderson, H. Newesely, J. Althoff, P. Quint, Y. Kitano, K. Simkiss, H. Fleisch.
Seated: D.S. Howell, G.H. Nancollas, P. Bennema, M.A. Crenshaw, A.E. Nielsen.

Biological Mineralization and Demineralization, ed. G.H. Nancollas, pp. 351-363.
Dahlem Konferenzen 1982. Berlin, Heidelberg, New York: Springer-Verlag.

Mechanisms of Mineralization (Normal) Group Report

K. Simkiss, Rapporteur
J. Althoff, C. Anderson, P. Bennema, A. Boyde,
M. A. Crenshaw, H. Fleisch, H.-J. Höhling, D. S. Howell,
Y. Kitano, G. H. Nancollas, H. Newesely, A. E. Nielsen,
P. Quint, J. D. Termine

INTRODUCTION

The word "biomineralization" encapsulates very well the goals of this workshop. It instantly recognizes that one is dealing with a process (i.e., a system that is time dependent, that it is a phenomenon that involves a phase change and a system that, because it is biological, is based in a complex saline medium. From the start, however, we recognized that whereas biomineralization might imply a recognizable phenomenon to a physical chemist, to a biologist there are a variety of biomineralizations. Our first task was to decide whether we would deal with generalizations based on all these systems or with specific points relevant only to particular tissues. It was felt that at the biological level generalizations would only lead to superficiality. This determined the direction of our second task, namely, how we would structure our discussion. The biological view of this complexity would precede the physicochemical approach to the problems.

THE BIOLOGICAL VIEW

Biologists involved with biomineralization concern themselves with questions such as: Is it an open or a closed system and is it enclosed by cells? What are the distances from the various sources of inputs (cells, blood vessels, etc.)? What are the rates of formation of the various components of the system (matrix, mineral, etc.)? What is the control over these processes (hormonal, etc.)? What is the uniqueness of the cellular contributions (enzymes, vesicles, ions, etc.), and how do these vary with time? The time relationship in biology is complex since it includes concepts such as the "maturation" time of osteoid or matrix vesicles (several days), the "ageing" process, and even the "life span" of vesicles or cells which may all be within the time scale of the mineralization process. The wish of all biologists is to identify, isolate, and purify the systems involved in the transport, delivery, and acceptance of lattice ions at sites of biomineralization. Much of the data which is relevant to many of these questions is summarized in Table 1 which illustrates the complexity of the systems being studied and enables one to discuss particular issues.

Spatial considerations are provocative, for the tissues shown in Table 1 fall into two groups. In most, new layers of tissue are formed by a continuous layer of cells which isolate a separate compartment in which matrix is deposited and mineralized. In some systems, e.g., enamel, the primary purpose of these cells appears to be to regulate the ingress of lattice ions, and if the integrity of this layer is interrupted, mineralization occurs rapidly and in a disorderly fashion. In the second type of system, of which cartilage is the sole example among mammals, the cells which make the matrix are isolated and do not define the anatomical compartment in which they elaborate their product. Numerous examples of this type occur among the invertebrates.

An interesting technical approach is that in order for biologists to study the temporal sequence of mineralization, they have to

TABLE 1 - Various types of phosphate mineralizing systems in mammals

QUESTIONS		Woven bone	Lamellar bone	Sharpey-fiber bone	Cemen-tum ⊕	Dentine (ivory)	Enamel	Calcifying growth cartilage
		TISSUES						
What is the typical distance of the mineralizable matrix from blood vessels (range in μm)?		20	20	?>100	50	50	200-1000	200
Are formative cells in contact with general extracellular fluid?		Yes	Yes	Yes	Yes	Yes	No	Yes
Is matrix sealed off from extracellular fluid?		Yes?	Yes?	No	No	Yes	Yes	No
Matrix-chemistry of main components in mineralized tissue	a) Collagen as a % of organic matter	90%	95%	>95%	>95%	>90%	0	20-50%
	b) Collagen type	1	1	1	1	1	nil	11
	c) Collagen fiber diameter (μm)	mostly 0.1	1-3	EF 5-12	EF 5-7	0.1	nil	0.02
	d) Relative amount of proteoglycan (+ = some to ++++ = much)	+	+	+	+	+	-	++++
Is there a separate non-collagenous, proteoglycan-rich phase in a pericellular location?		No	Yes	No	Yes	Yes	No	No
What is the rate of bulk matrix formation and mineralization? (μm/day)		4-10	0.5-3	?	?	16 (rat) 4 (man)	16 (rat) 4 (man)	>100
What is the thickness of the unmineralized matrix layer? (μm)		10	<10	EF→ IF<10	IF<5	20	0.05	large
What percentage of matrix components are lost during mineralization?		?∼2	?∼1	?0	?0	?∼1	>95	?∼10
Is there a "maturation" or continued mineralization process?		No	Yes	No	No	No	Yes	No
Are matrix vesicles demonstrable?		Yes	No?	No	No	Yes	No	Yes
What are the typical dimensions of single crystals? a) by X-ray diffraction (nm)		5 x 40	5 x 40	5 x 40	5 x 40	5 x 40	50 x 150	?
b) by electron microscopy	i) minimum thickness (nm)	2 - 6	2 - 6	2 - 6	2 - 6	2 - 6	40 - 70	2 - 10
	ii) maximum length (nm)	40→	40→	40→	40→	40→	>>1000	>100
What is the arrangement of the primary crystal clusters (e.g., calcospherites)?		*	//	//	//	*	///	*
Is there a sensory innervation in the complete tissue?		No	No	No	No	Yes	No	No
What volume of the completed, mineralized matrix is occupied by cells (%)?		>10	1-2	<1	0-<1	10	0	>50

⊕ Cementum may resemble bone, Sharpey-fiber bone, or cartilage, and it may be acellular.
ABBREVIATIONS: EF = extrinsic fibers; IF = intrinsic fibers; ? = controversial; * = roughly spherical and not specifically collagen centered; // = centered in and parallel with the collagen; /// = C and a axes parallel in groups; → = upper limit not defined.

use the spatial separation of events in the cartilage growth plate. There are assumptions in this approach which should be stressed; for example, are the matrix vesicles found in the different locations actually similar? (We would wish to see some method of identifying them.) The results obtained from cartilage growth plate are, however, interesting. There are microprobe results which claim that the extracellular calcium content remains relatively constant as far as the lower region of the columnar zone but with an increase in the hypertrophic zone (6). This zone also shows an increase in phosphorus content, the level of which is relatively low in the upper regions of the growth plate cartilage. These results correlate well with "topochemical" analyses (10). Quantitative electron probe microanalysis estimates a calcium content of about 40 to 60 m mole. kg wet wt^{-1} in this region, which raises the critical questions of whether cellular reservoirs (e.g., mitochondria) or extracellular systems are the sources of these ions.

Assuming that these flux problems can be resolved, the next problem to study is the identification of the nucleation mechanisms in biomineralization. Some systems appear to have matrix vesicles while others involve collagen (Table 1), and whichever system one considers, it has to be related to the production of a new mineral phase from some biological fluid.

According to the vesicle concept a) the vesicle may interact with the matrix via membrane-bound proteoglycans, b) the vesicle membrane not only separates an extra- and intravesicular compartment but may regulate the interaction of the two by enzyme activities, and c) the components of the vesicle contents are specialized phospholipids and phosphate-rich compounds (1). The number of possible biological domains that could interact in this situation is extensive, and it is controversial as to whether the final effect is to produce crystal nucleation on the inside or the outside of the vesicle membrane, and who would arrange the complexity of all these interactions into a temporal sequence?

A similar problem exists for those who favor collagen and the role of collagen holes in the nucleation process. On the one hand, it was eloquently argued that the hole in a collagen fiber was $100A^2$ in cross section area, and therefore only one apatite unit cell crystal (10 x 10 x 7Å) could fit in it. One would not be able to explain the diffraction patterns obtained from mineralizing bone on this basis, since these patterns require several crystal unit cells. On the other hand, the periodicity effects within the hollow space arrangement of the collagen fibers could provide a basis for this effect. We are still in a situation where we have to consider the whole question of the organization of collagen microfibrils (11). There is sufficient space for crystallites of mineral to grow in channels whose diameter is initially about 1.8nm wide but which will be widened by crystal growth compressing the protein microstructure.

One is forced, therefore, to accept a more complex organizational basis for the role of collagen. Purified collagens can only be made to mineralize by first allowing them to bind to other non-collagenous proteins such as osteonectin (3,13). This raises the possibility of a temporally regulated nucleation system controlled by several mechanisms and acting over the lifetime of the mineralization site.

The study of non-collagenous proteins in other mineralizing systems is equally provocative. In enamel very large and co-oriented apatite crystals grow in a milieu rich in hydrophobic amelogenin protein aggregates with acidic glycoproteins (enamelin) attached to their overall surfaces (3,13). This is a system that a) is a gel, b) has fixed polyelectrolyte charges, c) could be modelled in terms of an unstirred layer, and d) contains hydrophobic amelogenins. The role of hydrophobicity and the effects of protein layers over crystals are also of interest in a number of invertebrate systems, and it clearly represents an interesting challenge to existing theories of biomineralization.

At this stage of our discussion we felt that we had posed the problem of biological complexity. Even within the mammals, different tissues appear to form the same mineral through different routes while invertebrates invoke an additional range of intracellular and extracellular deposits of widely different composition. We had looked at the problem of how the ions entered the system, passed through the cell layers, and entered the mineralizing environment. We had seen the role of the matrix a) as a mixture of soluble and insoluble proteins, b) as a compartment, c) as a possible regulator system for the whole phenomenon of nucleation and crystal induction, d) as a regulatory system controlling growth, and e) as a modifier of crystal form. It was therefore time to approach the problem from a physicochemical viewpoint.

THE PHYSICOCHEMICAL VIEW

Micropuncture studies of cartilage can produce 15 nl of fluid which represents a sample drawn from around 600 cells (7). We started therefore by considering the thermodynamics of mineral formation in an isolated small drop of liquid. The case was argued for the application of nonequilibrium thermodynamics to show how the free energy in such a system was related to the formation of a solid with a subsequent reorganization and "sharpening up" of the crystal lattice (4) (Fig. 1). The biological implications were immediately obvious. If one felt that the normal biomineralization phenomenon occurred in an open system in which ions continually entered the site of crystal nucleation, then the additional refinements of this model were unnecessary. If, however, biomineralization occurred in enclosed unstirred spaces (e.g., vesicles) which were also closed in terms of influxes, the model was exciting. Furthermore, it may provide a basis for the growth of some of the multinucleated systems of invertebrates such as coccolithophorids.

If the analysis of this system could be carried further, it might provide estimates that are relevant not only to the

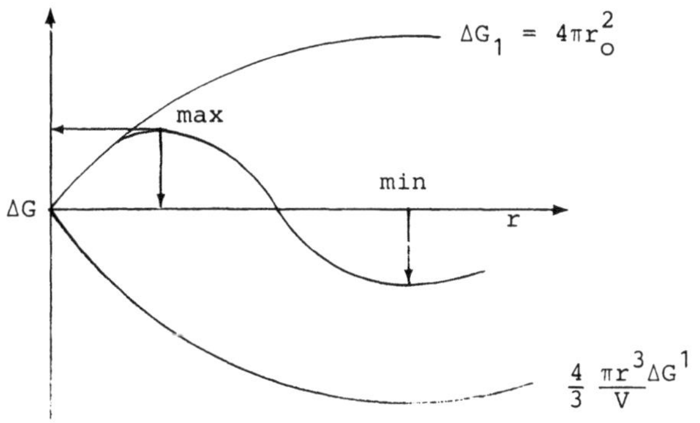

FIG. 1 - Bennema's application of classical thermodynamics to the particular problems of the formation of a crystal in a small (few μm) closed space (see also (2,5)).

With a small closed space instead of simply using the classical derivation for nucleus formation (see Garside, this volume, for further details), we have to consider the loss of supersaturation as the nucleus grows,

$$\text{i.e., } \Delta G = \frac{4}{3}\frac{\pi r^3}{V} \quad \Delta G^1 = \frac{4}{3}\frac{\pi r^3}{V} RT \ln \frac{\Pi_i(\gamma, C_i)\gamma_1}{\Pi_i(\gamma, C_i)\gamma_2},$$

where the product, Π, is that for all species i forming the crystal.

If we simplify and assume that $\gamma_1 = \gamma_1 eq$ and that all γ_1 terms are equal, we get

$$\Delta G = \frac{4}{3}\frac{\pi r^3}{V} RT \ln \frac{C}{C_{eq}} = \frac{4}{3}\frac{\pi r^3}{V} RT \ln \frac{n^{int} - n^{solid}}{n_\infty},$$

where n^{int} = the total number of particles in the original cavity before nucleation starts, n^{solid} = the number of particles in the crystal (nucleus), and n_∞ = the number of solute particles in the cavity at saturation.

As the crystal grows, we may reach a point where $n^{int} - n^{solid} = n_\infty$, i.e., the solution around the crystal is now saturated and ΔG is zero because lnI equals zero.

If r becomes larger, supersaturation decreases, and the ΔG(r) curve goes up. This means that we get a minimum in the ΔG(r) curve where the nucleus becomes a stable crystal with a radius r_{eq} in equilibrium with a supersaturated solution.

The experimental formation of equilibrium forms of NaCl cubes in closed droplets, that are only μm in size, may take hours or days to occur. Initially they may be dendritic, but they then reorganize themselves crystallographically. An equilibrium form is the crystalline analogue of the spherical critical nucleus.

mineralization process but also to the experimental artifacts that may be occurring during the study of biological samples. Answers to these problems are possible, but the calculations have not been undertaken.

The problem of nucleation in systems that remain supersaturated is easier to study. Homogenous nucleation is rare because of the high concentrations required and can probably be ignored by biologists. Heterogenous nucleation provides the most likely model, but secondary nucleation is usually dependent upon stirring, which can cause particle splintering. The possibility of its occurring cannot be ignored, however, and there was some discussion as to whether cell mobility might have this effect when cells retreat before a mineralization front (as in, e.g., dentine). Heterogenous nucleation is more likely, however, because of the large numbers of preexisting surfaces that occur in all biological systems.

Lattice information from a heterogenous nucleation surface could probably only be transmitted over a range of 1 or 2 atomic radii, and these distances would probably be hydration layers in most systems. It is therefore highly unlikely, for example, that dentine could nucleate enamel in teeth, since the distances, although microscopic, are extensive in atomic terms. On the other hand, proteins, especially if they formed a membrane with critically spaced charges, could transmit this information and also provide a way of reducing the energy requirements for nucleation. The possibility of β-pleated sheets of proteins acting in this manner is obviously an exciting prospect for investigation, especially if one could obtain such systems with varying degrees of hydration (12). The ultimate extension of this concept would be to relate nucleation to conformational changes on proteins. Thus dentine phosphoryn acts as an inhibitor of crystal growth until its β-pleating content is increased.

The ionotropic model for the role of proteins in the formation of molluscan nacre predicts that calcium and carbonate ions are concentrated locally by the hydrophilic, calcium-binding portion of the matrix. Then the β-pleated sheets in the underlaying proteins provide a substrate for oriented overgrowth of the crystals (epitaxy). Another interesting aspect of protein nucleating sites is that by bending they may reduce crystal strain in the nucleation process. Finally, we looked at the influence of other ion species on the nucleation and growth of crystals and chose the work of Kitano (8) to relate this to invertebrate skeletons. Calcium carbonate can exist as 3 polymorphic forms: calcite, aragonite, and vaterite, and these occur in a wide variety of layers in such structures as molluscan shells. When calcium carbonate is formed in vitro, different crystallographic forms may be stabilized or destabilized by the presence of various cations and anions which may not themselves be significantly incorporated into the lattice (Table 2). Thus magnesium ions favor the formation of aragonite but do not become incorporated into that crystal (although they may be incorporated into calcite). Barium ions favor the formation of calcite but mainly occur in aragonite deposits. It is interesting to note that the abnormally high amounts of magnesium and barium that are incorporated during these in vitro experiments occur only during the very early stages of crystal formation. Many organic metabolites, e.g., citrate and pyruvate ions, act similarly, and the importance of this in biological systems is in providing a clue as to the possible mechanisms whereby crystals, such as high magnesium carbonates, could form in sponge spicules. There must clearly be several conflicting influences between the ions in the fluids which form such crystals, and the subsequent study of the solubility of such materials is clearly problematical, since they appear to exist as solid solutions.

CONCLUSIONS

Studying any object from two viewpoints should provide one with a greater appreciation of depth, but it can also produce a squint, and the group was unsure as to which effect it was experiencing. The biologist studying the mineralization process

TABLE 2 - Influence of foreign ions on the stabilization of the various calcium carbonate crystals compared with the crystal forms in which these ions mainly accumulate.

Foreign ion	Crystal form that is stabilized	Crystal form that accumulates this ion
Mg^{2+}	Aragonite	Calcite
Ba^{2+}	Calcite	Aragonite

is in an almost impossible situation. He is trying to investigate a system which, by definition, is a) of barely stable molecular dimensions, b) unstable when taken into any other fluid system, c) transient, and d) in an ionically unique interfacial situation. Any attempt to measure a component of this system is likely to disrupt it. The criticisms of investigative techniques in this area have therefore been very severe, and the biologist clings to the concept that he must retain normal physiological conditions at all costs. It may be that some of the physiological conditions about which he is concerned may not be critical to the process being studied and may therefore actually restrain further physicochemical investigations in this system.

As a first example, one can consider in vitro models of crystal growth. This process involves the addition of single ions at growth sites which are "kinks" in steps on the crystal surface. Their arrival at the crystal surface requires either convection or diffusion as a transport process. At the crystal surface there may be a) adsorption, b) surface migration, c) adsorption at a step, d) migration along a step, or e) integration into a "kink." Most of these processes can be identified through the kinetics that are observed, but in a biological situation this will require their study at different supersaturation levels.

A second example where physicochemical methods could enormously simplify the study of biomineralization is in the design of in vitro mineralization tests. If a "mineralizable" system is added to a metastable supersaturated solution, it is possible to study the subsequent reaction at constant ion activities by the controlled addition of crystal lattice ions to the solution (9). The absolute levels of supersaturation may not be critical if the mechanism is understood (c.f. physiologically "normal"), providing that the driving forces are not excessive. With this system the rate of mineralization could be studied in relation to pH, inhibitors, tissue components, etc. The reproducibility and value of such test systems is, of course, enormously improved if one can ensure that only one crystal phase is involved in the system. This again could be achieved by using solubility isotherms to select conditions which precluded the occurrence of multiple crystal types. The biologists argue that this is too simple and that the solubility isotherms will have to be at least three-dimensional in order to begin to cope with the additional variables of the biological situation. But at least the concept is established and both groups can work around a range of values which are within physiological limitations. The most difficult problem in applying this approach to biological systems is that thermodynamics depends upon defining the driving forces that are present. If a biological system can release energy at a small localized site, then that site becomes transiently different from the thermodynamic influences in the bulk solution. This is clearly what happens in many enzyme-catalyzed situations, and it may be the whole basis of biomineralization. Certainly if we can identify that this is so, then the physical chemists will be able to study the processes by mimicking the phenomenon and scaling it up.

As with other groups, we kept a "wish list" of scientific aspirations, but in the area of our discussions, the majority of these resolved into the following requirements: 1) the need to work with activities of free ion species (possible), 2) the need to define mineralization test systems specific for precise mineral phases (possible), and 3) the need to achieve electrode systems capable of operating in vivo (requires a considerable technical advance).

In the meanwhile, it would be a help to have a mechanistic model for calcium and phosphate enrichment that meets the known data for biomineralization and which is in such a form that it can be refined as new results become available.

REFERENCES

(1) Ascenzi, A.; Bunucii, E.; and Bernard, B. de, eds. 1981. Proceedings of the Third International Conference on Matrix Vesicles, Montebico, Spoleto. Milano: Wichtig Editore srl.

(2) Bienfait, M., and Kern, R. 1964. Determination of the equilibrium form of a crystal (Lemmlein-Klija method). Bull. Soc. France. Mineral. Crist. 87: 604-611.

(3) Fincham, A.G.; Belcourt, A.B.; and Termine, J.D. 1981. The molecular composition of bovine fetal enamel matrix. In Chemistry and Biology of Mineralized Connective Tissues, ed. A. Veis, pp. 523-529. New York: Elsevier-North Holland.

(4) Glansdorff, P., and Prigogine, I. 1971. Thermodynamic theory of structure stability and fluctuations. New York: Wiley Interscience.

(5) Hartman, P. 1973. The equilibrium form in a phase of small dimensions. In Crystal Growth: An Introduction, ed P. Hartman. Amsterdam: North Holland.

(6) Höhling, H.J.; Althoff, J.; Barckhaus, R.H.; Krefting, E.-R.; Lissner, G.; and Quint, P. 1981. Early stages of crystal nucleation in hard tissue formation. In International Cell Biology 1980-81, ed. H.G. Schweiger. Berlin: Springer.

(7) Howell, D.S., and Picta, J.C. 1976. Calcification of growth plate cartilage with special reference to studies on micropuncture fluids. Clin. Orthop. 118: 208-229.

(8) Kitano, Y. 1979. Carbonate sediments. Recent Prog. Nat. Sci. (Japan) 4: 11-18.

(9) Koutsoukos, P.; Amjad, Z.; Tomson, M.B.; and Nancollas, G.H. 1980. Crystallization of calcium phosphates. A constant composition study. J. Phys. Chem. 102: 1553-1557.

(10) Krefting, E.-R.; Libner, G.; and Höhling, H.J. 1981. Quantitative Elektronenstrahl-Mikroanalyse der Wachstumsfuge. Beitr. elektronenmikroskop. Direktabb. Oberfl. 14: 373-384.

(11) Newesely, H.; Hoseman, R.; and Uther, B. 1980. Kollagen-Einschlußverbindungen. Z. Naturforsch. 35: 177-187.

(12) Termine, J.D.; Eanes, E.D.; and Conn, K.M. 1980. Phosphoprotein modulation of apatite crystallization. Calc. Tiss. Intern. 31: 247-251.

(13) Termine, J.D.; Kleinman, H.K.; Whitson, S.W.; Conn, K.M.; McGarvey, M.L.; and Martin, G.R. 1981. Osteonectin, a bone-specific protein linking mineral to collagen. Cell 26: 99-105.

Group on Mechanisms of Pathological Mineralization

Standing, left to right: D.W. Bach, A. Hesse, J. Garside, W.E. Klee, E.C. Moreno, M.D. Grynpas, E.W. Vahlensieck.
Seated: B. Finlayson, L.H. Smith, W.G. Robertson, M.D. Francis, D.J. McCarty.

Biological Mineralization and Demineralization, ed. G.H. Nancollas, pp. 367-386.
Dahlem Konferenzen 1982. Berlin, Heidelberg, New York: Springer-Verlag.

Mechanisms of Pathological Mineralization Group Report

W. G. Robertson, Rapporteur
D. W. Bach, B. Finlayson, M. D. Francis, J. Garside,
M. D. Grynpas, A. Hesse, W. E. Klee, D. J. McCarty,
E. C. Moreno, L. H. Smith, E. W. Vahlensieck

INTRODUCTION

Pathological mineralization may complicate a broad spectrum of clinical conditions and take such diverse forms as urinary stones, dental calculus, salivary calculi, biliary calculi, pancreatic stones, crystal deposits in joints, and various forms of ectopic calcification (Smith, this volume). It was decided, mainly because of the structure and interests of the group, to discuss only four of these in depth. These were considered under two main headings: (a) dental and urinary calculus formation, and (b) abnormal tissue mineralization (with emphasis on pseudogout and ectopic calcification).

DENTAL AND URINARY CALCULUS FORMATION

Although it was recognized that there are two fundamental aspects of calculus formation, physicochemical and metabolic, it was decided to limit discussion only to mechanisms involving the former. These included the factors involved in crystallization ((Nielsen and Christoffersen, this volume), with particular reference to supersaturation, nucleation, crystal growth, aggregation, and phase transformation), the processes of crystal dissolution and disaggregation, the site and mechanism of crystal retention in the urinary tract, the role of matrix, and the role of crystallization inhibitors in calculus formation (Finlayson, this volume).

Factors Important in Crystallization

Supersaturation. Several methods are currently available for estimating the degree of supersaturation with respect to the various crystalline phases found in calculi (through the use of activity products) in both saliva and urine. These take into account complexation and ionic strength in the calculation of free ion activities (Robertson, this volume). Some of these free ion activities (e.g., that of Ca^{2+}) may be checked, with some degree of certainty, directly in the medium concerned by means of ion-selective electrodes. Others may be checked indirectly by equilibration studies with crystals involving the ion concerned (e.g., HPO_4^{2-} by equilibration with brushite and $C_2O_4{}^{2-}$ by equilibration with calcium oxalate monohydrate).

The supersaturation of both saliva and urine is known to vary throughout the day. It was suggested that short-term peaks in supersaturation leading to nucleation, crystal growth, and phase transformation may be important and that collections of urine or saliva over longer periods may mask these short-term peaks (28). It is not clear, however, how important such short-term peaks are in the initiation of urinary calculi. It may be that the frequency of exposure to such peaks is also important. In the case of dental calculus, the main factor which causes the laying down of calcium phosphates on enamel is the presence of dental plaque in the mouth. Phase transformation leading to calcium apatite is brought about at a later stage by hydrolysis of the more acidic calcium phosphates which usually precipitate first.

Nucleation. The probabilities of the various types of crystal nucleation occurring in urine and saliva were considered. It was generally agreed that homogeneous nucleation of any of the urinary stone salts or acids is unlikely, although not entirely impossible, in urine. *There is, however, a dearth of reliable information in this area, and it would be worthwhile to study this process in more detail.* One of the major problems is the detection and measurement of primary nucleation (Garside, this volume). Continuous crystallizers almost certainly do not specifically measure homogeneous nucleation but probably some combination of secondary or heterogeneous nucleation. Furthermore, nucleation rates obtained by extrapolation of the data obtained from continuous crystallizers to zero particle size may be in error owing to two effects, first, the Gibbs-Thomson effect on the solubility of crystals less than about 1 μm in diameter, and second, size-dependent growth taking place under diffusion control. *One technique suggested for consideration was that of Garten and Head* (10) *employing the crystalloluminescence of salts doped with certain metal ions.*

The question was considered as to whether or not homogeneous nucleation is likely to be influenced by inhibitors at the stage of subcritical "embryos." Inhibition of further crystal growth once the critical cluster size has been reached would prevent the subsequent detection of crystals and so would also be interpreted as inhibition of nucleation. *There was not a consensus, however, on this point which deserves further consideration.*

The nucleation of both urinary and dental calculi was considered to be probably heterogeneous in nature and as such likely to be influenced by various inhibitors of crystallization. *In this connection, it is important to know what is the nature of the catalytic surfaces involved, what is the role of epitaxy, and what are the kinetics of the heterogeneous nucleation process.* Although epitaxial growth is an attractive theory, in practice there is not much evidence for it in stones. The failure of biologists to include considerations of lattice strain energy induced by lattice mismatch may explain the failure of epitaxial theory to have predictive value in their systems. Epitaxy and heterogeneous nucleation may be distinguished by employing a specific inhibitor either of the catalytic surface or of the salt being nucleated.

Crystal growth. The rates of crystal growth of calcium oxalate and various calcium phosphates are known to be affected by certain substances. These may act by complexing one (or more) of the precipitating ions, thereby reducing the level of supersaturation, or by adsorbing to the crystal surface and interfering with the growth mechanism (i.e., acting as true inhibitors (9,22)). In this latter case the reduction in growth rate is proportional to the fraction of the growth sites on the crystal surface covered by inhibitor as defined by some adsorption isotherm (8) (usually that of Langmuir (Nielsen and Christoffersen, this volume)). *To understand this completely it would be necessary to know what stage in the growth process is dominant and what stage is affected by the inhibitor. In this connection it would be desirable to know the growth rates and effects of inhibitors on individual crystal faces of biological crystals.*

The naturally occurring inhibitors of calcium phosphate crystal growth so far identified in urine include citrate, pyrophosphate, and magnesium, and in saliva, the peptide statherin and certain acidic proline-rich phosphoproteins (19). For calcium oxalate the list includes, in addition to citrate and pyrophosphate, certain glycoproteins, glycosaminoglycans, RNA-like material, and acidic glycopeptides. *Further work should be carried out to identify all the urinary inhibitors possibly using calcium oxalate column chromatography followed by elution with an oxalate concentration gradient.*

There was some discussion regarding the construction of a system for measuring total inhibitory activity in mixtures of inhibitors, particularly in whole urine. One possibility is the development of a "constant composition" system such as that employed by Nancollas and his co-workers (15,26). Another is a system involving seeding of urine and a comparable inorganic solution of the same supersaturation with increasing amounts of crystals of the salt of interest and comparing the initial growth rates in both systems. *This is of considerable importance in both fields of urinary and of dental calculus and ought to be pursued.*

Crystal aggregation. It was agreed that little is known about the aggregation of crystals of most calculus constituents, although a few studies have been carried out on calcium oxalate and calcium phosphate systems. The kinetics of aggregation have not been studied at all, and there is some question as to whether or not the methods used for the study of this phenomenon in vitro are relevant to the stone-forming process in vivo. However, even if not important in the initial retention of crystals in the urinary tract, particle accretion by aggregation with the stone nucleus may be necessary for the development of the final calculus. From in vitro studies it would appear that the aggregation of both calcium oxalate and calcium phosphate may be greatly retarded by highly charged anions such as pyrophosphate, diphosphonates, chondroitin-4-sulphate, heparin, Tamm-Horsfall mucoprotein and certain RNA-like materials.

Phase transformations. Phase transformations have been noted in salts involved in both dental and urinary calculi, since often the salt first precipitated is not the most thermodynamically stable under conditions existing in saliva and urine (Nancollas, this volume). In the case of dental calculus, the first salt laid down is probably dicalcium phosphate dihydrate which is converted, under the influence of saliva, to calcium apatite by solid → liquid → solid transformation. The sequence of phases from "young" to "old" calculi is in the order: dicalcium phosphate dihydrate → octacalcium phosphate → β-tricalcium phosphate → hydroxyapatite (19). In the case of urinary salts there are not only examples of the conversion of various metastable or amorphous calcium phosphates to calcium apatite but also of calcium oxalate trihydrate or dihydrate to monohydrate (Nancollas, this volume). The main importance of these transformations is that they result in the formation of a phase which is less soluble than that first precipitated and therefore more difficult to redissolve. *It is important to know whether these processes are dominated by kinetic or thermodynamic factors and further work is necessary. It was also suggested that the role of calcium oxalate trihydrate in stone formation should be investigated*, although it has not yet actually been found in stones or in the urinary tract.

Dissolution and Disaggregation of Crystals

The reverse processes of crystal dissolution and disaggregation were considered briefly. In the case of urinary salts and acids the most potent methods for undersaturating urine so as to produce dissolution of calculi containing these materials are to alter urinary pH by dietary or therapeutic means and to increase urinary volume. Thus it is possible to dissolve cystine, uric acid, magnesium ammonium phosphate, and, in some instances, calcium phosphate stones (16). Owing to their extremely low solubility, calcium oxalate stones (particularly those consisting primarily of the monohydrate) are resistant to dissolution because urine cannot be readily made undersaturated with respect to this salt. Some stones can be dissolved in situ by irrigation of the urinary tract over many hours with special solutions designed to dissolve the particular type of stone in question, but this procedure is not without some risk to the patient. It should be emphasized, however, that irrigation of the urinary tract by percutaneous nephrostomy is successful for the dissolution of some of the types of urinary stones mentioned above and may prove to be a useful alternative to surgical removal of the stone.

There was some question as to whether or not inhibitors or stone matrix might retard the rate of dissolution of stones. Evidence was reported that both may do so. Indeed, one of the roles of matrix may be to slow down dissolution and thereby cause stone embryos or stones themselves, which otherwise might dissolve, to persist (21). Some ions may stabilize certain forms of salts (e.g., magnesium stabilizes whitlockite formation in some dental calculi).

There was a brief discussion of disaggregation, and examples were reported in which diphosphonates and pyrophosphate have been noted to cause peptization of aggregates of both calcium phosphate and calcium oxalate. A similar phenomenon has been observed in diphosphonate-coated dental calculus. *It was felt that the effect of specific inhibitors and of stone matrix on the rates of both dissolution and disaggregation of stone salts should be determined.*

Site and Mechanism of Crystal Retention

This extremely important aspect of stone initiation was discussed at some length. From a consideration of urinary saturation with calcium oxalate, the linear growth rate of that salt, and the flow rate and average retention time of urine in various compartments of the urinary tract, it was argued that the initiation of stone formation was unlikely to occur by a "free-particle" mechanism in either the collecting tubules or the renal pelvis. It could, however, theoretically occur in the bladder because of the longer retention times there. It was therefore suggested that stone formation can only be initiated in the upper urinary tract if the retention time of crystals is increased sufficiently by some means to allow the crystallites to grow or aggregate to dimensions which will limit the probability of their expulsion from the urinary system (Finlayson, this volume). This may be achieved by two possible mechanisms. The first involves "fixation" of a crystal (or aggregate of crystals) to the epithelium of the renal tubule or pelvis. Evidence of this "fixed particle" mechanism of stone initiation has been seen in rats given large intraperitoneal injections of oxalate (14). The second type of particle retention may be predicted from the same model described above if allowance is made for the effects of gravity on calcium oxalate crystals of about 1 μm diameter or more which are still growing as they pass through a collecting tubule which might be draining against gravity. In this situation, the crystals are being increasingly retarded by gravity and eventually may stop or even drop back down the tubule where they may grow and possibly aggregate more freely under continuing conditions of supersaturation.

It should be noted that in the oxalate-loading study in rats, crystals were seen sticking to the luminal walls of the renal tubules. However, a degree of tubular necrosis was observed in these studies, and it was not clear which had come first - the necrosis or the sticking of the crystals to the wall. Papillary concretions akin to Randall's plaques (Type II) were also observed in some animals. Since crystals elsewhere are known to increase prostaglandin production when they come in contact with cells, *it was suggested that the attachment of crystals to tubular cell walls might lead to prostaglandin release and that this should be looked for in these experiments. In general, however, a desire was expressed for improved animal models of stone formation.*

Results from studies of calcium and oxalate levels in human kidney tissue removed at operation (partial nephrectomy) were reported showing a gradient of increasing concentration through the kidney in the order cortex $<$ medulla $<$ papilla. This may account for the formation of crystal deposits often found in the papillar region. *The need was emphasized to carry out further micropuncture studies on calcium and oxalate transport in rat kidneys.*

Role of Matrix in Calculus Formation

All urinary calculi contain an organic matrix which constitutes on average about 2 to 3% by weight of the stone but can be as high as 60% (2). This is basically a mucoprotein whose protein content has a high proportion of aspartic and glutamic acids. The main matrix component, designated "matrix substance A" by Boyce and King, is soluble in aqueous systems and was at first thought to be excreted in greater amounts by stone-formers than normals but later shown to be excreted in increased amounts by any patient with one of a number of renal disorders not usually associated with stone formation (13). It has never been clear whether this material has a primary role in the initiation of stones or in the structural integrity of stones or whether it is adsorbed adventitiously to the surfaces of growing crystals. When stones are decalcified and the matrix washed and suspended in aqueous solution, part is insoluble and this fraction promotes aggregation of crystals of calcium oxalate in vitro. The soluble component, on the other hand, is a very strong inhibitor of aggregation such that when the two matrix components are re-mixed in the original proportions, the net reaction is to cause inhibition of aggregation (24).

Matrix is not identical from one stone type to another, and *it was felt by some of the group that the relationships between matrix type and crystal type and between matrix type and stone morphology should be investigated. It was also suggested that it would be worthwhile to restudy the immunochemical properties of the various types of matrix.*

Role of Inhibitors in Calculus Formation

It is known that saliva is supersaturated with respect to the components of both enamel and dental calculus. A series of acidic proline-rich inhibitors (11,000 to 16,000 daltons), which are constituents of tooth pellicle but are also present in saliva, prevent the growth of enamel crystals detrimental to the normal tooth morphology (19). They probably also play some role in limiting dental calculus formation. Thus the present view is that the natural salivary inhibitors are degraded by the action of bacteria and this leads to the precipitation of the mineral phase.

In terms of urinary stone disease, a number of inhibitors of heterogeneous nucleation and/or crystal growth and/or aggregation have been identified as outlined earlier in this report. Differences in net overall inhibitory activity between stone formers and normals have been reported (6,23). However, there may be a major problem with the test systems used in these studies, each of which employs diluted urine (1 to 5% in a standard metastable solution of calcium oxalate in order to standardize the supersaturation driving force). The main problem is that urine contains a mixture of inhibitors with different potencies. By diluting urine, this may eliminate some inhibitors altogether which are significant in whole urine. The answer may be to use whole urine and one of the test systems suggested earlier in the section on "Crystal Growth."

Crystal growth inhibition can occur in a number of different ways, but all appear to involve adsorption on the crystal surface. For example, adsorption at or near the center of a screw dislocation may prevent further propagation of steps from that point; adsorption of ions or molecules on the edge of a step will prevent advance of the step if the adsorbed species are closer together than the radius of a surface nucleus. Adsorption over the whole surface will, of course, hinder growth. *It was suggested that the adsorption isotherms of the various important inhibitors should be determined.*

ABNORMAL TISSUE MINERALIZATION

There is a number of types of crystal deposition disease and forms of ectopic calcification associated with joint tissue and with various soft tissues in the body. The group decided to concentrate on two of these - pseudogout, which involves the deposition of calcium pyrophosphate dihydrate (CPPD), calcium apatite, or both, in joint tissue and ectopic calcification with particular reference to those types which may follow myocardial infarction, cell necrosis, calciphylaxis, and calcergy, and those which may result from increases in systemic supersaturation with respect to calcium salts.

Pseudogout

This is one of several types of crystal deposition diseases affecting joints and involving a number of salts, e.g., monosodium urate (as in gout), calcium pyrophosphate dihydrate (as in pseudogout or chondrocalcinosis), dicalcium phosphate dihydrate, calcium apatite, and mixed crystal deposition (calcium apatite + CPPD) (7). The crystals of CPPD occur in two polymorphic forms - monoclinic and triclinic - of which the latter is the dominant form. They may be up to 10-20 μm in length, and some appear to have "holes" as demonstrated by transmission electron microscopy. The origin, content, and significance of these "holes" is not known, but they are probably artifacts produced by the electron beam.

There are three groups of patients who develop CPPD crystal deposition disease. First, there is a hereditary form, which is autosomal dominant and so far only found in Caucasian families. Second, there is a form that may be associated with various metabolic conditions including hypercalcemia, such as in hyperparathyroidism (although this association has not yet been statistically proven), hypomagnesemia, hemochromatosis/hemosiderosis, hypothyroidism, and hypophosphatasia (a genetic disorder in which there is a deficiency of the enzyme alkaline phosphatase, a pyrophosphatase, and this allows the tissue levels of pyrophosphate to increase). Finally, there is a sporadic form of the disorder which is also the most common. The prevalence of this form increases rapidly in later life with about 5% of the population having the disorder by age 75 and about 50% by age 90.

Biochemical studies show that both urinary and plasma levels of pyrophosphate are normal in patients with CPPD disease (except those with hypophosphatasia). Pyrophosphate levels in joint fluid are generally higher than in plasma, but such levels do not correlate with the deposition of CPPD (18). There is, however, a correlation between joint fluid pyrophosphate levels and the degree of joint degeneration measured radiologically. Pyrophosphate levels in cartilage have not been measured but are probably higher than those in synovial fluid. In addition, the production of pyrophosphate by chondrocytes and its consequent diffusion gradient into the cartilage tissue may account for the appearance of crystalline CPPD in the cartilage joints. Cartilage slices, incubated in organ culture, release pyrophosphate into the ambient medium. The amount released is correlated with the metabolic activity of the chondrocytes, e.g., hexuronic acid release as an index of glycosaminoglycan synthesis. Both hyaline and fibro-cartilages secrete pyrophosphate. This may be significant since cartilage is the site of initial CPPD crystal formation. Recent, but as yet unconfirmed, data from Seegmiller's group (17) suggest that intracellular levels of pyrophosphate in cultured skin fibroblasts from patients from one family with familial CPPD crystal deposition are about twice those found in fibroblasts from unaffected adult members of the same family or from control subjects. *This may be extremely significant, since it may be evidence for a generalized cellular biochemical abnormality phenotypically expressed only in chondrocytes and ought to be followed up.*

Chemically it is not known whether or not the level of supersaturation of synovial fluid with CPPD is sufficiently high to permit precipitation of the salt. In vitro studies by Pritzker's group employing normal synovial fluid levels of pH, calcium, and magnesium, but levels of pyrophosphate some 5 to 20 times higher than those in the synovial fluid, suggest that the level of supersaturation in cartilage interstitial fluid may be approaching that which permits spontaneous precipitation of CPPD (5). It is possible, therefore, that precipitation might be brought about in vivo by localized increases in pyrophosphate concentration generated in the neighborhood of the chondrocytes. Increases in the enzyme, ATP pyrophosphate phosphohydrolase, which promotes the destruction of ATP to AMP and pyrophosphate, have been found in Triton X100 extracts of cartilage from patients with CPPD deposition compared with extracts of cartilage from patients with osteoarthritis (27). Alkaline phosphatase activity in these extracts was decreased and 5'-nucleotidase activity increased compared to the corresponding activities in osteoarthritis. Another possible source of pyrophosphate is that it is produced as a by-product of glycosaminoglycan synthesis or sulphation. In vitro studies indicate that the type of salt produced is highly dependent on the composition of the crystallizing medium. *It was suggested that further studies are necessary on the solubility of CPPD and the factors involved in the supersaturation and precipitation of that salt.*

CPPD deposits in cartilage have an associated organic matrix which is different from that in adjacent normal cartilage. These differences are mainly due to increased levels of glycosaminoglycans, in particular keratan sulphate. It is not clear whether the matrix plays an active or passive role in the production of CPPD deposits. Early CPPD crystal formation appears to occur at the junction between territorial matrix (found immediately around chondrocytes and which contains thin collagen fibrils and no proteoglycan aggregates) and inter-territorial matrix which forms the bulk of the cartilage extracellular matrix and contains thicker collagen fibrils and proteoglycan aggregates. *It was suggested that studies should be carried out to determine the structure of the matrix in which CPPD and calcium apatite is deposited.*

Little is known about the influence and importance of crystallization inhibitors in retarding the formation and growth of CPPD deposits. Some in vitro studies were reported which suggest that low concentrations of heparin (around 1 nmol/l) are able to inhibit the nucleation or growth of CPPD crystals (Robertson, unpublished manuscript). *It was thought that this aspect of the problem ought to be pursued and that specific inhibitors of calcium apatite and CPPD ought to be looked for in synovial fluid and cartilage.*

The events leading to an acute attack of inflammatory arthritis were outlined. The pseudogout attack corresponds with an acute drop in ionized calcium or pyrophosphate level in synovial fluid. This may reduce the supersaturation level of CPPD in the synovia and lead to the loosening and shedding of crystals of CPPD from cartilage into the fluid. The crystals coated with immunoglobulin G (IgG) are then taken up by macrophages and/or polymorphonuclear leucocytes. This causes the release of proteases which may damage tissue (thereby causing degenerative joint disease) and "strip-mine" additional crystals from the matrix. There is also a release of the prostaglandins PGE_2 and, in some cell types, $PGF_{2\alpha}$. The CPPD crystals are removed from human joints with a half-time of 1 to 3 months. All dissolution appears to take place within phagocytic synovial cells. Thus "crystal traffic" is set up as a net result of release and removal mechanisms. *The processes involved in the dissolution are not known and are worthy of future study.*

Several wishes were expressed by the group, including the development of animal and cell or organ culture models to study CPPD deposition disease and the development of a direct method for assaying pyrophosphate. It is also not yet clear whether the sequence of the disorder is (a) metabolic abnormality tissue degeneration (+ altered matrix) → CPPD deposition, (b) metabolic abnormality → simultaneous degeneration and deposition of CPPD, or (c) metabolic abnormality → CPPD deposition → degeneration. This ought to be clarified.

Ectopic Calcification

Three main types of ectopic calcification were discussed: first, that which often, but by no means uniformly, takes place following cell damage or necrosis; second, that which follows localized increases in saturation, such as in nephrocalcinosis, or in muscle and skin calcification which are similar to the experimental models of calciphylaxis and calcergy; finally, that due to systemic supersaturation such as, for example, in conditions resulting in general hypercalcemia or hyperoxalemia (oxalosis) (Smith, this volume).

Calcification secondary to cell necrosis. The main example discussed was the calcification which follows myocardial infarction. This is said to occur in 95% of the patients with this disorder. Following myocardial infarction, a number of biochemical changes combine to produce a high risk of calcification within the ischemic boundary region of the infarct. These include, first, a marked fall in intracellular magnesium (11); second, a rapid and marked increase in intracellular calcium concentration from between 10^{-7} to 10^{-6} mol/l to between 10^{-4} and 10^{-3} mol/l (4); third, a total loss of membrane ATP (20); and finally, a large efflux of amorphous calcium phosphate from the mitochondria of myocardial cells into the cell and, if the cell is ruptured, into the extracellular space (3). There is also a concomitant rise in the iron concentration following hemolysis (12). There is, therefore, a marked increase both in the level of supersaturation, with respect to calcium phosphate, and in the presence of a number of calcium phosphate nuclei and decreases in the levels of crystal growth inhibitors such as magnesium and ATP.

There was some suggestion that phospholipids might be involved in some forms of soft-tissue calcification by causing the release of pyrophosphatase with tissue damage, thereby increasing the destruction of polyphosphates. *It was considered that the role of both crystalline and non-crystalline lipid components in these calcification processes should be studied. Furthermore, it is important to define the role of Fe deposition in calcification associated with necrosis.*

The amount of calcium deposition appears to be maximal about 6 to 8 hours after the infarction in the rat. Thereafter the calcium level falls to normal at between 12 and 48 hours. In man, however, the entire process may take longer. *It is not known what causes the removal of calcification after infarction and this should be investigated.*

In necrosis of other types of soft tissue following trauma, e.g., in muscle, calcification is not as frequent or as marked as in myocardial tissue. *It is not clear why necrosis does not always result in calcification in these tissues, or if it always occurs, why has it not been detected in the past? This ought to be examined. There is also a question as to whether or not the ability to calcify after trauma is related to the vascular supply of the tissue concerned.*

Local ectopic calcification. This may be caused by tissue injury and may be studied in two experimental models in animals. The first of these is calciphylaxis ((1,25), and Lian et al., in preparation), which involves challenging the animal with dihydrotachysterol or vitamin D_3 (thereby causing hypercalcemia) and subsequently injuring the tissue. In time, apatite crystals are deposited in intimate relationship with the collagen fibrils. The second method of causing local ectopic calcification is called calcergy ((29), and Lian et al., in preparation). This requires local injection of a chemical agent (usually a heavy metal compound) which causes apatite crystal deposition at a distance from the collagen fibrils. In both of these models calcification is a highly ordered process, cellularly controlled, and accompanied by increased concentrations of various macromolecules found in normal calcification, e.g., bone Gla-protein.

Systemic calcification. There was only brief discussion on this topic. It is clear that there is a great variability in response to different levels of plasma calcium. The occurrence of calcification would appear to depend on more than just the saturation levels of the extracellular fluid with respect to calcium phosphate. It can result, in some instances, in the general deposition of crystals throughout the body.

For example, in oxalosis, calcium oxalate crystal deposits may be found in tissues throughout the body, even in the osteoid of bone. This is almost certainly due to increased supersaturation of tissue fluids with calcium oxalate as a result of hyperoxalemia. *It was suggested that the possible role of matrix vesicles in ectopic calcification and the effect of cellular inhibitors of crystal growth and enzyme inhibitors on calcification be investigated.*

In summary, it would appear that, in general, pathological mineralization may be accounted for in terms of the thermodynamic driving force (supersaturation) and the factors which affect the kinetics of crystallization (inhibitors, promoters, and supersaturation) in a given system. *However, not all the factors nor their relative importance are known for each form of abnormal calcification, and these must be determined in order to enable us to understand completely why these calcifications occur in some tissues and not in others and in some individuals and not in others.* The ultimate aim of research in this area must be to permit us to prevent abnormal mineralization occurring or recurring in various parts of the human body.

REFERENCES

(1) Boivin, G. 1975. Étude chez le rat d'une calcinosis cutanée induite par calciphylaxie locale. I. Aspects ultrastructuraux. Arch. Anat. Microsc. Morphol. Exp. 64: 183-205.

(2) Boyce, W.H. 1968. Organic matrix of human urinary concretions. Am. J. Med. 45: 673-683.

(3) Buja, L.M.; Parkey, R.W.; Deef, J.H.; Stokley, E.M.; Harris, R.A.; Bonte, F.J.; and Willerson, J.T. 1975. Morphological correlates of technetium-99m stannous pyrophosphate imaging of acute myocardial infarcts in dogs. Circulation 52: 596-607.

(4) Buja, L.M.; Tofe, A.J.; Kulkarni, P.V.; Mukherjee, A.; Parkey, R.W.; Francis, M.D.; Bonte, F.J.; and Willerson, J.T. 1977. Sites and mechanisms of localization of technetium-99m phosphorus radiopharmaceuticals in acute myocardial infarcts and other tissues. J. Clin. Invest. 60: 724-740.

(5) Cheng, P.K.; Pritzker, K.P.H.; Adams, M.E.; Nyburg, S.C.; and Omar, S.A. 1980. Calcium pyrophosphate crystal formation in aqueous solutions. J. Rheumatol. 7: 609-616.

(6) Coe, F.L.; Margolis, H.G.; Deutsch, L.H.; and Strauss, A.L. 1980. Urinary macromolecular crystal growth inhibitors in calcium nephrolithiasis. Min. Electrolyte Metab. 3: 268-275.

(7) Dieppe, P.A.; Doyle, D.V.; Huskisson, E.C.; Willoughby, D.A.; and Crocker, P.R. 1978. Mixed crystal deposition disease and osteoarthritis. Brit. Med. J. 1: 150.

(8) Francis, M.D.; Davis, T.L.; Benedict, J.J.; and Tofe, A.J. 1980. Diphosphonates: in vitro adsorption and desorption studies on hydroxyapatite and diffusion in bone. In: Etidronate, ed. A. Caniggia, pp. 33-50. Pisa: Gentili.

(9) Francis, M.D., and Martodam, R.R. 1982. Chemical, biochemical and medicinal properties of diphosphonates. In The Role of Phosphonates in Living Systems, ed. R.L. Hildebrand. Boca Raton: CRC Press, in press.

(10) Garten, V.A., and Head, R.B. 1973. Nucleation in salt solutions. J. Chem. Soc. (Perkin I) 69: 514-520.

(11) Jennings, R.B.; Ganote, C.E.; and Reimer, K.A. 1975. Ischemic tissue injury. Am. J. Pathol. 81: 179-194.

(12) Jones, A.G.; Francis, M.D.; and Davis, M.A. 1976. Bone scanning: radionucleidic reaction mechanisms. Seminars Nucl. Med. 6: 3-18.

(13) Keutel, H.; King, J.S.; and Boyce, W.H. 1964. Further studies of uromucoid in normal and stone urine. Urol. Int. 17: 324-341.

(14) Khan, S.R.; Finlayson, B.; and Hackett, R.L. 1979. Histologic study of the early events in oxalate induced intranephronic calculosis. Invest. Urol. 17: 199-202.

(15) Koutsoukos, P.; Amjad, Z.; Tomson, M.B.; and Nancollas, G.H. 1980. Crystallization of calcium phosphates. A constant composition study. J. Am. Chem. Soc. 102: 1553-1557.

(16) Leading article. 1979. Dissolution of renal stones. Brit. Med. J. 1: 1746-1747.

(17) Lust, G.; Faure, G.; Netter, P.; Gaucher, A.; and Seegmiller, J.E. 1981. Evidence of generalized metabolic defect in patients with hereditary chondrocalcinosis: increased inorganic pyrophosphate in cultured fibroblasts and lymphoblasts. Arthrit. Rheum., in press.

(18) McCarty, D.J. 1979. Calcium pyrophosphate dihydrate crystal deposition disease; pseudogout; articular chondrocalcinosis. In Arthritis and Allied Conditions, 9th ed., ed. D.J. McCarty, pp. 1276-1297. Philadelphia: Lea & Febiger.

(19) Moreno, E.C., and Varughese, K. 1981. Growth of calcium apatites from dilute solutions. J. Cryst. Growth 53: 20-30.

(20) Neely, J.R.; Rovetto, M.J.; Whitmer, J.T.; Morgan, H.E. 1973. Effects of ischemia on function and metabolism of the isolated working rat heart. Am. J. Physiol. 225: 651-658.

(21) Pawelchak, J.; Flanagan, D.R.; and Simonelli, A.P. 1981. Rates and mechanisms of dissolution of renal calculi. III. Mechanisms and rates of dissolution of simulated oxalate calculi in acid (and EDTA solutions). In Urolithiasis - Clinical and Basic Research, eds. L.H. Smith, W.G. Robertson, and B. Finlayson, pp. 551-556. New York: Plenum.

(22) Randolph, A.D., and Drach, G.W. 1981. A proposal of standardized nomenclature in the study of crystallization in biological systems, e.g., Urolithiasis. In Urolithiasis - Clinical and Basic Research, eds. L.H. Smith, W.G. Robertson, and B. Finlayson, pp. 383-390. New York: Plenum.

(23) Robertson, W.G.; Peacock, M.; Marshall, R.W.; Marshall, D.H.; and Nordin, B.E.C. 1976. Saturation-inhibitor index as a measure of the risk of calcium oxalate stone formation in the urinary tract. New Eng. J. Med. 294: 249-252.

(24) Robertson, W.G.; Scurr, D.S.; and Bridge, C.M. 1981. Factors influencing the crystallisation of calcium oxalate in urine - critique. J. Cryst. Growth 53: 182-194.

(25) Selye, H. 1962. Calciphylaxis. Chicago: University Press.

(26) Sheehan, M.E., and Nancollas, G.H. 1980. Calcium oxalate crystal growth. A new constant composition method for modelling urinary stone formation. Invest. Urol. 17: 446-450.

(27) Tenenbaum, J.; Muniz, O.; Schumacher, H.R.; Good, A.E.; and Howell, D.S. 1981. Comparison of phosphohydrolase activities from human cartilage in calcium pyrophosphate deposition disease and primary osteoarthritis. Arthrit. Rheum. 24: 492-500.

(28) Vahlensieck, W.; Bach, D.; Hesse, A.; and Schneeberger, W. 1981. Metabolic investigations in 407 recurrent stone-formers and healthy controls on condition of individual as well as standard diet. In Urolithiasis - Clinical and Basic Research, eds. L.H. Smith, W.G. Robertson, and B. Finlayson, pp. 29-34. New York: Plenum.

(29) Walzer, C.; Boivin, G.; Schonborner, A.A.; and Baud, C.A. 1980. Ultrastructural and cytochemical aspects of the initial phases of an experimental cutaneous calcinosis (calcergy) in the rat. Cell Tissue Res. 212: 185-202.

Group on Mechanisms of Normal and Pathological Demineralization

Standing, left to right: K.H. Schmidt, J. Christoffersen, M.A.H. Gebhardt, J.C. Elliott.
Seated: H. Schäfer, L.G. Raisz, J. Arends, J.J. Reynolds.

Biological Mineralization and Demineralization, ed. G.H. Nancollas, pp. 389-398.
Dahlem Konferenzen 1982. Berlin, Heidelberg, New York: Springer-Verlag.

Mechanisms of Normal and Pathological Demineralization
Group Report

J. J. Reynolds, Rapporteur
J. Arends, J. Christoffersen, J. C. Elliot,
M. A. H. Gebhardt, L. G. Raisz, H. Schäfer, K. H. Schmidt

INTRODUCTION

Scientists from a wide variety of disciplines were brought together in this group in the hope that useful exchanges of ideas would in the long-term help in our understanding of the problem of demineralization, in terms of cell biology and chemistry. A better understanding of this problem has important consequences for many areas of research. In medicine and dentistry bone demineralization is an important pathological mechanism in common and serious disorders, for examples, periodontal disease and malignant neoplasms: the fertilizer and phosphate industry, for which apatites form the raw materials, would also benefit from the application of new information.

Because of the small number in the group, agreement was first reached that only a relatively few issues would be tackled. Broad questions were formulated from the preliminary discussions to serve as a basis for more detailed discussions, and in general the questions are used as the subject matter headings below. Some related problems were discussed in the final session and a few comments included in this report.

No attempt has been made either to discuss or to summarize the papers prepared as specific background material (Arends and Raisz, both this volume), but they serve as excellent guides for anyone wishing to understand more fully our discussions.

ROLE OF HYDROGEN ION AND OTHER COMPOUNDS IN DEMINERALIZATION

The concept of the overall importance of hydrogen ions was first reviewed by several discussants, and primary stress was placed on the interaction of hydrogen ions with phosphate groups in the solid surface. It can be envisaged that when the inorganic part of mineralized tissue is dissolved two types of processes operate, one by which anions are transformed into more soluble forms by complexing with hydrogen ions, and in the other cations are transformed into more soluble forms by complexation. There is sound thermodynamic and experimental evidence to suggest that it is the action of hydrogen ions on the phosphate groups of the hydroxyapatite lattice that is the most important consideration. The role of anions, particularly the Ca^{2+}-complexing agents such as citrate and lactate, were thought to be of less importance. Other compounds that could be involved in demineralization were not discussed, and it was thought that hydrogen ions might be unique.

In stressing the importance of the solid state structure it was generally agreed that crystal defects could locally enhance dissolution and cause destruction of crystals. Minor changes in the lattice parameters per se were not considered to be of major importance. The presence of carbonate in the crystals is also an important determining factor in dissolution, and this has been well characterized in enamel research (Arends, this volume). It was stressed that biological minerals are often particularly rich in carbonate, and so their properties are different from pure hydroxyapatite. The ultrastructural evidence suggests that there is a carbonate controlled mechanism in enamel dissolution, which might also occur in bone, but precisely what kind of crystal defect and how it influences the rate of dissolution is unknown. Despite the uncertainties it was considered possible that "carbonate-induced strain" could be the explanation as to how dissolution produces a "pipe formation" effect in a large percentage of the crystallites (Arends, this volume).

Although there was the general agreement on the importance of hydrogen ions in crystal dissolution, it was also agreed very early in the discussions that we know little about bone demineralization by such a process. Such chemical processes as those discussed above must, however, be taken into account when exploring how osteoclasts act in bone resorption, particularly because the distance over which the cells act is about 1 µm from the inorganic crystal. Although hydrogen ions could be the end agent at the crystal surface, how they could be generated could only be conjectured. It was agreed that new evidence from the analysis of the dissolving surface in both chemical and biological model systems could be very valuable. Two open questions were formulated:

"How can microanalyses of the resorbing surface be made?"
"How can data on the ion activities of calcium, phosphate, and other relevant species and pH be obtained?"

To try and answer these questions it was suggested that new physical methods might soon be at hand, and the following open question was posed:

"Can the development and application of physical microtechniques such as secondary ion mass spectrometry (SIMS) yield new information on mineralized tissues, cells, and crystals?"

All participants wished that this question could soon be answered. Although techniques have already been developed (3,5), the equipment is expensive: the major need of most investigators seems to be money!

The final open question that everyone wished could be answered was:

"Can we identify the agents that are responsible for bone demineralization?"

It now seems worthwhile to go back to bone research with the ideas and methods that have been developed in other areas.

ROLE OF INHIBITORS IN DEMINERALIZATION

A further discussion of the physical model for crystal dissolution centered on the source of "holes" or etch pits where dissolution begins. It was again stressed that these were indeed related to impurities in the crystal structure and that further work on carbonate would be very fruitful. The surface covering physicochemical model mentioned above ((1), and Nielsen and Christoffersen, this volume) also gives a good explanation for how fluoride acts as an inhibitor at the solid interface. It was also pointed out, however, that much of the effect of fluoride in biological systems could be via cell-mediated processes. The discussion led to a very exciting period when it was realized that methods now exist for investigating the problem of what natural biological inhibitors there might be. It is known from the physicochemical studies that polyphosphates, phosphoserine, and phosphopolycarboxylic acids are potent inhibitors of dissolution. Also, the fact that γ-carboxy-glutamic acid (Gla) inhibits dissolution in physicochemical systems suggests that a very fruitful area would be to study isolated materials from mineralized tissues, especially the bone Gla-protein. The next open question was:

"Can we characterize organic inhibitors of demineralization and action?"

This problem is clearly now open for investigation. The production of biological materials for physicochemical testing is currently a difficult biochemical problem, but new techniques, such as genetic engineering, should be available to attack this problem in the not too distant future. The properties that such as yet uncharacterized biological materials should have include the ability to act at the crystal surface at an acid pH so that they are not stripped off. Clearly acid groups and hydroxyl groups will be involved, and much speculative discussion took place on likely organic components of bone to test.

Evidence was discussed as to the role, if any, of the organic matrix in either inhibiting dissolution directly at the crystal surface or being involved indirectly via cell-mediated processes. The coupled regulation of matrix degradation and mineral dissolution is clearly important for the whole process of biological demineralization (4), but it is poorly defined and needs further study. On the one hand, it was considered unlikely that structural macromolecular components per se were important and that these were probably resorbed at a later stage to complete the whole process of resorption in biological systems. On the other hand, there was general agreement that there could be important crystal-associated proteins whose prior enzymatic digestion could precede crystal dissolution. The open question was therefore formulated:

"How can we increase knowledge about the sequence of events in bone and dentine resorption? Are there extracellular proteins that must be digested prior to mineral dissolution?"

The investigation of this problem is, of course, closely related to the one discussed above for organic inhibitors and could soon be a fruitful area of research. Many groups are now at work at isolating macromolecules from hard tissues. There was also a discussion on the possibility that one could define crystal changes near the osteoclast ruffled border by use of physical methods, and this seems a likely area for research in the immediate future.

ROLE OF CELLS AND TISSUES IN DEMINERALIZATION

A useful outline was given of the steps now considered important for the dissolution of enamel crystals, thus leading to dental caries. The first phase concerns the adherence and growth of bacteria and then the fermentation of food to give acids, principally lactic acid. The physical chemistry of the dissolution of enamel crystals is probably now the best understood of the biological systems (Arends, this volume), but it is also clear that the role of bacteria, except for acid formation, does not give useful analogies for other resorptive processes. Further enamel research could, however, be exceedingly useful in bridging the gap between the most complex biological resorptive process (bone resorption) with the studies on pure crystals.

An extensive discussion arose from considerations concerning the processes in ectopic calcification, where it seems that the site of calcification is very important in determining whether resorption later takes place or not. Much emphasis was placed on the possible role of macrophages in the dissolution of such mineralized deposits. It was clear that many useful experiments are now possible since there may be a major role for intracellular digestion where phagocytosing macrophages are concerned. These open questions were formulated:

"How can use be made of microelectrodes for analyzing macrophages (and giant cells derived from them) and their intracellular vacuoles with ingested particles?"
"Can use be made of the developing physical methods (such as SIMS) for analyzing the dissolution of ingested inorganic materials?"

Work with macrophages looks very promising but the question of whether there is a significant role for intracellular demineralization in osteoclastic bone resorption was left unanswered. The morphological evidence suggests that this is unlikely to be a significant dissolution process in bone, but any new knowledge is likely to be helpful in the long run.

PATHOLOGICAL DEMINERALIZATION AND MODEL SYSTEMS

An extensive discussion of the interaction of various cells and factors outlined the physiology and pathology of bone resorption (Raisz, this volume). It was suggested that chemotaxis (the reaction of cells to chemical stimuli) would be a very important factor in the recruitment of osteoclast precursor cells and in determining the site of resorption. It was envisaged that the products of degradation of the organic matrix of bone, especially those derived from collagen, could lead to the increased recruitment of cells involved in resorption, and that it would be a regenerative system. Although these pieces of evidence are central issues to biomedical research, it was thought that they were probably not processes of primary significance in the actual mechanism of crystal dissolution. What was interesting was that many ideas were put forward about what the gradient(s) between the osteoclast and bone mineral surfaces could be. One suggestion was that it could actually be the H^+ ions themselves that attracted osteoclasts to bone surfaces. It has been suggested on morphological grounds that osteoclasts may be attracted to bone surfaces without any cell covering. Perhaps the reaction of the bone surface with hydrogen ions could produce a chemotactic signal. Another suggestion was that piezo electric effects generating positive and negative changes on bone surfaces may influence the direction and magnitude of osteoclastic bone resorption. Consequently, this question was formulated:

"Can we define any of the events when intercellular channels on bone surfaces open?"

It was suggested that model systems with synthetic implants, possibly coated with biological materials, could form useful model systems to try to gain new data on this problem.

Potential model systems (at all levels of complexity) were discussed at this and other times, and these suggestions included not only ectopic calcification and implants but also giant cells that could be derived from macrophages in cell culture. It was also suggested that animal models of hydrogen ion excess (e.g., NH_4Cl loading) could also be useful in studying the role of H^+ ions. All of these models still do not help directly in the issue concerning bone resorption and the discussions led to the open question:

"Can methods be developed to give better functioning cell preparations of osteoclasts and precursors?"

Progress in this area is conceivable in the near future and could provide new insights into understanding the major overall question from the biological point of view:

"Can we gain a much better understanding of the origin and differentiation of cells resorbing mineralized tissues and their interaction with their extracellular environment?"

Current research in many laboratories indicates that a positive reply can soon be given to this question.

Finally, under the topic of pathology, a useful discussion concerned such agents as heparin, which is known to increase bone resorption. It seemed quite possible to test out such materials in physicochemical systems to see if they have important effects on dissolution. Disaggregated crystals would clearly be resorbed faster because of the increase in surface area. The question was posed to the physical chemists:

"Can experiments now be devised to determine if disaggregation of crystals is important in dissolution?"

It was thought that the answer was positive.

DEMINERALIZATION IN NONMAMMALIAN SYSTEMS

Since most of the discussions on biological problems were related to mammalian tissues, some time was taken to learn of other information on birds, fishes, and other lower species. Most of the problems concerned the dissolution of calcium carbonate crystals, and it was clear that hydrogen ions, albeit often with different anions, were the major factor in dissolution. Discussions on the dissolution of calcium carbonate from the eggshell were particularly interesting in that the enzyme carbonic anhydrase seems to play an important role. The enzyme has been localized near the site of crystal dissolution in the shells of developing chick embryos, and it seemed natural and important to pose the open question:

"Can new information be obtained on whether carbonic anhydrases are involved in the generation of hydrogen ions at the site of crystal dissolution in bone?"

This question could be tackled by available techniques (2) based on current knowledge of the enzyme. It is already known that osteoclasts have relatively more of this enzyme than other bone cells, but precisely how this enzyme is controlled is not known.

It was also of interest from these discussions on nonmammalian systems that the osteoid-like material in fish scales seems to be resorbed by cells that may be akin to mononuclear osteoclasts. Clearly more work with fishes and avian embryos could be of great usefulness in understanding the cell biology of resorption.

A discussion was also organized on the possible role of phosphate and bi- and polyphosphonates in demineralization, and two open questions were formulated:

"What is the role of phosphate in calcium carbonate mineralization and demineralization?"
"What are the mechanisms of action of bi- and polyphosphonates in inhibiting demineralization?"

Phosphate is known to slow down the precipitation of carbonate (Crenshaw, this volume) and could perhaps influence dissolution. The relevance to bone crystals was not clear nor was the evidence for the precise mechanisms of action of polyphosphonates. It should be possible and interesting to study the mode of action of all these substances on both bone powder and synthetic crystals containing carbonate that would be more nearly related to natural minerals than pure hydroxyapatite.

The final open question was:

"Can new methods for studying the measurement of bone resorption in vivo be devised?"

This question arose out of the desire to develop and exploit new physical techniques for medical practice and clearly would be a significant advance. There was, however, not a great deal that could be suggested.

CONCLUSIONS

The important role of H^+ ions in the dissolution of apatitic minerals was clearly accepted as a general concept that could be taken from the physicochemical systems and applied to many biological problems. The concept is based on sound chemical evidence and theory, and many useful analogies were drawn from a wide variety of biological systems. Moreover, many new ideas for experimentation at both the physical and biological levels were suggested, and some of these have been indicated above. Perhaps the largest challenges now are for biologists to identify cellular mechanisms that could be responsible for the production of H^+ ions and for physical chemists to define whether any organic materials isolated from mineralized tissues could play important roles in crystal dissolution. The discussions were an extremely useful forum and all the participants agreed that their awareness of unsolved problems had been widened and that possible applications of new physical techniques clarified. Finally, the whole group felt that the conference goal had been achieved in that a very useful interaction between scientists from many disciplines had taken place.

REFERENCES FOR SUGGESTED FURTHER READING

(1) Christoffersen, J. 1982. Dissolution of calcium hydroxyapatite. Calcif. Tissue Int., in press.

(2) Gay, C.V. 1980. Recent developments in carbonic anhydrase localization: occurrence of carbonic anhydrase in calcium - mobilizing tissues. In Biophysics and Physiology of Carbon Dioxide, eds. C. Bauer, G. Gros, and H. Bartels, pp. 399-405. Berlin: Springer-Verlag.

(3) Lefèvre, R. 1980. Secondary ion microscopy of biomineralisations. Scanning 3: 90-97.

(4) Sellers, A.; Meikle, M.C.; and Reynolds, J.J. 1980. Collagenase and collagenase inhibitor levels following changes in bone resorption in vitro. Calcif. Tissue Int. 31: 35-43.

(5) Spurr, A.R. 1980. Applications of SIMS in biology and medicine. Scanning 3: 97-109.

List of Participants

ALTHOFF, J.
Institut für Medizinische Physik
der Universität Münster
4400 Münster, German Federal Republic

Field of research: Alkaline phosphatases of hard tissue formation, Ca + P metabolism

ANDERSON, C.
Dept. of Pathology
University of Western Ontario
London, Ontario N6A 5C1, Canada

Field of research: Metabolic bone disease, the effect of environmental pollutants on bone turnover rates

ARENDS, J.
Lab. voor Materia Technika
University of Groningen
9713 AV Groningen, Netherlands

Field of research: Apatites, de- and remineralization of hard mineralized tissues, fluoridation, di- and polyphosphonates

ARNOTT, H.J.
Dept. of Biology, University of Texas
Arlington, TX 76019, USA

Field of research: Cytology, ultrastructure and biophysics of biomineralization in plants

BACH, D.W.
Urologische Universitätsklinik
der Universität Bonn
5300 Bonn-Venusberg
German Federal Republic

Field of research: Pathogenesis and therapy of kidney stone disease

BENNEMA, P.
RIM Laboratory of Solid State Chemistry
Catholic University Nijmegen
Nijmegen, Netherlands

Field of research: Crystal growth theory and experiment

BOYDE, A.
Dept. of Anatomy and Embryology
University College London
London WC1E 6BT, England

Field of research: Development and structure of normal and pathological mineralized tissues (SEM)

BROWN, W.E.
American Dental Association Health Foundation Research Unit
National Bureau of Standards
Washington, DC 20234, USA

Field of research: Crystal chemistry of calcium phosphates

CHRISTOFFERSEN, J.
Medicinsk-Kemisk Institut
Panum Instituttet
University of Copenhagen
2200 Copenhagen N, Denmark

Field of research: Kinetics and inhibition of dissolution of calcium hydroxyapatite

CRENSHAW, M.A.
Dental Research Center and Dept. of Pedodontics
University of North Carolina
Chapel Hill, NC 27514, USA

Field of research: Biomineralization

DOSCH, W.
Institut für Geowissenschaften
der Universität Mainz
6500 Mainz, German Federal Republic

Field of research: Biomineralization - crystal chemistry of urinary and biliary calculi

ELLIOT, J.C.
Dept. of Biochemistry
The London Hospital Medical College
London E1 2AD, England

Field of research: Structure of biological calcium phosphates and physical chemistry of dental caries

FINLAYSON, B.
Division of Urology, Dept. of Surgery
University of Florida
Gainesville, FL 32607, USA

Field of research: Urolithiasis

FLEISCH, H.
Pathophysiologisches Institut
der Universität Bern
3010 Bern, Switzerland

Field of research: Bone, calcium and phosphate metabolism

FRANCIS, M.D.
The Proctor & Gamble Co.
2N142, Miami Valley Laboratories
Cincinnati, OH 45247, USA

Field of research: Heterotopic ossification, arthritis and inflammation, nuclear medicine bone scanning mechanisms

GARSIDE, J.
Dept. of Chemical Engineering
University of Manchester
Institute of Science and Technology
Manchester M60 1QD, England

Field of research: Mechanisms of nucleation (primary and secondary) and of crystal growth, modelling of size distribution in crystallizers

GEBHARDT, M.A.H.
Mineralogisch-Petrologisches Institut
der Universität Bonn
5300 Bonn, German Federal Republic

Field of research: Urinary calculi, bone mineral content measurement by X-ray absorptiometry

GRYNPAS, M.D.
Children's Hospital
Medical Center
Orthopaedics Research
Boston, MA 02115, USA

Field of research: Biophysics of calcified and connective tissues

HASCALL, V.C.
Bldg. 30, Rm. 106
National Institute of Dental Research
NIH
Bethesda, MD 20205, USA

Field of research: Proteoglycan biochemistry

HAUTMANN, R.
Abteilung für Urologie
der Medizinischen Fakultät
R.W.T.H. Aachen
5100 Aachen, German Federal Republic

Field of research: Urolithiasis, renal handling of oxalate (pharmacokinetic, micropuncture), oxalate and calcium concentration in the tissue (papilla)

HESSE, A.
Harnsteinsforschungsstelle der
Urologische Universitätsklinik Bonn
5300 Bonn-Venusberg
German Federal Republic

Field of research: Urolithiasis

HÖHLING, H.-J.
Institut für Medizinische Physik
der Universität Münster
4400 Münster, German Federal Republic

Field of research: Hard tissue formation, bone disease, and pathological mineralization

HOWELL, D.S.
Dept. of Medicine D8-7
University of Miami
Miami, FL 33101, USA

Field of research: Osteoarthritis, related biological mineralization

KITANO, Y.
Water Research Institute
Nagoya University
Nagoya 464, Japan

Field of research: Carbonate geochemistry, water geochemistry, environmental chemistry

KLEE, W.E.
Institut für Kristallographie
der Universität Karlsruhe
7500 Karlsruhe
German Federal Republic

Field of research: Composition of biliary and urinary calculi, structure of carbonate apatites

KRAMPITZ, G.P.
Abteilung für Biochemie
Institut für Anatomie, Physiologie
und Hygiene der Haustiere
der Universität Bonn
5300 Bonn-Venusberg
German Federal Republic

Field of research: Biochemistry of biological calcification, protein chemistry

MATTHEWS, J.L.
A. Webb Roberts Center for Continuing
Medical Education
Baylor University Medical Center
Dallas, TX 75246, USA

Field of research: Ultrastructural studies of mineralized tissues, factors influencing mineralization and resorption

McCARTY, D.J.
Dept. of Medicine
University of Wisconsin
Milwaukee, WI 53226, USA

Field of research: Rheumatology

MORENO, E.C.
Forsyth Dental Center
Boston, MA 02115, USA

Field of research: Biological mineralization, crystal growth, cariogenesis

MÜNZENBERG, K.J.
Orthopädische Universitätsklinik
der Universität Bonn
5300 Bonn-Venusberg
German Federal Republic

Field of research: Mineralization of bone

NANCOLLAS, G.H.
Dept. of Chemistry
State University of New York
Buffalo, NY 14214, USA

Field of research: Physical chemistry of crystallization

NEWESELY, H.
Institut für Klinisch-theoretische
Zahn-, Mund- und Kieferheilkunde
der Freien Universität
1000 Berlin 33
German Federal Republic

Field of research: Mechanisms of mineralization (normal)

NIELSEN, A.E.
Medicinsk-Kemisk Institut
Panum Instituttet
University of Copenhagen
2200 Copenhagen N, Denmark

Field of research: Nucleation and growth of crystals during electrolyte precipitation in aqueous solution

PRITZKER, K.P.H.
Dept. of Laboratories
Mount Sinai Hospital
Toronto, Ontario M5G 1X5, Canada

Field of research: Pathology of crystal-associated arthritis, calcium pyrophosphate dihydrate crystal formation, crystal localization and deposition in disease processes

PROCKOP, D.J.
Dept. of Biochemistry, College of Medicine and Dentistry of New Jersey
Rutgers Medical School
Piscataway, NJ 08854, USA

Field of research: Collagen structure, biosynthesis, genes, diseases involving

QUINT, P.
Institut für Medizinische Physik der Universität Münster
4400 Münster, German Federal Republic

Field of research: Investigation of hard tissue formation (e.g., porcine epiphyseal growth plate) by chemical analysis

RAISZ, L.G.
Dept. of Medicine, University of Connecticut Health Center
Farmington, CT 06032, USA

Field of research: Calcium regulation and bone metabolism

REYNOLDS, J.J.
Dept. of Cell Physiology
Strangeways Research Laboratory
Cambridge CB1 4RN, England

Field of research: Connective tissue metabolism, tumor invasion

ROBERTSON, W.G.
M.R.C. Mineral Metabolism Unit
The General Infirmary
Leeds LS1 3EX, England

Field of research: The cause and treatment of urinary stone disease

SCHÄFER, H.
Institut für Pathologie der Universität Hamburg
2000 Hamburg 20
German Federal Republic

Field of research: Ultrastructural calcium distribution in soft tissues, morphology of ectopic calcification

SCHENK, R.K.
Anatomisches Institut der Universität Bern
3012 Bern, Switzerland

Field of research: Ultrastructure of cartilage and bone

SCHMIDT, K.H.
Chirurgische Abteilung der Universität Tübingen
7400 Tübingen
German Federal Republic

Field of research: Morphogenesis of bone, osteoinduction and osteoblastic differentiation, cultivation of osteoblasts

SIMKISS, K.
Dept. of Zoology
University of Reading
Reading RG6 2AJ, England

Field of research: Metal ions in cells, biomineralization

SMITH, L.H.
Dept. of Nephrology and Internal Medicine
Mayo Clinic and Foundation
Rochester, MN 55905, USA

Field of research: Urolithiasis

TERMINE, J.D.
Bldg. 30, Rm. 211
National Institute of Dental Research
NIH
Bethesda, MD 20205, USA

Field of research: Mineralization biochemistry; non-collagenous protein chemistry of bone, dentin, and enamel extracellular matrices

VAHLENSIECK, E.W.
Urologische Universitätsklinik
der Universität Bonn
5300 Bonn-Venusberg
German Federal Republic

Field of research: Abnormal mineralization, urolithiasis

WATABE, N.
Electron Microscopy Center
University of South Carolina
Columbia, SC 29208, USA

Field of research: Mechanisms of mineralization in the invertebrates and fish

YOUNG, R.A.
School of Physics and Engineering
Experiment Station
Georgia Institute of Technology
Atlanta, GA 30332, USA

Field of research: Atomic-scale mechanisms and detail in apatites, tooth enamel, and related materials

Subject Index

Author Index

MIX
Papier aus verantwortungsvollen Quellen
Paper from responsible sources
FSC® C105338

If you have any concerns about our products,
you can contact us on
ProductSafety@springernature.com

In case Publisher is established outside the EU,
the EU authorized representative is:
Springer Nature Customer Service Center GmbH
Europaplatz 3, 69115 Heidelberg, Germany

Printed by Libri Plureos GmbH
in Hamburg, Germany